ENDANGERED AND THREATENED SPECIES OF THE PLATTE RIVER

Committee on Endangered and Threatened
Species in the Platte River Basin

Board on Environmental Studies and Toxicology
Water Science and Technology Board

Division on Earth and Life Studies

NATIONAL RESEARCH COUNCIL
OF THE NATIONAL ACADEMIES

THE NATIONAL ACADEMIES PRESS
Washington, DC
www.nap.edu

THE NATIONAL ACADEMIES PRESS 500 Fifth Street, NW Washington, DC 20001

NOTICE: The project that is the subject of this report was approved by the Governing Board of the National Research Council, whose members are drawn from the councils of the National Academy of Sciences, the National Academy of Engineering, and the Institute of Medicine. The members of the committee responsible for the report were chosen for their special competences and with regard for appropriate balance.

This project was supported by Grant 98210-3-G-483 between the National Academy of Sciences and the Fish and Wildlife Service and Bureau of Reclamation. Any opinions, findings, conclusions, or recommendations expressed in this publication are those of the authors and do not necessarily reflect the view of the organizations or agencies that provided support for this project.

Library of Congress Control Number 2004116614

International Standard Book Number 0-309-09230-2 (Book)
International Standard Book Number 0-309-53263-9 (PDF)

Additional copies of this report are available from

The National Academies Press
500 Fifth Street, NW
Box 285
Washington, DC 20055

800-624-6242
202-334-3313 (in the Washington metropolitan area)
http://www.nap.edu

Printed in the United States of America.

THE NATIONAL ACADEMIES

Advisers to the Nation on Science, Engineering, and Medicine

The **National Academy of Sciences** is a private, nonprofit, self-perpetuating society of distinguished scholars engaged in scientific and engineering research, dedicated to the furtherance of science and technology and to their use for the general welfare. Upon the authority of the charter granted to it by the Congress in 1863, the Academy has a mandate that requires it to advise the federal government on scientific and technical matters. Dr. Bruce M. Alberts is president of the National Academy of Sciences.

The **National Academy of Engineering** was established in 1964, under the charter of the National Academy of Sciences, as a parallel organization of outstanding engineers. It is autonomous in its administration and in the selection of its members, sharing with the National Academy of Sciences the responsibility for advising the federal government. The National Academy of Engineering also sponsors engineering programs aimed at meeting national needs, encourages education and research, and recognizes the superior achievements of engineers. Dr. Wm. A. Wulf is president of the National Academy of Engineering.

The **Institute of Medicine** was established in 1970 by the National Academy of Sciences to secure the services of eminent members of appropriate professions in the examination of policy matters pertaining to the health of the public. The Institute acts under the responsibility given to the National Academy of Sciences by its congressional charter to be an adviser to the federal government and, upon its own initiative, to identify issues of medical care, research, and education. Dr. Harvey V. Fineberg is president of the Institute of Medicine.

The **National Research Council** was organized by the National Academy of Sciences in 1916 to associate the broad community of science and technology with the Academy's purposes of furthering knowledge and advising the federal government. Functioning in accordance with general policies determined by the Academy, the Council has become the principal operating agency of both the National Academy of Sciences and the National Academy of Engineering in providing services to the government, the public, and the scientific and engineering communities. The Council is administered jointly by both Academies and the Institute of Medicine. Dr. Bruce M. Alberts and Dr. Wm. A. Wulf are chair and vice chair, respectively, of the National Research Council.

www.national-academies.org

COMMITTEE ON ENDANGERED AND THREATENED SPECIES IN THE PLATTE RIVER BASIN

WILLIAM L. GRAF (*Chair*), University of South Carolina, Columbia
JOHN A. BARZEN, International Crane Foundation, Baraboo, WI
FRANCESCA CUTHBERT, University of Minnesota, St. Paul
HOLLY DOREMUS, University of California, Davis
LISA M. BUTLER HARRINGTON, Kansas State University, Manhattan
EDWIN E. HERRICKS, University of Illinois, Urbana
KATHARINE L. JACOBS, University of Arizona, Tucson
W. CARTER JOHNSON, South Dakota State University, Brookings
FRANK LUPI, Michigan State University, East Lansing
DENNIS D. MURPHY, University of Nevada, Reno
RICHARD N. PALMER, University of Washington, Seattle
EDWARD J. PETERS, University of Nebraska, Lincoln
HSIEH W. SHEN, University of California, Berkeley
JAMES ANTHONY THOMPSON, Willow Lake Farm, Windom, MN

Staff
SUZANNE VAN DRUNICK, Project Director
LAUREN ALEXANDER, Program Officer
NORMAN GROSSBLATT, Senior Editor
KELLY CLARK, Assistant Editor
MIRSADA KARALIC-LONCAREVIC, Research Associate
BRYAN P. SHIPLEY, Research Associate
LIZA R. HAMILTON, Program Assistant
SAMMY BARDLEY, Library Assistant

With best regards,
Suzanne

Sponsors
U.S. DEPARTMENT OF THE INTERIOR
BUREAU OF RECLAMATION
FISH AND WILDLIFE SERVICE

WATER SCIENCE AND TECHNOLOGY BOARD

OTHER REPORTS OF THE BOARD ON ENVIRONMENTAL STUDIES AND TOXICOLOGY

Air Quality Management in the United States (2004)
Atlantic Salmon in Maine (2004)
Endangered and Threatened Fishes in the Klamath River Basin (2004)
Cumulative Environmental Effects of Alaska North Slope Oil and Gas Development (2003)
Estimating the Public Health Benefits of Proposed Air Pollution Regulations (2002)
Biosolids Applied to Land: Advancing Standards and Practices (2002)
The Airliner Cabin Environment and Health of Passengers and Crew (2002)
Arsenic in Drinking Water: 2001 Update (2001)
Evaluating Vehicle Emissions Inspection and Maintenance Programs (2001)
Compensating for Wetland Losses Under the Clean Water Act (2001)
A Risk-Management Strategy for PCB-Contaminated Sediments (2001)
Acute Exposure Guideline Levels for Selected Airborne Chemicals (4 volumes, 2000-2004)
Toxicological Effects of Methylmercury (2000)
Strengthening Science at the U.S. Environmental Protection Agency (2000)
Scientific Frontiers in Developmental Toxicology and Risk Assessment (2000)
Ecological Indicators for the Nation (2000)
Waste Incineration and Public Health (1999)
Hormonally Active Agents in the Environment (1999)
Research Priorities for Airborne Particulate Matter (4 volumes, 1998-2004)
The National Research Council's Committee on Toxicology: The First 50 Years (1997)
Carcinogens and Anticarcinogens in the Human Diet (1996)
Upstream: Salmon and Society in the Pacific Northwest (1996)
Science and the Endangered Species Act (1995)
Wetlands: Characteristics and Boundaries (1995)
Biologic Markers (5 volumes, 1989-1995)
Review of EPA's Environmental Monitoring and Assessment Program (3 volumes, 1994-1995)
Science and Judgment in Risk Assessment (1994)
Pesticides in the Diets of Infants and Children (1993)
Dolphins and the Tuna Industry (1992)
Science and the National Parks (1992)
Human Exposure Assessment for Airborne Pollutants (1991)
Rethinking the Ozone Problem in Urban and Regional Air Pollution (1991)
Decline of the Sea Turtles (1990)

Copies of these reports may be ordered from the National Academies Press
(800) 624-6242 or (202) 334-3313
www.nap.edu

OTHER REPORTS OF THE WATER SCIENCE AND TECHNOLOGY BOARD

Managing the Columbia River: Instream Flows, Water Withdrawals, and Salmon Survival (2004)
Review of the U.S. Army Corps of Engineers Upper Mississippi-Illinois Waterway Restructured Feasibility Study: Interim Report (2004)
Review of the Desalination and Water Purification Technology Roadmap (2004)
A Review of the EPA Water Security Research and Technical Support Action Plan (2004)
Groundwater Fluxes Across Interfaces (2004)
Riparian Areas: Functions and Strategies for Management (2003)
Bioavailability of Contaminants in Soils and Sediments: Processes, Tools, and Applications (2003)
Environmental Cleanup at Navy Facilities: Adaptive Site Management (2003)
Review Procedures for Water Resources Planning (2002)
Privatization of Water Services in the United States: An Assessment of Issues and Experiences (2002)
Opportunities to Improve the U.S. Geological Survey National Water Quality Assessment Program (2002)
Predictability and Limits-to-Prediction in Hydrologic Systems (2002)
Estimating Water Use in the United States: A New Paradigm for the National Water-Use Information Program (2002)
Missouri River Ecosystem: Exploring the Prospects for Recovery (2002)
Review of USGCRP Plan for a New Science Initiative on the Global Water Cycle (2002)
Assessing the TMDL Approach to Water Quality Management (2001)
Classifying Drinking Water Contaminants for Regulatory Consideration (2001)
Envisioning the Agenda for Water Resources Research in the Twenty-first Century (2001)
Inland Navigation System Planning: The Upper Mississippi River-Illinois Waterway (2001)
Investigating Groundwater Systems on Regional and National Scales (2000)
Risk Analysis and Uncertainty in Flood Damage Reduction Studies (2000)
Clean Coastal Waters: Understanding and Reducing the Effects of Nutrient Pollution (2000)

Natural Attenuation for Groundwater Remediation (2000)
Watershed Management for Potable Water Supply: Assessing the New York City Strategy (2000)

Copies of these reports may be ordered from the National Academies Press
(800) 624-6242 or (202) 334-3313
www.nap.edu

Acknowledgments

We are appreciative of the generous support provided by the Fish and Wildlife Service and the Bureau of Reclamation. We are especially grateful to the outstanding assistance provided by R. Thomas Weimer, U.S. Department of the Interior, and Larry Schulz, Bureau of Reclamation.

Many people assisted the committee and National Research Council by providing data and reports, and assisting with committee hearings and field trips. We are grateful for the information and support provided by the following:

J. David Aiken, University of Nebraska
Steven Anschutz, Fish and Wildlife Service
Jane Austin, Northern Prairie Wildlife Research Center, U.S. Geological Survey
Maryanne Bach, Bureau of Reclamation
Edward Bartell, Water for Life Foundation
John Bartholow, U.S. Geological Survey
Curtis A. Brown, Bureau of Reclamation
Rick Brown, Colorado Water Conservation Board
Mark Butler, Fish and Wildlife Service
David Carlson, Fish and Wildlife Service
Robert Cox, Northern Prairie Wildlife Research Center, U.S. Geological Survey
Mark Czaplewski, Central Platte Natural Resources District
Betsy Didrickson, International Crane Foundation

John Dinan, Nebraska Game and Parks Commission
Kenny Dinan, Fish and Wildlife Service
Jeffrey L. Drahota, Fish and Wildlife Service
Sara Gavney-Moore, International Crane Foundation
James Harris, International Crane Foundation
Mathew Hayes, International Crane Foundation
Lynn Holt, Bureau of Reclamation
James Jenniges, Nebraska Public Power District
J. Michael Jess, University of Nebraska
Wallace Jobman, Fish and Wildlife Service
Douglas H. Johnson, Northern Prairie Wildlife Research Center, U.S. Geological Survey
Kenneth Jones, Dyersburg State Community College
John W. Keys III, Bureau of Reclamation
Gary Krapu, Northern Prairie Wildlife Research Center, U.S. Geological Survey
Steven Krentz, Fish and Wildlife Service
Kammie L. Kruse, Fish and Wildlife Service
Frank Kwapnioski, Nebraska Public Power District
Anne Lacy, International Crane Foundation
Robert C. Lacy, Chicago Zoological Society/IUCN SSC Conservation Breeding Specialist Group
Brent Lathrop, The Nature Conservancy
John H. Lawson, Bureau of Reclamation
Gary L. Lewis, Parsons
Gary Lingle, University of Nebraska
Steven Lydick, Fish and Wildlife Service
Jeremiah L. Maher, Kleinschmidts Associates
Robert McCue, Fish and Wildlife Service
Ted Melis, U.S. Geological Survey
Ralph O. Morgenweck, Fish and Wildlife Service
Peter J. Murphy, Bureau of Reclamation
James E. Parham, University of Nebraska
Thomas R. Payne, Thomas R. Payne & Associates
Mark M. Peyton, Central Nebraska Public Power & Irrigation District
Timothy J. Randle, Bureau of Reclamation
William E. Rinne, Bureau of Reclamation
Jeff Runge, Fish and Wildlife Service
John J. Shadle, Nebraska Public Power District
David E. Sharp, Fish and Wildlife Service
Hal Simpson, Colorado State Engineer
Tammy S. Snyder, Nebraska Game and Parks Commission

Thomas Stehn, Fish and Wildlife Service
Dale Strickland, Western EcoSystems Technology, Inc.
Raymond J. Supalla, University of Nebraska
Martha Tacha, Fish and Wildlife Service
Paul Tebbel, National Audubon Society
Sharon Whitmore, Fish and Wildlife Service
Erika Wilson, Fish and Wildlife Service
Duane Woodward, Central Platte Natural Resources District
Margot Zallen, U.S. Department of the Interior

The committee's work also benefited from written and oral testimony submitted by the public, whose participation is much appreciated.

Review Participants

This report has been reviewed in draft form by people chosen for their diverse perspectives and technical expertise in accordance with procedures approved by the National Research Council Report Review Committee. The purpose of this independent review is to provide candid and critical comments that will assist the institution in making its published report as sound as possible and to ensure that the report meets institutional standards of objectivity, evidence, and responsiveness to the study charge. The review comments and draft manuscript remain confidential to protect the integrity of the deliberative process. The committee and the National Research Council thank the following for their review of this report:

Jacob Bendix, Syracuse University
Leo M. Eisel, Brown and Caldwell
Paul J. Goossen, Canadian Wildlife Service
Peter Kareiva, The Nature Conservancy
James C. Lewis, (Retired) Fish and Wildlife Biologist
Richard Marston, Oklahoma State University
Steven A. Nesbitt, Florida Wildlife Conservation Commission
Gordon Orians, University of Washington
Bruce Rhoads, University of Illinois
J.B. Ruhl, Florida State University
Ernest T. Smerdon, University of Arizona
Vince Travnichek, Missouri Department of Conservation
Peter Wilcock, The Johns Hopkins University

Although the reviewers listed above have provided many constructive comments and suggestions, they were not asked to endorse the conclusions or recommendations, nor did they see the final draft of the report before its release. The review of this report was overseen by Stanley V. Gregory, Oregon State University, and Frank H. Stillinger, Princeton University. Appointed by the National Research Council, they were responsible for making certain that an independent examination of this report was carried out in accordance with institutional procedures and that all review comments were carefully considered. Responsibility for the final content of this report rests entirely with the committee and the National Research Council.

Cover Art

Mouth of the Platte River, 1833,
watercolor and pencil on paper, 10 5/8 × 16 5/8, Karl Bodmer.

The image on the cover of this book is one of the earliest accurate depictions of the Platte River. The view, constructed from direct observation on May 3, 1833, is from the channel of the Missouri River looking westward to the two channels separated by an island that formed the mouth of the Platte River. The artist was Karl Bodmer, a Swiss view painter and portrait artist born in Zurich in 1809. By the time Prince Maxmillian of Wied-Neuwied began organizing an expedition to explore the American West in 1832, he was already aware of the talents of the young Swiss artist, and he invited him to join a trip to the then little known plains of North America to collect natural specimens and artifacts of Native American cultures. Also in 1832, the American artist George Catlin ascended the Missouri, but Bodmer's superior work reached the public first.

In the spring of 1833, Maxmillian—along with his hunting companion and servant, David Dreidoppel, and Bodmer—began a 2-year journey up the Missouri River, examining the landscapes and cultures along the way. The three explorers traveled by keelboat and the *Yellow Stone*, a steamboat operated by the American Fur Company to supply its far-flung trading empire and the first steamboat to ply the upper Missouri. Eventually, they reached Fort McKenzie, near the present day Great Falls, Montana, before turning downstream for the return. Maxmillian and Bodmer wintered at Fort Clark, a company trading post associated with the Mandan villages near what is now Bismarck, North Dakota. In 1834 they continued downstream to St. Louis aboard the *Yellow Stone* on its return trip. They had traveled to the far reaches of the upper Missouri, collecting a treasure-trove of natural specimens and scores of Bodmer's paintings depicting a changing landscape and a vanishing people. They also collected two black bears that hibernated through the winter. Unfortunately, much of the specimen collection was lost after being transferred to another steamboat, the *Assiniboine*, which promptly sank. The most important collections survived the trip, however, as did all of Bodmer's works. In the summer of 1834, the three explorers and their specimens and paintings (as well as the two bears) departed for Europe, never to return to North America.

Bodmer's astonishing paintings educated a generation of Europeans about the American plains. In 1836 he exhibited most of his works, five years before George Catlin's work reached Europe and a year before Catlin's work appeared in public in the United States. Bodmer's work is meticulous and was executed with great skill. Those parts of the Missouri River landscape that remain unchanged from his day are still exactly recognizable in his paintings, and his renditions of Native Americans and their artifacts are confirmable through comparison with the few remaining items in collections. His river scenes provide insights to river geomorphology and ecology from a period that predated photographs of the region by more than two decades.

After the Missouri River excursion, Bodmer never again traveled widely. He painted mostly woodland scenes from Barbizon, France, and created illustrations for books and magazines. When he died in 1893, his estate sale dispersed much of his work, but the Joslyn Art Museum of Omaha, Nebraska, has recollected almost all of his paintings, along with Maxmillian's journals. The University of Nebraska has published them in the volume *Karl Bodmer's America*.

Preface

The Platte River of central Nebraska has undergone great transformations during the last two centuries. The installation of water-control infrastructures and influences of climatic change have altered the river's biophysical characteristics, and cultural perceptions of the river as a resource have undergone dramatic changes. As a nation, we have viewed the river as a pioneer trail, as a commodity, and finally as an ecosystem. American Indians and early European settlers saw the river as part of a primary transportation route that eventually became the Oregon Trail, one of several connective threads that bound an expanding nation together. When agricultural development transformed the Nebraska landscape, the river was viewed as a conduit for the economically and legally defined commodity of water. More recently, the Platte River has come to be perceived as an ecosystem, not only supplying water for human use but also providing important habitat for many plant and animal species that are part of our natural heritage.

The Platte River ecosystem is enormously complex from a resource perspective. The physical, chemical, and biological aspects of the river are complicated because the river flows west to east through a transition zone from nearly arid to more humid conditions that typify the Great Plains in the midsection of the North American continent. The river is part of a vast system of dams, diversions, and canals that distribute water across the landscape and that are connected to the groundwater system of the region. The river crosses the Central Flyway, a primary north-south corridor for migratory birds, and the river's riparian zones provide valuable habitat for these and a variety of other birds. The shallow waters of the river interact with a complex series of islands and bars to create unique habitats for birds and fish.

Although many public policies govern the Platte River resources, ranging from legally defined water rights to nationally specified goals for restoration under the Clean Water Act, some of the most pressing issues for river managers on the Platte emerge from the Endangered Species Act (ESA). Under the provisions of the ESA, federal officials have listed three birds—the whooping crane, the piping plover, and the interior least tern—as requiring special protection. They have also listed one fish, the pallid sturgeon. Management of the river is inextricably bound to mandated efforts to restore the populations of those species to viable, self-sustaining sizes. Such efforts inevitably focus on manipulating the water flow in the river, potentially affecting the management and use of that water for other purposes. The problem of reconciling the management of water for species and for other beneficial uses is typical of many rivers, and the Platte is not an unusual case in this respect. Similar debates occur regarding the Rio Grande, Snake, Klamath, Trinity, Truckee, Sacramento, Missouri, and Colorado Rivers.

Reconciliation of apparently competing uses lies in administrative decisions and political and legal processes, but science also plays an important role. The best decisions for public policy are likely to be the best informed, and considerable research is now available to explain biophysical processes associated with the Platte River and its listed species. Decision makers for the Platte River, particularly those participating in a cooperative agreement among state and federal agencies responsible for the river, rely on scientific data, information, and interpretations related to the species, their habitat, and the behavior of the river. They asked the National Research Council to determine whether current central and lower Platte habitat conditions affect the likelihood of the listed species' survival and recovery and to assess the validity of the science supporting the designation of critical habitat, descriptions of habitat-suitability guidelines, and management of river processes.

This report presents the findings of the National Research Council Committee on Threatened and Endangered Species in the Platte River Basin. The committee addressed specific questions about the quality of the science that decision makers have used in administering the Platte River to meet requirements of the ESA for the four listed species. The committee investigated only the scientific aspects of species and river management and sought to evaluate the quality of the research objectively. It adhered to the highest scientific principles in its evaluations.

The committee's work was greatly aided by the hospitality of many Nebraskans during our two visits to the Platte River. People from Nebraska, Colorado, and Wyoming provided their views and experiences with the river and its resources in valuable public hearings. Federal, state, and privately supported researchers were generous in sharing with the committee the fruits of their professional labor, and they took valuable time from their own schedules to help us with their testimony and to supply us with necessary but elusive documents and data. The committee benefited appreciably from a report, produced at its request by J. Michael Reed, on viability issues for listed bird species in the Platte River Basin.

The committee's work was immeasurably enhanced by the marvelous support of the National Research Council staff. James Reisa (director of the Board on Environmental Studies and Toxicology), David Policansky (scholar of the Board on Environmental Studies and Toxicology), and Stephen Parker (director of the Water Science and Technology Board) created a vision for the committee, and their guidance and wise council were exceptionally important. Suzanne van Drunick (project director and senior program officer) was a central figure in the deliberations and the production of this report, which would not have reached fruition without her good judgment and hard work. Lauren Alexander (program officer) was a helpful participant in committee deliberations. Staff members Bryan Shipley (research associate) and Liza Hamilton (program assistant) were true partners in the study processes, and their skills, from arranging initial public hearings to assembling the final report, were pivotal in our success. Our report benefited from important help from Norman Grossblatt (senior editor), Mirsada Karalic-Loncarevic (research assistant), and Sammy Bardley (library assistant). To all the fine Research Council personnel, a sincere thank you.

This report is not only the product of the efforts of committee members and National Research Council staff members; it reflects the input of the Board on Environmental Studies and Toxicology, especially Patrick O'Brien of Chevron Research and Technology who had the responsibility of report oversight; 13 independent external reviewers, listed in the acknowledgments; and Stanley V. Gregory, Oregon State University, and Frank H. Stillinger, Princeton University, who oversaw the external review. Those scientists and professionals provided us with sage reflections and remarkable insights into the complexities of the research underpinning decisions for the Platte River and its listed species.

The committee is under no illusions about the use of this report. We will not end all the controversies surrounding the Platte River Basin and its listed species, but we hope to contribute to resolving some of the questions related to the science of the matter. In this process, committee members share an overriding vision with decision makers and citizens: to have a sustainable river ecosystem that is a social, economic, and environmental bequest for future generations.

William L. Graf, *Chair*
Committee on Endangered and Threatened Species
in the Platte River Basin

Abbreviations and Acronyms

ANWR: Aransas National Wildlife Refuge
ATV: all-terrain vehicle
AWP: Aransas-Wood Buffalo migratory population of whooping cranes
COHYST: Cooperative Hydrology Study
CNPPID: Central Nebraska Public Power & Irrigation District
DOI: U.S. Department of the Interior
EPA: Environmental Protection Agency
ESA: Endangered Species Act
FERC: Federal Energy Regulatory Commission
GLO: General Land Office
GPS: Global Positioning System
HCP: Habitat Conservation Plan
IFIM: Instream Flow Incremental Methodology
MCL: maximum contaminant level
msl: mean sea level
NAS: National Academy of Sciences
NESCA: Nebraska Endangered Species Conservation Act
NGP: Northern Great Plains
NGPC: Nebraska Game and Parks Commission
PCE: Primary Constituent Elements
PHABSIM: Physical Habitat Simulation System
PVA: population viability analysis
RK: river kilometer
RPAs: reasonable and prudent alternatives

SEDVEG: sediment-vegetation model
USBR: U.S. Bureau of Reclamation
USFWS: U.S. Fish and Wildlife Service
USGS: U.S. Geological Survey
WBNP: Wood Buffalo National Park
WUA: weighted usable area

Contents

Boxes, Figures, and Tables

BOXES

FIGURES

TABLES

ENDANGERED AND THREATENED SPECIES OF THE PLATTE RIVER

SUMMARY

The North Platte River and the South Platte River rise in the Rocky Mountains of Colorado and flow through Wyoming and Colorado, respectively, to join in western Nebraska to form the Platte River, which continues eastward to its confluence with the Missouri River. The central Platte River and the lower Platte River are the focus of this report. The central Platte River (as defined in this report) includes the reach from Lexington to Columbus, Nebraska, and the lower Platte River is the segment from Columbus to the confluence with the Missouri River (Figure S-1).

A portion of the Platte River corridor is within the North American Central Flyway and provides habitat for migratory and breeding birds, including three endangered or threatened species: the whooping crane (*Grus americana*), the northern Great Plains population of the piping plover (*Charadrius melodus*), and the interior least tern (*Sterna antillarum athalassos*). Most of the interest related to habitat areas for these listed birds extends from Lexington to Chapman (Figure S-1). The broad, shallow waters of the lower Platte River provide important habitat for the endangered pallid sturgeon (*Scaphirhynchus albus*).

Changing landscape and ecological conditions well beyond the Platte River are responsible for the declines in populations of those four species that resulted in their listings under the Endangered Species Act (ESA) or, in the case of the cranes, prior legislation. The decline in whooping crane populations began many years ago with overhunting and widespread habitat destruction. Whooping cranes, the rarest species of crane in the world,

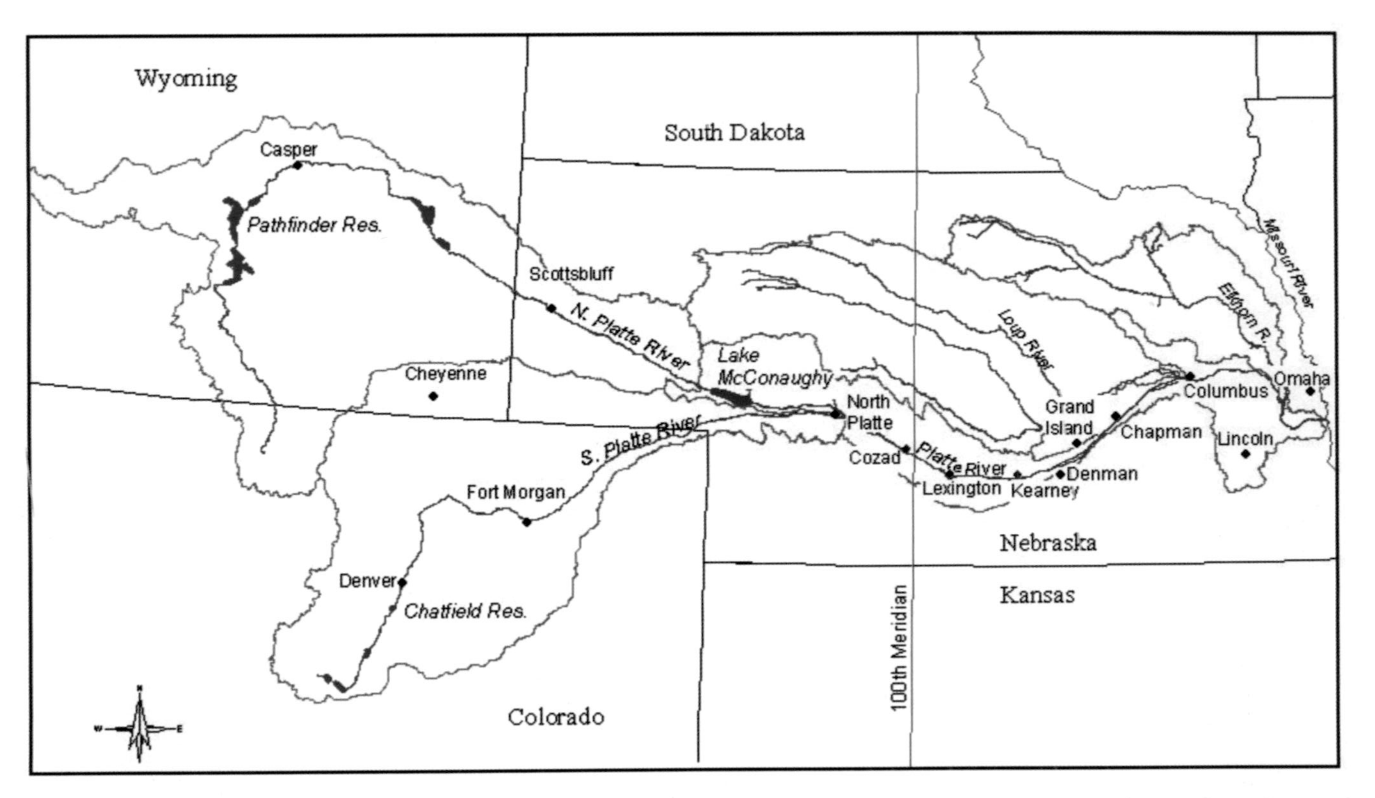

FIGURE S-1 General location and features of Platte River Basin, including its position across 100th meridian. Source: Adapted from DOI 2003.

were federally listed as endangered in 1967 under the Endangered Species Preservation Act. Critical habitat for the whooping crane was designated in 1978. Only about 185 wild birds remain, and another 118 are in captivity.

The northern Great Plains population of the piping plover was federally listed as threatened in 1986. Critical habitat for the piping plover was designated in 2002. The population on the Platte River was estimated in 2001 at about 85 nesting pairs. The number and extent of suitable nesting sites have declined with changes in magnitudes and frequency of river flows, flooding from local runoff, changes in vegetation, and human interference during nesting.

Interior least terns were federally listed as endangered in 1985. Observations of the interior least tern are rare in the central Platte River. The estimated total number of birds in the lower Platte River area is now less than 500. Their population decline results from the loss of open sandy areas in and along rivers, a byproduct of inundation by reservoirs, channelization, large-scale changes in flow regimes, and replacement of open areas with woodlands, sand and gravel mines, housing, and roadways.

The pallid sturgeon was federally listed as endangered in 1990 in the lower Platte River. Populations of pallid sturgeon have declined throughout its range; 500 observations per year in the 1960s declined to about seven per year in the 1980s. Pallid sturgeon seem to prefer warm, turbid waters with annually variable flows and firm, sandy channel bottoms; however, extensive damming has disrupted fish passage and resulted in cooler stream flows, less turbid waters, and inconsistent flow regimes. Commercial harvesting, now illegal, also contributed to the decline of the pallid sturgeon.

The Platte River delivers water, mostly from precipitation in the Rocky Mountains, to an extensive water-control system for irrigated agriculture and urban water in all three states. This system of large dams with storage reservoirs and diversion works with canals provides such benefits as water supply, flood control, electrical power generation, and recreation; it also has substantially altered the river's hydrology and geomorphology. Additional hydrological alterations occur with additions to groundwater through seepage from canals and irrigation and subtractions from wells. The geomorphic and hydrological alterations have caused changes in wildlife habitat and may affect species that depend on particular types of habitat. For example, altered stream flow has resulted in the expansion of woodlands and narrowing of river channels, but the endangered and threatened birds that breed or stop over in the central Platte River appear to prefer sparsely vegetated, open, sandy areas near shallow water.

Protection of federally listed species has been in tension with water management in the Platte River Basin for more than 25 years. Dam construction, new diversions, and federal relicensing of power projects have all been complicated by conflicts with the perceived needs of endangered and threatened species. The conflicts were sharpened by the ongoing litigation among the basin states over division of the waters of the North Platte River, which is not governed by an interstate compact. In 1997, in an effort to find a nonadversarial means of resolving listed-species disputes in the Platte River Basin, the basin states and the federal government entered into a cooperative agreement that established a Governance Committee representing state, federal, environmental, and water-user interests. The committee was charged with developing and implementing a recovery program for the listed species of the basin. Progress toward a recovery program proved slower than the parties had hoped. Meanwhile, implementation of the ESA in the Platte River Basin was increasingly controversial as the U.S. Fish and Wildlife Service (USFWS) issued a series of "jeopardy opinions," finding that any new depletions of the Platte River would have to be compensated by mitigation measures, and a lawsuit forced the designation of "critical habitat" for the northern Great Plains population of the piping plover. Members of the Governance Committee, the interests they represent, and others whose interests would be affected by any recovery program began to question the science supporting current management of the basin's listed species and sought an outside review of the science before the recovery program was made final.

In 2003, the U.S. Department of the Interior (DOI) asked the National Academies to direct its investigative arm, the National Research Council, to evaluate independently the habitat requirements for the whooping crane, piping plover, interior least tern, and pallid sturgeon; to examine the scientific aspects of USFWS's instream-flow recommendations and habitat suitability guidelines; and to assess the scientific support for the connections among the physical systems of the river related to the habitat as explained and modeled by the U.S. Bureau of Reclamation (USBR) (Box S-1). To help focus the National Research Council's task, the Governance Committee offered 10 specific questions related to science and policy for the four threatened and endangered species (Box S-2).

The National Research Council formed the Committee on Endangered and Threatened Species in the Platte River Basin to address the charge described in Boxes S-1 and S-2. The 14-member committee includes biologists specializing in the study of cranes, plovers, terns, and sturgeon; ecologists; engineers specializing in hydraulics, hydrology, and civil-environmental topics; a geomorphologist; a geographer; legal, economic, and water-policy experts; and a farmer.

The committee met three times. Its first two meetings (held in Kearney and Grand Island, Nebraska) were open to the public and included invited presentations from researchers and decision makers and a public-comment session. During those two meetings, the committee participated

BOX S-1
Statement of Task for the National Research Council

A multidisciplinary committee will be established to evaluate the central Platte River habitat needs of the federally listed whooping crane, Northern Great Plains breeding population of the piping plover, interior least tern, and the Lower Platte River habitat needs of the pallid sturgeon. The committee will review the government's assessments of how current Platte River operations and resulting hydrogeomorphological and ecological habitat conditions affect the likelihood of survival of and/or limit the recovery of these species, and whether other Platte River habitats do or can provide the same values that are essential to the survival and/or recovery of these species. The committee will consider the scientific foundations for the current federal designation of central Platte habitat as "critical habitat" for the whooping crane and Northern Great Plains breeding population of the piping plover.

The study will also examine the scientific aspects of (1) the processes and methods used by the U.S. Fish and Wildlife Service in developing its Central Platte River instream-flow recommendations, taking the needs of the listed species into account (i.e., annual pulse flows, and peak flows); (2) characteristics described in the U.S. Fish and Wildlife Service habitat suitability guidelines for the central Platte River; and (3) the U.S. Department of Interior's conclusions about the interrelationships among sediment movement, hydrologic flow, vegetation, and channel morphology in the central Platte River.

in an observational flight over the Platte River from Lake McConaughy to Chapman, Nebraska, and visited the Rowe Sanctuary and Shelton Cottonwood Demography Site. The third meeting (held in Boulder, Colorado) was not open to the public, so the committee could complete its report. Members of the committee visited DOI researchers at their installations in Denver and Grand Island. The committee also reviewed documents describing the methods and procedures used by DOI investigators in reaching their determinations and other written documentation provided by experts and the public.

The focus of the committee's review is the habitat needs of the Platte River endangered and threatened species. The ESA protects *critical habitat*, defined as the specific areas that contain physical or biological features essential to the conservation of the species and that may require special management considerations or protection (ESA § 3(5)). This report uses the term only to refer to areas that have been formally designated under the ESA. Other key terms in the statement of task, defined by the committee for the purpose of this report, are *limit,* which was interpreted by the committee to mean adversely affect or influence; *recovery*, interpreted to mean improvement in the status of listed species to the point at which they would

BOX S-2
Governance Committee's NAS Review Questions
(October 31, 2002)

The Governance Committee offers these questions to focus NAS in their scientific review. Not all members of the GC agree with all of the questions. However, we are unanimous that the NAS not review the Program, but stay focused on the science related to the questions. During the implementation of the review, individual GC members expect that they will have the opportunity to provide the NAS with their views on the specific issues and areas of concern to be reviewed. In reviewing the government's assessments, the committee should consider how the following 10 questions apply to them.

1. Do current Central Platte habitat conditions affect the likelihood of survival of the whooping crane? Do they limit its recovery?
2. Is the current designation of Central Platte River habitat as "critical habitat" for the whooping crane supported by the existing science?
3. Do current Central Platte habitat conditions affect the likelihood of survival of the piping plover? Do they limit its recovery?
4. Is the current designation of Central Platte River habitat as "critical habitat" for the piping plover supported by the existing science?
5. Do current Central Platte habitat conditions affect the likelihood of survival of the interior least tern? Do they limit its recovery?
6. Do current habitat conditions in the Lower Platte (below the mouth of the Elkhorn River) affect the likelihood of survival of the pallid sturgeon? Do they limit its recovery?
7. Were the processes and methodologies used by the USFWS in developing its Central Platte River Instream Flow Recommendations (i.e. species, annual pulse flows, & peak flows) scientifically valid?
8. Are the characteristics described in the USFWS habitat suitability guidelines for the Central Platte River supported by the existing science and are they essential to the survival of the listed avian species? To the recovery of those species? Are there other Platte River habitats that provide the same values that are essential to the survival of the listed avian species and their recovery?
9. Are the conclusions of the Department of the Interior about the interrelationship of sediment, flow, vegetation, and channel morphology in the Central Platte River supported by the existing science?
10. What were the key information and data gaps that the NAS identified during their review?

no longer be designated as endangered or threatened; and *survival,* interpreted to mean the persistence of the listed entity.

This report represents the unanimous consensus of all members of the National Research Council committee. It is limited to the specific charge as agreed on by the Research Council, USFWS, USBR, and the Governance Committee (Box S-1).

To address its charge, the committee considered the extent of the data available for each question and whether the data were generated according to standard scientific methods that included, when feasible, empirical testing. The committee also considered whether those methods were sufficiently documented and whether and to what extent they had been replicated, whether either the data or the methods used had been published and subject to public comment or been formally peer reviewed, whether the data were consistent with accepted understanding of how the systems function, and whether they were explained by a coherent theory or model of the system. To assess the scientific validity of the methods used to develop instream-flow recommendations, the committee applied the criteria listed above, but focused more directly on the methods. For example, the committee considered whether the methods used were in wide use or generally accepted in the relevant field and whether sources of potential error in the methods have been or can be identified and the extent of potential error estimated. The committee acknowledges that none of the above criteria is decisive, but taken together they provide a good sense of the extent to which any conclusion or decision is supported by science. Because some of the decisions in question were made many years ago, the committee felt that it was important to ask whether they were supported by the existing science at the time they were made. For that purpose, the committee asked, in addition to the questions above, whether the decision makers had access to and made use of state-of-the-art knowledge at the time of the decision.

The study committee did not evaluate four items that are closely related to, but not part of, its charge: (1) USBR's draft environmental impact statement, which was completed and released after the committee finished its deliberations on this report, (2) an advanced computer model, SEDVEG, to evaluate the interactions among hydrology, river hydraulics, sediment transport, and vegetation being developed, but not yet completed or tested, by USBR for application on the Platte River, (3) an evaluation of the models and data used by USFWS to set flow recommendations for whooping cranes being developed, but not yet completed, by USGS, and (4) the Central Platte River Recovery Implementation Program proposed in the cooperative agreement by the Governance Committee.

Principal Findings of the Committee

1. Do current central Platte habitat conditions affect the likelihood of survival of the whooping crane? Do they limit (adversely affect) its recovery?

The committee concluded that, given available knowledge, current central Platte habitat conditions adversely affect the likelihood of survival of the whooping crane, but to an unknown degree. The Platte River is important to whooping cranes: about 7% of the total whooping crane population stop on the central Platte River in any one year, and many, if not all, cranes stop over on the central Platte at some point in their lifetimes. Population viability analyses show that if mortality were to increase by only 3%, the general population would likely become unstable. Thus, if the cranes using the Platte River were eliminated, population-wide effects would be likely. Resources acquired by whooping cranes during migratory stopovers contribute substantially to meeting nutrient needs and probably to ensuring survival and reproductive success. Because as much as 80% of crane mortality appears to occur during migration, and because the Platte River is in a central location for the birds' migration, the river takes on considerable importance. The committee concluded that current habitat conditions depend on river management in the central Platte River, but the population also depends on events in other areas along the migratory corridor. If habitat conditions on the central Platte River—that is, the physical circumstances and food resources required by cranes—decline substantially, recovery could be slowed or reversed. The Platte River is a consistent source of relatively well-watered habitat for whooping cranes, with its water source in distant mountain watersheds that are not subject to drought cycles that are as severe as those of the Northern Plains. There are no equally useful habitats for whooping cranes nearby: the Rainwater Basin dries completely about once a decade, and the Sandhills are inconsistent as crane habitat, while the Niobrara and other local streams are subject to the same variability as the surrounding plains. Future climatic changes may exacerbate conflicts between habitat availability and management and human land use. If the quality or quantity of other important habitats becomes less available to whooping cranes, the importance of the central Platte River could increase.

2. Is the current designation of central Platte River habitat as "critical habitat" for the whooping crane supported by the existing science?

An estimated 7% of the wild, migratory whooping crane population now uses the central Platte River on an annual basis and many, if not all, cranes stop over on the central Platte at some point in their lifetimes. The proportion of whooping cranes that use the central Platte River and the amount of time that they use it are increasing (with expected inter-annual variation). The designation of central Platte River migratory stopover habitat as critical to the species is therefore supported because the birds have specific requirements for roosting areas that include open grassy or sandy areas with few trees, separation from predators by water, and proximity to foraging areas such as wetlands or agricultural areas. The Platte River critical habitat area is the only area in Nebraska that satisfies these needs on a consistent basis. However, some habitats designated as critical in 1978 appear to be largely unused by whooping cranes in recent years, and the birds are using adjacent habitats that are not so designated.

Habitat selection (to the extent that it can be measured) on multiple geographic scales strongly suggests that Nebraska provides important habitat for whooping cranes during their spring migration. Riverine, palustrine, and wetland habitats serve as important foraging and roosting sites for whooping cranes that stop over on the central Platte River. Whooping cranes appear to be using parts of the central Platte River that have little woodland and long, open vistas, including such areas outside the zone classified as critical habitat. In some cases the cranes appear to be using areas that have been cleared of riparian woodland, perhaps partly explaining their distribution outside the critical habitat area.

3. Do current central Platte habitat conditions affect the likelihood of survival of the piping plover? Do they limit (adversely affect) its recovery?

Reliable data indicate that the northern Great Plains population of the piping plover declined by 15% from 1991 to 2001. The census population in Nebraska declined by 25% during the same period. Resident piping plovers have been virtually eliminated from natural riverine habitat on the central Platte River. No recruitment (addition of new individuals to the population by reproduction) has occurred there since 1999. The disappearance of the piping plover on the central Platte can be attributed to harassment caused by human activities, increased predation of nests, and losses of suitable habitat due to the encroachment of vegetation on previously unvegetated shorelines and gravel bars.

The committee concluded that current central Platte River habitat conditions adversely affect the likelihood of survival of the piping plover, and, on the basis of available understanding, those conditions have adversely affected the recovery of the piping plover. Changes in habitat along the river—including reductions in open, sandy areas that are not subject to flooding during crucial nesting periods—have been documented through aerial photography since the late 1930s and probably have adversely affected populations of the piping plover. Sandpits and reservoir edges with beaches may, under some circumstances, mitigate the reduction in riverine habitat areas. Because piping plovers are mobile and able to find alternative nesting sites, changes in habitat may not be as severe as they would be otherwise, but no studies have been conducted to support or reject this hypothesis.

4. Is the current designation of central Platte River habitat as "critical habitat" for the piping plover supported by the existing science?

The designation of central Platte habitat as critical habitat for the piping plover is scientifically supportable. Until the last several years, the central Platte supported substantial suitable habitat for the piping plover, including all "primary constituent elements" required for successful reproduction by the species. Accordingly, the central Platte River contributed an average of more than 2 dozen nesting pairs of plovers to the average of more than 100 pairs that nested each year in the Platte River Basin during the 1980s and 1990s. The critical habitat designation for the species explicitly recognizes that not all areas so designated will provide all necessary resources in all years and be continuously suitable for the species. It is also now understood that off-stream sand mines and reservoir beaches are not an adequate substitute for natural riverine habitat.

5. Do current central Platte habitat conditions affect the likelihood of survival of the interior least tern? Do they limit (adversely affect) its recovery?

The committee concluded that current habitat conditions on the central Platte River adversely affect the likelihood of survival of the interior least tern—in much the same fashion as they affect the likelihood of survival of the piping plover—and that on the basis of available information, current habitat conditions on the central Platte River adversely affect the likelihood of recovery of the interior least tern. Reliable population estimates indicate that the total (regional) population of interior least terns was at the recovery goal of 7,000 in 1995, but some breeding areas, including the central Platte River, were not at identified recovery levels. The central Platte subpopulation of least terns declined from 1991 to 2001. The number of terns using the Platte River is about two-thirds of the number needed to reach the interior least tern recovery goal for the Platte. The interior tern is nesting in substantial numbers on the adjacent lower Platte River, but numbers continue to decline on the central Platte, reflecting declining habitat conditions there. The decline in the tern population on the central Platte River has been coincidental with the loss of numerous bare sandbars and beaches along the river. Control of flows and diversion of water from the channel are the causes of these geomorphic changes. Woodland vegetation, unsuitable as tern habitat, has colonized some parts of the central Platte River. Alternative habitats, such as abandoned sand mines or sandy shores of Lake McConaughy, are not suitable substitutes for Platte River habitat because they are susceptible to disturbance by humans and natural predators. The shores of Lake McConaughy are available only at lower stages of the reservoir, and they disappear at high stages.

6. Do current habitat conditions in the lower Platte (below the mouth of the Elkhorn River) affect the likelihood of survival of the pallid sturgeon? Do they limit (adversely affect) its recovery?

Current habitat conditions on the lower Platte River (downstream of the mouth of the Elkhorn River) do not adversely affect the likelihood of survival and recovery of the pallid sturgeon because that reach of the river appears to retain several habitat characteristics apparently preferred by the species: a braided channel of shifting sandbars and islands; a sandy substrate; relatively warm, turbid waters; and a flow regime that is similar to conditions that were found in the upper Missouri River and its tributaries before the installation of large dams on the Missouri. Alterations of discharge patterns or channel features that modify those characteristics might irreparably alter this habitat for pallid sturgeon use. In addition, the lower Platte River is connected with a long undammed reach of the Missouri River, which allows access of the pallid sturgeon in the Platte River to other segments of the existing population. Channelization and damming of the Missouri River have depleted pallid sturgeon habitats throughout its former range, so the lower Platte may be even more important for its survival and recovery. The population of pallid sturgeon is so low in numbers, and habitat such as the lower Platte River that replicates the original undisturbed habitat of the species is so rare that the lower Platte River is pivotal in the management and recovery of the species.

7. Were the processes and methodologies used by the USFWS in developing its central Platte River instream-flow recommendations (i.e., species, annual pulse flows, and peak flows) scientifically valid?

USFWS used methods described in an extensive body of scientific and engineering literature. Reports of interagency working groups that addressed instream-flow recommendations cite more than 80 references that were in wide use and generally accepted in the river science and engineering community. The committee reviewed that information, as well as oral and written testimony critical of the research conducted by DOI agencies, and it concluded that the methods used during the calculations in the early 1990s were the most widely accepted at that time. Revisions were made as improved knowledge became available. Although the Instream Flow Incremental Method (IFIM) and Physical Habitat Simulation System (PHABSIM) were the best available science when DOI agencies reached their recommendations regarding instream flows, there are newer developments and approaches, and they should be internalized in DOI's decision processes for determining instream flows. The new approaches, centered on the river as an ecosystem rather than focused on individual species, are embodied in the concepts of the normative flow regime. Continued credibility of DOI instream-flow recommendations will depend on including the new approach.

The instream-flow recommendations rely on empirical and model-based approaches. Surveyed cross sections along the river provided DOI investigators with specific information on the morphology of the river and vegetation associated with the river's landforms. The portions of the cross sections likely to be inundated by flows of various depths were directly observed. Model calculations to simulate the dynamic interaction of water, geomorphology, and vegetation that formed habitat for species were handled with the prevailing standard software PHABSIM, which has seen wide use in other cases and has been accepted by the scientific community. The software was used by DOI researchers in a specific standard method, IFIM, which permits observations of the results as flow depths are incrementally increased.

The continuing DOI model developments, including the emerging SEDVEG model, are needed because of the braided, complex nature of the Platte River—a configuration that is unlike other streams to which existing models are often applied. The committee did not assess the newer models, because they have not yet been completed or tested, but it recommends that they be explored for their ability to improve decision making.

The committee also recognizes that there has been no substantial testing of the predictions resulting from DOI's previous modeling work,[1] and it recommends that calibration of the models be improved. Monitoring of the effects of recommended flows should be built into a continuing program of adaptive management to help to determine whether the recommendations are valid and to indicate further adjustments to the recommendations based on observations.

8. Are the characteristics described in the USFWS habitat suitability guidelines for the central Platte River supported by the existing science and are they (i.e., the habitat characteristics) essential to the survival of the listed avian species? To the recovery of those species? Are there other Platte River habitats that provide the same values that are essential to the survival of the listed avian species and their recovery?

The committee concluded that the habitat characteristics described in USFWS's habitat suitability guidelines for the central Platte River were supported by the science of the time of the original habitat description during the 1970s and 1980s. New ecological knowledge has since been developed. The new knowledge, largely from information gathered over the last 20 years, has not been systematically applied to the processes of designating or revising critical habitat, and the committee recommends that it be done.

---

[1]The committee did not consider USGS's in-progress evaluation of the models and data used by USFWS to set flow recommendations for whooping cranes.

The committee also concluded that suitable habitat characteristics along the central Platte River are essential to the survival and recovery of the piping plover and the interior least tern. No alternative habitat exists in the central Platte that provides the same values essential to the survival and recovery of piping plovers and least terns. Although both species use artificial habitat (such as shoreline areas of Lake McConaughy and sandpits), the quality and availability of sites are unpredictable from year to year. The committee further concluded that suitable habitat for the whooping crane along the central Platte River is essential for its survival and recovery because such alternatives as the Rainwater Basin and other, smaller rivers are used only intermittently, are not dependable from one year to the next, and appear to be inferior to habitats offered by the central Platte River.

9. Are the conclusions of the U.S. Department of the Interior about the interrelationship of sediment, flow, vegetation, and channel morphology in the central Platte River supported by the existing science?

The committee concluded that DOI conclusions about the interrelationships among sediment, flow, vegetation, and channel morphology in the central Platte River were supported by scientific theory, engineering practice, and data available at the time of those decisions. By the early 1990s, when DOI was reaching its conclusions, the community of geomorphologists concerned with dryland rivers had a general understanding of the role of fluctuating discharges in arranging the land forms of the channel, and DOI included this understanding in its conclusions about the river. In the early 1990s, engineering practice, combined with geomorphology and hydrology, commonly used IFIM and PHABSIM to make predictions and recommendations for flow patterns that shaped channels, and this resulted in adjustments in vegetation and habitat. In fact, despite some criticisms, IFIM and PHABSIM are still widely used in the professional community of river restorationists in 2004. In applying scientific theory and engineering practice, the DOI agencies used the most current data and made additional measurements to bolster the calculations and recommendations. Since the early 1990s, more data have become available, and the USBR has conducted considerable cutting-edge research on a new model (SEDVEG) that should update earlier calculations but is not yet in full operation (and was not reviewed by this committee).

Sediment data are obtained by sampling sediment concentrations and multiplying the concentrations by discharges and duration. For flow, gaging records on the Platte River are 50 years in duration or longer, and they are in greater density than on many American rivers; the gages provide quality data on water discharge for the Platte River. Murphy and Randle (2003) review the analyses and other sources of knowledge about the flows that provide a sound basis for DOI decisions. In addition to the review by Murphy et al. (2001) concerning vegetation, several studies over the last 20 years have provided an explanation of vegetation dynamics that the committee found to be correct and that is the basis of DOI decisions. Early work by USFWS (1981a) and Currier (1982) set the stage for an evolution of understanding of vegetation change on the river that was later expanded by Johnson (1994). For channel morphology, there is a long history of widely respected research to draw on, including early geomorphologic investigations by Williams (1978) and Eschner et al. (1983), continuing with the reviews by Simons and Associates (2000), and culminating in recent work by Murphy and Randle (2003).

10. What were the key information and data gaps that the NAS identified in the review?

The committee reached its conclusions for the preceding nine questions with reasonable confidence on the basis of the scientific evidence available. However, the committee identified the following gaps in key information related to threatened and endangered species on the central and lower Platte River, and it recommends that they be addressed to provide improved scientific support for decision making.

- *A multiple-species perspective is missing from research and management of threatened and endangered species on the central and lower Platte River.* The interactions of the protected species with each other and with unprotected species are poorly known. Efforts to enhance one species may be detrimental to another species, but these connections remain largely unknown because research has been focused on single species. One approach is to shift from the focus on single species to an ecosystem perspective that emphasizes the integration of biotic and abiotic processes supporting a natural assemblage of species and habitats.
- *There is no systemwide, integrated operation plan or data-collection plan for the combined hydrological system in the North Platte, South Platte, and central Platte Rivers that can inform researchers and managers on issues that underlie threatened and endangered species conservation.* Natural and engineered variations in flows in one part of the basin have unknown effects on other parts of the basin, especially with respect to reservoir storage, groundwater storage, and river flows.
- *A lack of a full understanding of the geographic extent of the populations of imperiled species that inhabit the central Platte River and a lack of reliable information on their population sizes and dynamics limit our ability to use demographic models to predict accurately their fates under different land-management and water-use scenarios.* Detailed population

viability analyses using the most recent data would improve understanding of the dynamics of the populations of at-risk bird species and would allow managers to explore a variety of options to learn about the probable outcomes of decisions. Continuation of population monitoring of at-risk species using the best available techniques, including color-banding of prefledged chicks and application of new telemetry techniques, is recommended.

• *There is no larger regional context for the central and lower Platte River in research and management.* Most of the research and decision making regarding threatened and endangered species in the Platte River Basin have restricted analysis to the basin itself, as though species used its habitats in isolation from other habitats outside the basin. There are substantial gaps in integrative scientific understanding of the connections between species that use the habitats of the central and lower Platte River and adjacent habitat areas, such as the Rainwater Basin of southern Nebraska and the Loup, Elkhorn, and Niobrara Rivers and other smaller northern Great Plains rivers.

The committee is confident that the central Platte River and lower Platte River are essential for the survival and recovery of the listed bird species and pallid sturgeon. However, in light of the habitat it provides and the perilously low numbers of the species, there is not enough information to assess the exact degree to which the Platte contributes to their survival and recovery.

• *Water-quality data are not integrated into knowledge about species responses to reservoir and groundwater management and are not integrated into habitat suitability guidelines.* Different waters are not necessarily equal, either from a human or a wildlife perspective, but there is little integration of water-quality data with physical or biological understanding of the habitats along the Platte River.

• *The cost effectiveness of conservation actions related to threatened and endangered species on the central and lower Platte River is not well known.* Neither the cost effectiveness nor the equitable allocation of measures for the benefit of Platte River species has been evaluated. The ESA does not impose or allow the implementing agencies to impose a cost-benefit test. Listed species must be protected no matter what the cost, unless the Endangered Species Committee grants an exemption. Cost effectiveness, however, is another matter. The ESA permits consideration of relative costs and benefits when choosing recovery actions, for example. USFWS has adopted a policy that calls for minimizing the social and economic costs of recovery actions, that is, choosing actions that will provide the greatest benefit to the species at the lowest societal cost (Fed. Regist. 59: 3472 [1994]). In addition, persons asked to make economic sacrifices for the sake of listed species understandably want assurances that their efforts will provide some tangible benefit. In the Platte, the direct economic costs of measures taken for the benefit of species appear reasonably well understood. The biological benefits are another matter. For example, the costs of channel-clearing and other river-restoration measures are readily estimated. Their precise value for cranes is more difficult to estimate, although their general use is fairly well established.

The allocation of conservation costs and responsibility also has not been systematically evaluated. USFWS has concentrated its efforts to protect listed species in the Platte system on federal actions, such as the operation of federal water projects. That focus is understandable. Water projects with a federal nexus account for a large and highly visible proportion of diversions from the system. In addition, those actions may be more readily susceptible to regulatory control than others because they are subject to ESA Section 7 consultation. But some nonfederal actions also affect the species. Water users that depend on irrigation water from the federal projects may well feel that they are being asked to bear an inordinate proportion of the costs of recovering the system. A systematic inventory of all actions contributing to the decline of the species could help the parties to the cooperative agreement channel their recovery efforts efficiently and equitably. The National Research Council committee charged with evaluating ESA actions in the Klamath River Basin recently reached a similar conclusion (NRC 2004a).

• *The effects of prescribed flows on river morphology and riparian vegetation have not been assessed.* Adaptive-management principles require that the outcomes of a management strategy be assessed and monitored and that the strategy be adjusted accordingly, but there has been no reporting of the outcomes of the 2002 prescribed flow, no analysis of vegetation effects of managed flows, no measurement of their geomorphic effects, and no assessment of their economic costs or benefits.

• *The connections between surface water and groundwater are not well accounted for in research or decision making for the central and lower Platte River.* The dynamics of and connections between surface water and groundwater remain poorly known, but they are important for understanding river behavior and economic development that uses the groundwater resource. The effects of groundwater pumping, recently accelerated, are unknown but important for understanding river flows.

• *Some of the basic facts of issues regarding threatened and endangered species in the central and lower Platte River are in dispute because of unequal access to research sites.* Free access to all data sources is a basic tenet of sound science, but DOI agencies and Nebraska corporations managing water and electric power do not enter discussions about threatened and endangered species on the central and lower Platte River with the same datasets for species and physical environmental characteristics. USFWS personnel are not permitted to collect data on some privately owned lands. As a result, there are substantial gaps between data used by DOI and data used by the companies, and resolution is impossible without improved cooperation and equal access to measurement sites.

- *Important environmental factors are not being monitored.* Monitoring, consistent from time to time and place to place, supports good science and good decision making, but monitoring of many aspects of the issues regarding threatened and endangered species on the central and lower Platte River remains haphazard or absent. Important gaps in knowledge result from a lack of adequate monitoring of sediment mobility, the pallid sturgeon population, and movement of listed birds. Responses of channel morphology and vegetation communities to prescribed flows and vegetation removal remain poorly known because the same set of river cross sections is not sampled repeatedly. Groundwater may play an important role in flows, but groundwater pumping is not monitored.
- *Long-term (multidecadal) analysis of climatic influences has not been used to generate a basis for interpretation of short-term change (change over just a few years).* The exact interactions between climate and the system are poorly known because only short-term analyses of climate factors have been accomplished so far. In addition, the relative importance of human and climatic controls remains to be explicitly defined by researchers, even though such knowledge is important in planning river restoration for habitat purposes.
- *Direct human influences are likely to be much more important than climate in determining conditions for the threatened and endangered species of the central and lower Platte River.* Potentially important localized controls on habitat for threatened and endangered species on the central and lower Platte River are likely to be related to urbanization, particularly near freeway exits and small cities and towns where housing is replacing other land uses more useful to the species. Off-road vehicle use threatens the nesting sites of piping plovers and interior least terns in many of the sandy reaches of the river. Sandy beaches and bars are inviting to both birds and recreationists. Illegal harvesting has unknown effects on the small remaining population of pallid sturgeon. In each of those cases, additional data are required to define the threats to the listed species.

Successful conservation in the Platte River Basin must begin with water management. The committee found that sufficient scientific knowledge and understanding exist and have been used to make informed decisions about the management of water resources, the Platte River, and the threatened and endangered species that use the river as habitat. Regarding the critical understanding and modeling that DOI has used to explain the connections among stream flow, sediment movement, vegetation, and habitats, the committee found that valid science was used when recommendations were made in the past but that future decisions must rely on the use of newer methods and perspectives, particularly the concept of normative flow regimes. The quality of the information upon which decisions are based could be further improved by publishing research findings in peer-reviewed journals or in externally reviewed synthesis volumes to increase accessibility and decrease the reliance on non-peer-reviewed literature. The committee found numerous gaps in knowledge. Addressing them could substantially improve science and management for the river, its human population, and its threatened and endangered species. Those gaps are mostly related to problems of integration of the various lines of scientific investigation, a focus on highly localized rather than more broadly based ecosystem perspectives, a lack of analysis of basinwide connections, a lack of standardized procedures for data collection among government and private agencies, and lack of understanding of the relative cost effectiveness and distributional consequences of alternative conservation measures.

1

INTRODUCTION

Experience teaches that it is not wise to depend upon rainfall where the amount is less than 20 inches annually. The isohyetal or mean rainfall line of 20 inches...in a general way...it may be represented by the one hundredth meridian. [In this region] agriculturalists will early resort to irrigation.

—John Wesley Powell, 1878

The 100th meridian is a defining geographic feature in the United States that marks important transitions for land, water, and life. In the Great Plains, this north-south line (immediately east of Cozad, Nebraska) transects the corridor of the central Platte River Valley of Nebraska, the subject of this report. The river is emblematic of the human and ecological complexities of policy, science, and management for a dryland river (Figure 1-1). The river was named the Platte River, French for "flat river," on June 2, 1739, by two French explorers who were searching for a route from Illinois to Santa Fe, New Mexico (Sheldon 1913).

The Platte River's two great branches, the North Platte and the South Platte, rise in the Rocky Mountains in the West and join in western Nebraska to the west of the 100th meridian (Figure 1-2). The river flows 310 mi (about 500 km) through Nebraska across the meridian to a confluence with the Missouri River. The Platte River delivers the runoff of its 86,000-mi^2 (223,000-km^2) drainage area, largely the result of precipitation in the high western mountains, to an extensive system of water control for the highly productive agricultural area in the plains. The river is more than a conduit for water, however; along its course, it creates an aquatic and riparian ecosystem that provides wildlife habitat unlike habitats found outside the river valley. The river corridor, within the Central Flyway of North America, provides habitat for migratory and breeding birds, including three endangered or threatened species: the whooping crane (*Grus americana*), the piping plover (*Charadrius melodus*), and the interior least tern (*Sterna*

FIGURE 1-1 South channel of central Platte River at Rowe Sanctuary near Kearney. Channel is periodically dry, as in this view. Riparian vegetation includes cottonwood-dominated forests and more open areas. Source: Photograph by W.L. Graf, August 2003.

antillarum athalassos). The broad shallow waters of the Platte near its confluence with the Missouri River constitute an important habitat for at least one endangered species of fish: the pallid sturgeon (*Scaphirhynchus albus*).

The infrastructure investments that made irrigated agriculture and urban water supply possible in the Platte River Basin have substantially altered the hydrologic regime of the Platte River. During the twentieth century in the Platte River Basin, construction of storage reservoirs and diversion dams and installation of wells to tap groundwater supported the economic vitality of the region. The structures continue to provide flood reduction, water supply, hydroelectricity, and recreational benefits. By controlling and diverting water flows, however, the dams altered the stream flows, and that caused widespread environmental changes. A major habitat change involved the expansion of woodland and the narrowing of river channels. Whooping cranes, piping plovers, and interior least terns, whose populations were already declining because of other factors, prefer more sparsely vegetated, open, sandy areas near shallow water.

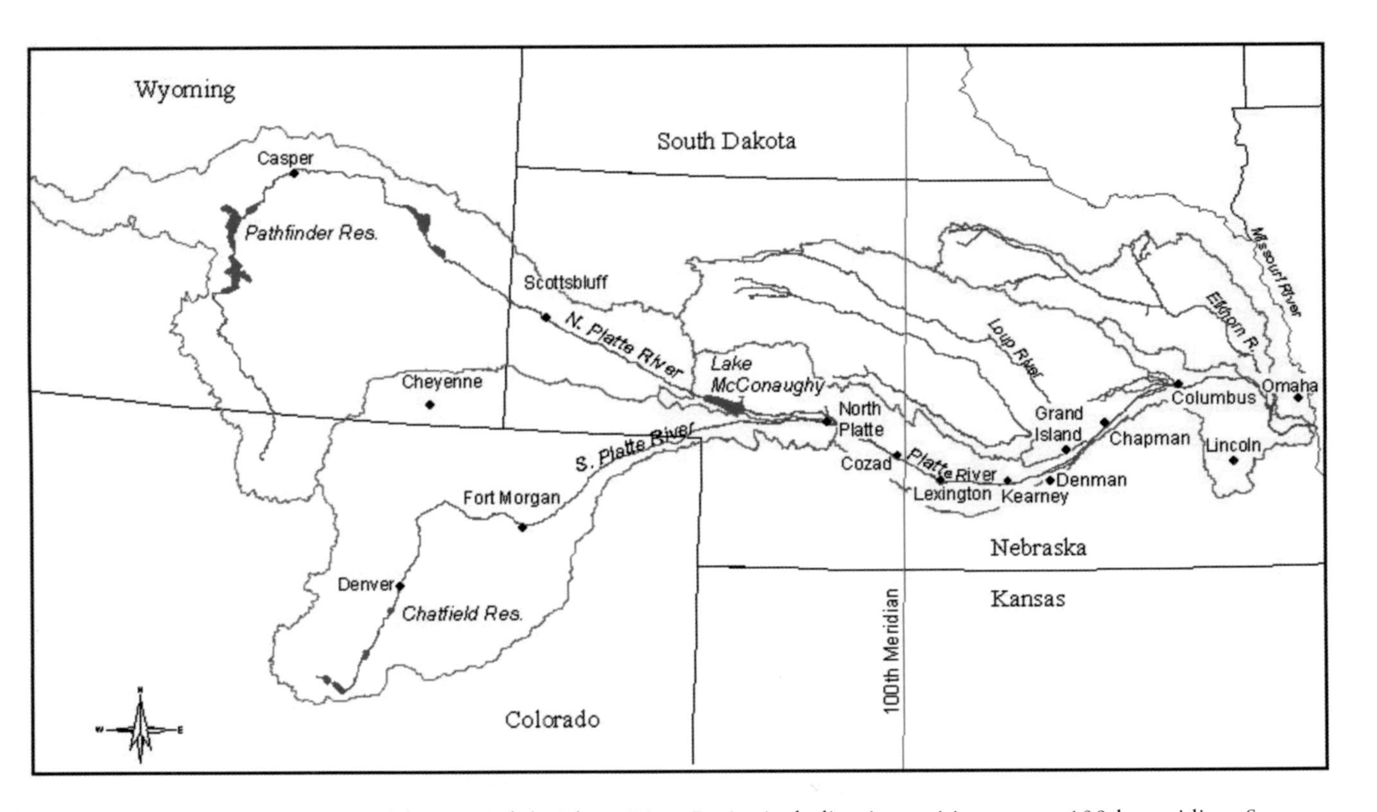

FIGURE 1-2 General location and features of the Platte River Basin, including its position across 100th meridian. Source: Adapted from DOI 2003.

Between 1967 and 1990, the whooping crane, piping plover, interior least tern, and pallid sturgeon were listed under federal legislation related to threatened and endangered species. Rather than address the individual species in isolation from each other, Nebraska, Colorado, Wyoming, and agencies of the U.S. Department of the Interior (DOI) signed the Platte River Cooperative Agreement in July 1997 to provide Endangered Species Act (ESA) compliance for all four species simultaneously. The signatories created the Platte River Endangered Species Partnership—which includes water users, environmental groups, and others—to implement the cooperative agreement. The agreement seeks to develop a recovery implementation program to improve and conserve habitat for the four listed species and seeks to enable existing and new water uses in the basin to proceed without additional regulatory actions related to the species. In 2003, DOI, with input from the Governance Committee of the state-federal partnership, asked the National Research Council to evaluate the scientific validity of the instream flow recommendations, habitat requirements for the species, and connections among the physical systems of the river related to the habitat as specified by the U.S. Fish and Wildlife Service (USFWS). The present volume is the final report of the committee.

This chapter provides a brief overview of the Platte River Basin, the primary management issues that are related to the river and the endangered and threatened species, and the governing policies that define potential approaches to these issues. It also explains the specific charge given to the Research Council's Committee on Endangered and Threatened Species in the Platte River Basin.

Endangered and Threatened Species in the Central and Lower Platte River

Among the consequences of extensive economic development of the river resources and the associated changes in hydrology, geomorphology, and riparian vegetation were the changes brought about in wildlife populations. The physical changes in the river are reflected in habitat changes. In the central Platte River, more than 40 mi^2 (104 km^2) of river channel has been altered. Declines in the populations of piping plovers and interior least terns have been particularly notable. The whooping crane declined to a low of 15 birds in the late 1800s and early 1900s and is now rebounding. Nationally, the pallid sturgeon also has become very rare. All three bird species prefer open, sandy areas near shallow water. Such areas are precisely the type of habitat that has shrunk in response to the hydrological, geomorphic, and vegetation changes. Continent-wide impacts caused the four species to become listed. Whooping crane populations were substantially decreased by overhunting and habitat degradation. Piping plovers and interior least tern populations were greatly reduced by dam and reservoir construction and human interference during nesting. Pallid sturgeon populations were adversely affected by large dams and channel works built on the mainstem Missouri River. Ecological changes in the central Platte River also have had habitat implications for other species, such as neotropical migrant birds that have been favored by expansion of riparian woodlands.

Whooping cranes, federally listed in 1967 under the Endangered Species Preservation Act, are the rarest species of crane in the world (Figure 1-3). Although exact numbers are not known, historical accounts indicate that the central Platte River was a stopping area for whooping cranes, which prefer its long vistas, shallow waters, and food sources in adjacent meadows and grasslands; sightings have been documented since 1820. Piping plovers, federally listed as threatened in 1986 under the ESA, are small migratory shorebirds that breed in three regions of North America: the Atlantic Coast, the Great Lakes, and the northern Great Plains (Figure 1-4). About 1% of the northern Great Plains breeding population uses the central Platte River as an area for nesting sites—almost exclusively the bare sandy areas associated with the active channel and in some cases alternative sites with open sandy areas, such as sand and gravel mines. In 2001, the population was estimated at about 85 nesting pairs on the Platte River (J. Dinan, Nebraska Game and Parks Commission, pers. comm., 2003). Interior least terns, federally listed as endangered under the ESA in 1985, are also small migratory birds that use open sandy islands, bars, and beaches

FIGURE 1-3 Whooping crane. Source: Photograph by George Archibald, International Crane Foundation, 1997. Reprinted with permission; copyright 1997, International Crane Foundation.

FIGURE 1-4 Piping plover. Source: USFWS 2004a.

on inland rivers for breeding sites (Figure 1-5). By the 1980s, the range for the interior least tern in the central Platte River had shrunk to include only a portion of the Platte River Valley between Kearney and Grand Island, and the bird was a rare migrant and an infrequent nester. The estimated total number of birds in the lower Platte River area is now lower than 500 (Kirsch and Sidle 1999). The pallid sturgeon was federally listed in 1990 under the ESA as endangered on the lower Platte River (Figure 1-6). The

FIGURE 1-5 Interior least tern. Source: USFWS 2004b.

FIGURE 1-6 Pallid sturgeon. Source: Photograph by Jason Olnes, University of Nebraska, May 2, 2001.

fish favors warm, turbid waters with annually variable flows and firm sandy channel bottoms with dunes and pockets, where it feeds on small fish and aquatic insects. Its population has declined throughout its range over recent decades; 500 observations per year in the 1960s declined to about seven per year in the 1980s (Fed. Regist. 55 (173): 36641 [1990]).

Policy Responses to the Species Issues

Nationwide declines in the populations of whooping cranes, interior least terns, piping plovers, and pallid sturgeon led to the listing of each under the provisions of the ESA or, in the case of the cranes, a prior act. By 1990, all four species had been listed (Table 1-1), and there was a continuing series of consultations and opinions by USFWS related to projects in the Platte River Basin. Water withdrawals were especially at issue because of the importance of water and flow regimes to the habitat of the species. Since 1978, the history of public decision making for water projects along the Platte River has been littered with lawsuits, negotiations, contentious debates over jeopardy opinions, and occasional agreements that allowed some projects to move forward with mitigation strategies. An example of the complications is the relicensing of hydroelectric projects by the Federal Energy Regulatory Commission. The applications of the Central Nebraska

TABLE 1-1 Dates of Federal Listings Under Endangered Species Act for Threatened and Endangered Species in Central Platte River

Species	Status	Date Listed	Date Critical Habitat Designated	Date of Latest Recovery Plan
Whooping crane	Endangered	Mar. 11, 1967	Aug. 17, 1978	Feb. 11, 1994
Piping plover (northern Great Plains breeding population)	Threatened	Jan. 10, 1986	Sep. 11, 2002	May 12, 1988
Interior least tern	Endangered	May 28, 1985	None	Sep. 9, 1990
Pallid sturgeon	Endangered	Sep. 6, 1990	None	Nov. 7, 1993

Sources: EA Engineering Science 1985, 1988; Lutey 2002.

Public Power and Irrigation District and the Nebraska Public Power District required 15 years for the approval of their licenses for their water and power operations. The USFWS recommendations for instream flows were also problematic for water managers. The instream flows, peak discharges, and pulses that USFWS determined to be essential to maintaining habitat for at-risk species restricted withdrawals and limited the use of river flow for other purposes.

Many of the data and nearly all the explanations of causal connections among the physical aspects of the Platte River, the ecology of habitats, and the biology of the four listed species have been questioned by commentators outside USFWS. In particular, the data related to whooping cranes, their use of the riverine habitats, and the importance of the Platte River in their ecology have attracted critical comment (G. Lingle, University of Nebraska, unpublished material, March 22, 2000). M.M. Czaplewski, J.J. Shadle, J.J. Jenniges, and M.M. Peyton (unpublished material, June 12, 2003) also have raised important questions about the science supporting the designation of the species as threatened or endangered and the science used to define instream flow requirements. The National Research Council committee has explored those and other criticisms of USFWS's scientific work and has considered them with agency replies.

The complex nature of water-resource development and endangered-species management in the Platte River Basin became an increasingly difficult issue beginning in 1994 when six water-related projects in the Colorado Front Range became the subject of negotiations between project proponents and USFWS. The settlement of the negotiations included permission for the projects to continue and support for a recovery effort that encompassed areas downstream along the Platte River. During the 1990s, recovery efforts included work by the Whooping Crane Trust and the National Audubon Society to modify habitat in the river corridor. In recognition of the complicated policy issues involving Wyoming, Colorado, Nebraska, and the federal government (represented by DOI agencies, such as USFWS and the U.S. Bureau of Reclamation [USBR]), a new cooperative agreement was created among these entities. On July 1, 1997, the governors of the three states and the secretary of the interior signed a cooperative agreement defining a Central Platte River Recovery Implementation Program (DOI 1997). The agreement established a 10-member Governance Committee whose members represented the full array of participants and were charged with designing and implementing the recovery program. The success of the recovery program depends on a sound understanding of the relationships among water flows, channels, control structures, vegetation, and wildlife and on decisions by the Governance Committee that are founded on the scientific explanations of those relationships. Discussions among the Governance Committee members and the various publics they represent revealed important questions about the scientific underpinnings of recovery plans for the species in question and of the explanations of environmental changes along the river that have affected wildlife habitat.

Committee Charge and Response

In January 2003, DOI, with the concurrence of and input from the Governance Committee, asked the National Academies to direct its investigative arm, the National Research Council, to assess the science underlying many of the decisions reached by agencies of the federal government related to endangered and threatened species in the Platte River Basin. Specifically, the Governance Committee asked for an evaluation of the habitat requirements of the listed species of the central and lower Platte River and for an evaluation of the scientific basis of USFWS's instream-flow recommendations and habitat-suitability guidelines and of DOI's conclusions about the relationships among sediment movement, hydrologic flow, vegetation, and channel morphology. See Boxes 1-1 and 1-2 for details of the charge to the committee and Box 1-3 for the committee's definition of terms included in the statement of task. In early 2003, the Research Council formed the

BOX 1-1
Statement of Task

A multidisciplinary committee will be established to evaluate habitat needs of four federally listed species: the whooping crane, Northern Great Plains breeding population of the piping plover, interior least tern of the central Platte River, and pallid sturgeon of the lower Platte River (below the mouth of the Elkhorn river). The committee will review the government's assessments of how current Platte River operations and resulting hydrogeomorphologic and ecological habitat conditions affect the likelihood of survival of and/or limit the recovery of these species, and whether other Platte River habitats do or can provide the same values that are essential to the survival and/or recovery of these species. The committee will consider the scientific foundations for the current federal designations of central Platte habitat as "critical habitat" for the whooping crane and Northern Great Plains breeding population of the piping plover.

The study will also examine the scientific aspects of (1) the processes and methods used by the U.S. Fish and Wildlife Service in developing its central Platte River instream flow recommendations, taking the needs of the listed species into account (i.e., annual pulse flows and peak flows); (2) characteristics described in the U.S. Fish and Wildlife Service habitat suitability guidelines for the central Platte River; and (3) the Department of Interior's conclusions about the interrelationships among sediment movement, hydrologic flow, vegetation, and channel morphology in the central Platte River. This plan of action is based on 10 specific questions that were offered to the National Research Council by the Governance committee to be addressed by the proposed study.

BOX 1-2
Governance Committee's NAS Review Questions
(October 31, 2002)

The Governance Committee offers these questions to focus the NAS in their scientific review. Not all members of the GC agree with all of the questions. However, we are unanimous that the NAS not review the Program, but stay focused on the science related to the questions. During the implementation of the review, individual GC members expect that they will have the opportunity to provide the NAS with their views on the specific issues and areas of concern to be reviewed. In reviewing the government's assessments, the committee should consider how the following 10 questions apply to them.

1. Do current central Platte habitat conditions affect the likelihood of survival of the whooping crane? Do they limit its recovery?
2. Is the current designation of central Platte River habitat as "critical habitat" for the whooping crane supported by existing science?
3. Do current central Platte habitat conditions affect the likelihood of survival of the piping plover? Do they limit its recovery?
4. Is the current designation of central Platte River habitat as "critical habitat" for the piping plover supported by existing science?
5. Do current central Platte habitat conditions affect the likelihood of survival of the interior least tern? Do they limit its recovery?
6. Do current habitat conditions in the lower Platte (below the mouth of the Elkhorn River) affect the likelihood of survival of the pallid sturgeon? Do they limit its recovery?
7. Were the processes and methodologies used by the USFWS in developing its central Platte River Instream Flow Recommendations (i.e., species, annual pulse flows, and peak flows) scientifically valid?
8. Are the characteristics described in the USFWS habitat suitability guidelines for the central Platte River supported by the existing science and are they essential to the survival of the listed avian species? To the recovery of those species? Are there other Platte River habitats that provide the same values that are essential to the survival of the listed avian species and their recovery?
9. Are the conclusions of the Department of Interior about the interrelationships of sediment, flow, vegetation, and channel morphology in the central Platte River supported by the existing science?
10. What were the key information and data gaps that the NAS identified in the review?

Committee on Endangered and Threatened Species in the Platte River Basin, a panel of 13 members (later expanded to 14) that included a specialist for each listed species in the charge; two ecologists; engineers specializing in hydraulics, hydrology, and civil-environmental topics; a geomorphologist; a geographer; legal, economic, and water-policy experts; and a farmer (see Appendix A for details).

BOX 1-3
Definitions of Terms Used in This Report

CRITICAL HABITAT: Defined in the Endangered Species Act to mean the specific areas within the range occupied by the species at the time of listing on which are found physical or biologic features essential to the conservation of the species that may require special management considerations or protection, and areas outside the range occupied at the time of listing if the secretary of the interior determines that the areas are essential to the conservation of the species (16 U.S.C. 1532(5)).
LIMIT: Adversely affect or influence.
RECOVERY: Not defined in the Endangered Species Act. Defined in regulations issued by USFWS to mean improvement in the status of listed species to the point at which they would no longer qualify as endangered or threatened.
SURVIVAL: The committee understands this to refer to the listed entity as a whole. As used in this report, means the persistence of the listed entity.
JEOPARDY: According to USFWS regulations, an action "jeopardizes the continued existence of" a listed species if it would be expected, directly or indirectly, to reduce appreciably the likelihood of survival and recovery of the species in the wild by reducing its reproduction, numbers, or distribution (50 C.F.R. 402.02).
DESTRUCTION OR ADVERSE MODIFICATION of critical habitat: Defined by USFWS regulations to mean direct or indirect alteration that appreciably diminishes the value of the critical habitat for both the survival and recovery of the species; not limited to alteration of the physical and biologic features that were the basis for designating the habitat as critical.
BEST PROFESSIONAL JUDGMENT: Not used in the Endangered Species Act or its implementing regulations. Used in some other laws, notably the Clean Water Act, which requires the use of best professional judgment to set limits in discharge permits when no general standards have been set.
SOUND PROFESSIONAL JUDGMENT: A finding, determination, or decision that is consistent with principles of sound fish and wildlife management and administration, available science and resources (National Wildlife Refuge System Improvement Act of 1997 [Public Law 105-57-Oct 9, 1997: 111 STAT]).
CURRENT: Refers to a few years (less than 10), broadly equivalent to the present.
EXISTING SCIENCE: The widely held understanding of natural and human processes at the time of the decision for which the science was used. For example, if a decision regarding a particular species was made in 1990, existing science refers to the available understanding in 1990.
INFORMATION AND DATA GAPS: Data (raw, unprocessed measurements) or information (data with added value of interpretation or explanation) unavailable at the time of preparation of this report (2003) but that could inform managers and thus lead to improved decisions.
SCIENTIFIC VALIDITY: The property of management decisions that are based on widely accepted scientific literature or that have the support of specially commissioned research that adheres to the principles of the scientific method. Actions or decisions are scientifically valid if they take into account, following generally accepted scientific practices, the available scientific evidence, published or unpublished, specifically commissioned or not. Where the evidence is thin, judgments can be scientifically valid even if they are only the opinions of the most knowledgeable experts in the field.

The committee held two public meetings in Nebraska—one in Kearney and the other in Grand Island—to collect information, meet with researchers and decision makers, and accept testimony from the public. The meetings included an extensive flight over the Platte River from Lake McConaughy to Chapman and visits to the Rowe Sanctuary and Shelton Cottonwood Demography Site. The committee met a third time, in executive session, in Boulder, Colorado, to complete its report. Subcommittees visited DOI researchers at their installations in Denver and Grand Island. The committee reviewed documents describing the methods and procedures used by DOI investigators in reaching their determinations.

The committee evaluated the literature relied on by the DOI agencies in formulating their decisions on threatened and endangered species and their understanding of the connections among water, sediment, river morphology, and habitat. The committee found that much of the literature pertaining to the Platte River for these applications was not in the refereed journal literature, the most common highly reliable sources. The refereed literature is significant because it represents the results of rigorous review and examination before publication, and it is the intellectual currency in the environmental sciences. Much of the literature of importance in dealing with threatened and endangered species along with the dynamics of the Platte River is in the "gray literature," publications that are products of governmental agencies and private foundations. Although these publications are sometimes reviewed, they are not as widely available nor are they as highly regarded in the scientific community as refereed journal articles. Nonetheless, the gray literature is an important conduit of scientific information, and in the absence of extensive journal literature it represents the only viable substitute. When judging whether or not DOI agencies relied on the best available science, the committee assessed the availability of background literature and determined whether or not DOI agencies relied on all the available sources, regardless of the type. In many instances, the committee found that at the time of their decisions, DOI agencies were forced to rely on in-house investigations, agency reports, and conference or working-group reports. The committee assessed these sources in reaching its conclusions. The committee did not evaluate four items that are closely related to but not part of its charge: (1) USBR's draft Environmental Impact Statement for the Platte River Recovery Program, which was completed and released when the committee had finished its deliberations and was close to finalizing its report; (2) an advanced computer model, SEDVEG, to evaluate the interactions among hydrology, river hydraulics, sediment transport, and vegetation for the Platte River—the model is being developed by USFWS but has not yet been completed or tested; (3) an evaluation of the models and data used by USFWS to set flow recommendations for whooping cranes that is being developed, but has not yet been completed, by USGS; and

(4) the Central Platte River Recovery Implementation Program proposed in the cooperative agreement by the Governance Committee, which, as stated in Box 1-2, was excluded from the committee's charge.

The present report is the product of the efforts of the entire Research Council committee and underwent extensive, independent external review overseen by the Research Council's Report Review Committee. It specifically addresses the statement of task as agreed on by the Research Council, USFWS, USBR, and the Governance Committee. The report is limited to that formal statement and to addressing 10 questions that the Governance Committee provided as guidance.

The remaining chapters of this report constitute the findings of the committee on Endangered and Threatened Species in the Platte River Basin. Chapter 2 provides a descriptive overview of the Platte River Basin and the historical background of the issues, including reviews of human, hydrological, and ecological history. Chapter 3 provides the foundation for understanding policy and science issues connected with the ESA. Chapter 4 evaluates the science behind decisions related to the physical systems of the Platte River. Chapters 5, 6, and 7 evaluate the validity of the biological science of the whooping crane, the piping plover and interior least tern, and the pallid sturgeon, respectively. Chapter 8 summarizes the committee's conclusions and recommendations and presents succinct answers to the specific questions outlined by the Governance Committee.

In the 1870s, John Wesley Powell correctly predicted the high degree of variability of water resource supplies along and west of the 100th meridian, but he could not foresee the changes that have occurred in general climate, land use, and land cover and in the fundamental physical and biological characteristics of the Platte River in the vicinity of the meridian. Powell's emphasis on change applies to all the components of the ecological systems in the central and lower Platte, but it is left to modern researchers to understand how those changes in water, land, and life interact with and influence each other. The primary underlying question that is a thread throughout the following pages is this: Are resource managers using the best available valid scientific information to support their decisions?

2

Regional Context for Water and Species

The validity of science for decision-making related to threatened and endangered species of the Platte River Basin depends on a fundamental understanding of the resources of the region, the nature of science, and the policies developed to address species concerns. Any consideration of environmental concerns must begin with the recognition that various natural processes and human activities interact to create complex conditions in the environment. The landscape and resources of the central Platte River Basin are no exception, and connections among land, water, wildlife, and human activities have shaped the present conditions. Understanding the central Platte ecosystem requires knowledge of habitat needs of the species and of the complexity of their interactions with natural and human controls. This chapter provides an overview of the central Platte region—its geography; human history; hydrological, geomorphic, and vegetation changes; and population trends in key groups of organisms. The overview in this chapter identifies several key issues that are threads through later chapters: interactions among components of the hydrological cycle, including surface water and groundwater; river behavior, including floods, droughts, and annual flow patterns; climatic and human influences on environmental change; and complex connections between the physical environment and vegetation and wildlife.

GEOGRAPHY

The Platte River Basin has three distinct geographic sub-basins: the North Platte, the South Platte, and the Platte River Valley of Nebraska. The North Platte River rises in the Medicine Bow Mountains of northern Colorado and flows northward and then eastward through Wyoming; it continues eastward into western Nebraska near Scottsbluff. The South Platte River originates in the Front Range of the Rocky Mountains around the valley of South Park, Colorado, and flows northeastward past Denver; it leaves the state at the northeast corner and joins the North Platte in west central Nebraska. The combined streams form the Platte River, which flows generally eastward through Nebraska; its valley has a distinctive southward bend, sometimes referred to locally as the "Big Bend." The Platte River joins the Missouri River south of Omaha at Plattsmouth and is its largest tributary (Figure 2-1).

For the purposes of this report, the central Platte River is the segment of the river extending from Lexington, Nebraska, downstream (eastward) to Columbus, Nebraska, although most of the interest related to habitat areas for the listed birds extends only to Denman, Nebraska (Figure 2-1). Also for the purposes of this report, the lower Platte River is the segment of the river from Columbus to the confluence with the Missouri River (Figure 2-1). The points along the Platte River that define those segments are related to

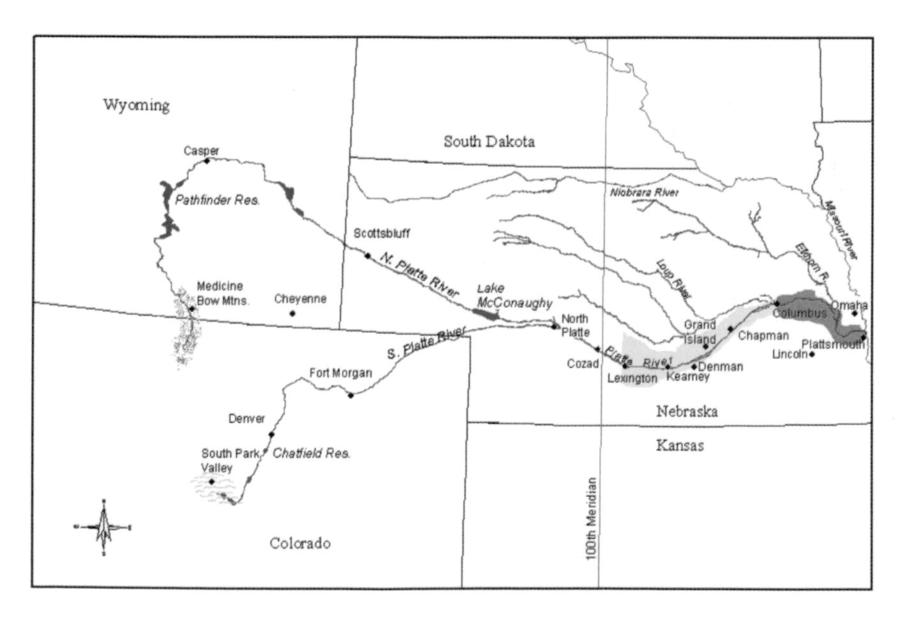

FIGURE 2-1 General location and places of interest in Platte River Basin, including its position across 100th meridian. Central Platte designated in yellow, lower Platte in Orange. Source: Adapted from DOI 2003.

the modified hydrology and geomorphology of the river. The upstream boundary of the central Platte River is the Johnson-2 (also known as J-2) return (the point at which water may either be returned to the south channel of the Platte River or be diverted into the Phelps County Canal). Upstream of that point, the channel is substantially dewatered by diversions. Downstream of that point, Johnson-2 return flows, local storm runoff, and returns from irrigated fields provide continuous flows even during nonflood periods. The division between the central and lower Platte River at Columbus corresponds to the entry of the Loup River.

The western mountains of the Platte River Basin have peaks above 14,000 ft (4,200 m), and the precipitation they intercept creates runoff, which, with groundwater base flow, provides most of the water that flows through the Platte River. The Great Plains form a ramp that slopes gently eastward from the base of the mountains, descending from almost 6,000 ft (1,800 m^2) on the Colorado Piedmont in the West to less than 1,000 ft (300 m) at the Missouri River. This plains region—characterized by sand and loess (fine silt) sheets, sand hills, and planar surfaces—is slightly dissected by the regional rivers (Figure 2-2). The Platte River Valley, bounded by bluffs of unconsolidated sands and silts, includes a flat floor and the river, a wide sandy ribbon of intersecting channels, and numerous vegetated islands (Figure 2-3).

HUMAN USE OF PLATTE RIVER BASIN

European perceptions of the geographic area encompassing the Platte River have always focused on its water. Until the twentieth century, the region east of the Rockies was described as high plains and prairie, with the division somewhere near the 96th to 98th meridian. However, much of the national perception of the region was based on the Great American Desert label derived from the reports of the expeditions of Zebulon Pike in 1806-1807, Stephen H. Long in 1819-1820, and others. Long's 1823 map labeled the central plains as the Great American Desert (Long and James 1823), helping to perpetuate a perception that had already become firmly established (Webb 1931).

Pre-European occupants of the western part of the region were nomadic plains tribes, including the Arapaho, Cheyenne, and Dakotah Sioux (Figure 2-4). These were hunting cultures that depended particularly on bison. In eastern and central parts of the region, tribes were more settled. The Pawnee, for example, were farmers; they occupied lodges for about 10 years at a time (Holmgren et al. 1993). They occupied the central Platte in Nebraska and other areas, including the Loup River, cultivating crops, gathering wild fruits and vegetables, and hunting bison and other wildlife. Early Euro-American reports indicated that Indians used cottonwood bark and leaves as food for horses and used wood for fuel, poles, and stakes.

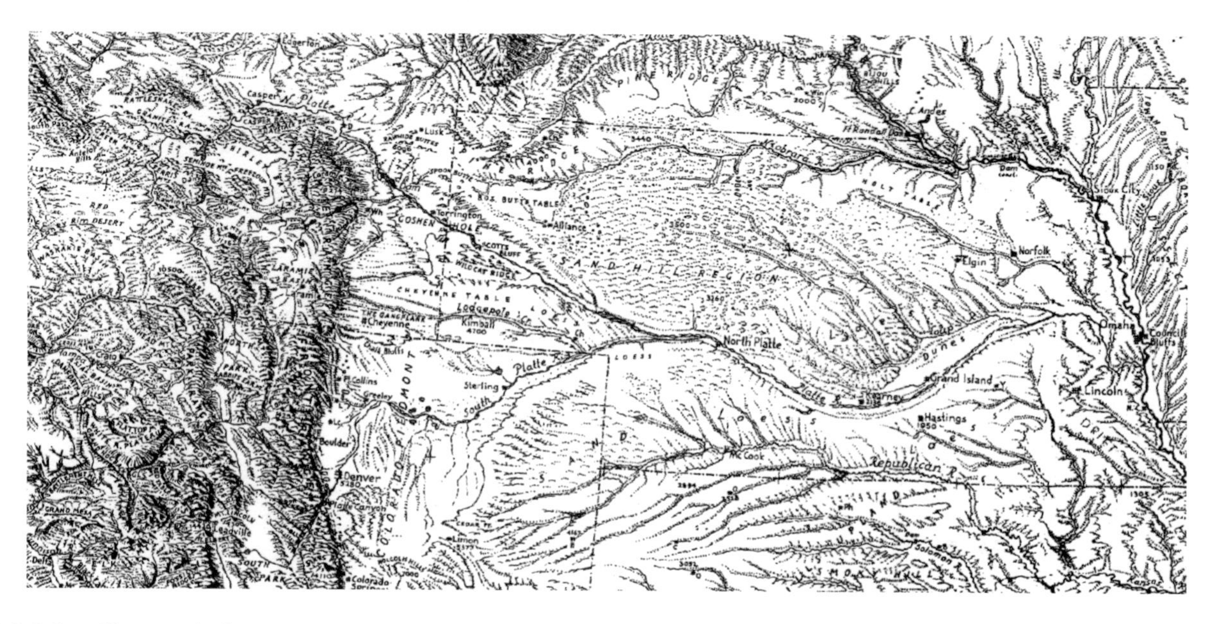

FIGURE 2-2 Landforms of Platte River Basin and nearby regions. Source: Raisz and Atwood 1957. Reprinted with permission; copyright 1957, E. Raisz, Cambridge, MA.

FIGURE 2-3 Channel of Platte River showing its broad, shallow, braided nature. Source: Photograph by W.L. Graf, May 2003.

Euro-Americans followed the Platte River westward, and much of the route remained in use as the Mormon Trail (on the north side of the Platte) and the Oregon Trail (on the south side) from the 1840s to the late 1860s (Brown and Whitaker 1948; Holmgren et al. 1993). The width of the river, its shifting channels, and its soft bed meant that there were few places where pioneers could easily cross. As westward movement of settlers proceeded, so did conflicts with aboriginal Americans. To protect pioneers and control Indian activities, a number of military encampments and forts were established along the Platte and its tributaries, including Fort Kearny, Fort Laramie, and Fort Collins. By 1855, there were violent conflicts between Plains Indians and Euro-Americans, which continued through the Civil War. After the Civil War, General Philip H. Sheridan ordered the destruction of their subsistence resources with the statement "Kill the buffalo and you kill the Indians." The federal government had removed the tribes from the Platte region by the end of the 1870s. Trapping of beaver and intensive hunting of bison for both strategic purposes and sport characterized the beginning of substantial Euro-American-induced changes in the fauna of the Platte River Basin.

Even before gaining complete control over the Platte River Basin, Euro-Americans were gaining an understanding of the region's environmental conditions and increasing the use of its resources. The earliest settlers followed the trails to areas in California and Oregon. Perceptions of the region changed as railroads made the high plains and prairies more accessible and more attractive for settlement. John C. Frémont's 1842 expedition along

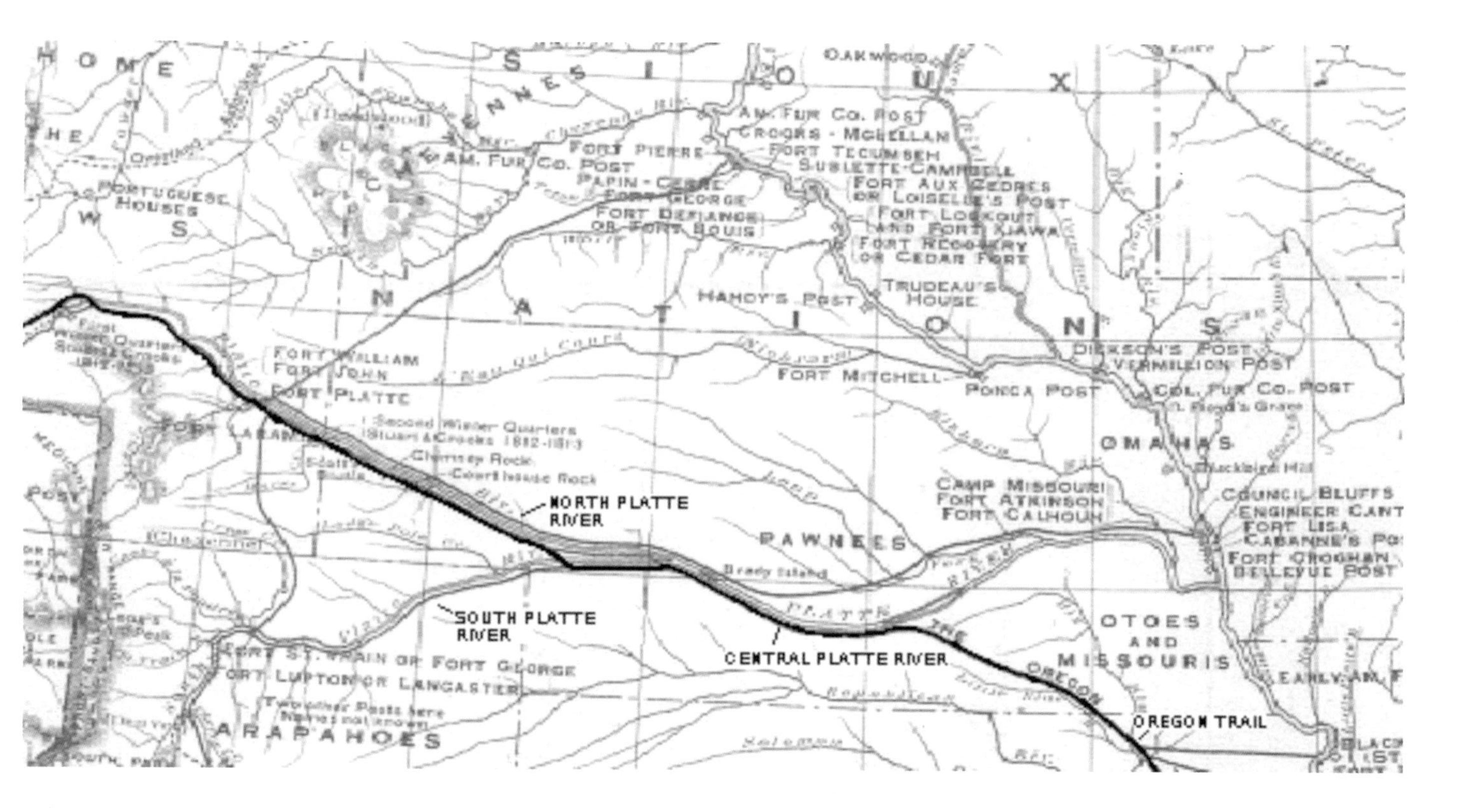

FIGURE 2-4 Portion of Platte River Basin as shown on *Map of the Trans-Mississippi of the United States During the Period of the American Fur Trade as Conducted from St. Louis Between the Years 1807-1843*. Oregon Trail alignment along central Platte River westward and then along North Platte River is shown by heavy black line. Source: Chittenden 1902.

the Oregon Trail coincided with a period of high rainfall, and settlers traveling in the 1840s experienced relatively wet conditions (Lawson 1974). Terminology shifted from "desert" to "subhumid" and "semiarid" to describe the region (Brown and Whitaker 1948). More settlers came to the Platte River Basin and stayed after the end of the Civil War in 1865 following passage of the Homestead Act of 1862. The Union Pacific Railroad, with a vast amount of granted land along the Platte, contributed to the settlement of the region. The Transcontinental Railroad was completed across Nebraska in 1867 and finished in May 1869. Railroad construction, housing, and fencing stimulated the rapid cutting of available timber.

Settlement and land use in the Platte region were affected by environmental perception and environmental reality. The Great American Desert came to be regarded as a potentially productive agricultural region, perhaps with trends toward increasing rainfall (Brown and Whitaker 1948). The idea that "rain follows the plow," popular in the 1870s and 1880s, seemed accurate when settlers experienced increasingly abundant rainfall (Glantz 1994). Central plains homestead entries climbed rapidly in the 1880s (see, e.g., Mock 2000). As available rainfall records became longer, however, they revealed alternating periods of relatively abundant moisture and drought. As Saarinen (1966) noted, "although Great Plains wheat farmers are aware of the drought hazard they appear to underestimate its frequency and to overestimate the number of very good years" in spite of being preoccupied with weather conditions and drought potential. Especially severe droughts in the 1930s Dust Bowl period and in the 1950s led to population declines, although many inhabitants held on through difficult periods.

HYDROLOGICAL CHANGES

Three of the most important hydrological changes that have occurred along the Platte River since the early 1880s are the development of reservoir storage in the Platte watershed, the development of extensive irrigated farming, and municipal and industrial water use in growing urban areas.

Reservoir Storage

Reservoir construction began in the Platte River Basin around the beginning of the 20th century and continued until the early 1980s (Table 2-1). The first relatively large dams built in the basin were constructed on the South Platte River: Jackson Lake (1900, 47,000 acre-ft), Lake Cheesman (1905, 87,227 acre-ft), and Antero Reservoir (1909, 115,000 acre-ft). Cheesman Dam was the world's highest dam at the time of its construction. In 1909, a much larger dam, Pathfinder Dam, was constructed on the North Platte River; its 1,016,500 acre-ft of storage made it one of the biggest dams in the world at that time. Dam construction during the period

TABLE 2-1 Dams in the Platte River Basin

Dam	River	Height (ft)	Storage (acre-ft)	Year Completed	Drainage Area (mi^2)	Owner
Eleven Mile DBWC	South Platte	150.5	128,000	1932	963	Canyon
Prewitt	South Platte (OS)	42.5	51,387	1912	105	LID
Cheesman	South Platte	221	87,227	1905	1,750	DBWC
Antero	South Fork South Platte	53	115,000	1909	185	DBWC
Empire	South Platte	38	52,280	1973	11.1	BIC
Jackson Lake	South Platte (TR)	38	47,000	1900	15.9	JLRC
Milton Lake	South Platte	55	39,660	1975	120	FRIC
Julesberg #4	South Platte (TR)	76	38,600	1910	6.5	JID
Spinney Mountain	South Platte	95	83,300	1982	772	City of Aurora
Chatfield Dam	South Platte	148	355,000	1973	3,018	CENWO
Alcova	North Platte	265	184,300	1938	10,376	USBR
Glendo	North Platte	190	1,118,653	1958	1,500	USBR
Guernsey	North Platte	135	45,228	1927	2,145	USBR
Pathfinder	North Platte	214	1,016,500	1909	14,600	USBR
Seminoe	North Platte	295	1,017,279	1939	7,210	USBR
Kingsley	North Platte	162	1,900,600	1941		CNPPID
Southerland	North and South Platte	30	65,000	1935		NPPD
Johnson Lake	Platte Canal	24	59,000	1941		CNPPID
Riverside	Sanborn Draw	41	94,500	1904	89.7	RID

Abbreviations: BIC, Bijou Irrigation Company; CENWO, Corps of Engineers Northwestern Division Omaha District; CNPPID, Central Nebraska Public Power and Irrigation District; DBWC, Denver Board of Water Commissioners; FRIC, Farmers Reservoir and Irrigation Company; JID, Julesburg Irrigation District; JLRC, Jackson Lake Reservoir Company; LID, Logan Irrigation District; NPPD, Nebraska Public Power District; OS, off stream; RID, Riverside Irrigation District; TR, tributary; USBR, U.S. Bureau of Reclamation.
Source: USACE 1996.

was primarily in response to the need for irrigation water and municipal water. From 1910 to 1939, a series of dams were constructed in the South Platte Basin (Julesburg, 1919, 38,600 acre-ft; Prewitt, 1912, 51,387 acre-ft; and Eleven Mile Canyon, 1932, 128,000 acre-ft) and in the North Platte Basin (Guernsey, 1927, 45,228 acre-ft; and Alcova, 1938, 184,300 acre-ft). From the late 1930s to the late 1950s, storage in the North Platte Basin increased dramatically with the construction of three dams that added over 4 million acre-ft of storage (Seminoe, 1939, 1,017,279 acre-ft; Kingsley, 1941, 1,900,600 acre-ft; and Glendo, 1958, 1,118,653 acre-ft). The largest dam in terms of storage quantity is Kingsley Dam. Its reservoir, Lake McConaughy, impounds a maximum of 1.743 million acre-ft (2.15 km^3) at elevation 3,265 on the North Platte River (Figure 2-5). The South Platte Basin in Colorado added a series of off-stream storage reservoirs during the 1970s and 1980s, although the volume added was less than 600,000 acre-ft (Empire, 1973, 52,280 acre-ft; Chatfield, 1973, 355,000 acre-ft; Milton, 1975, 39,660 acre-ft; and Spinney Mountain, 1982, 83,300 acre-ft).

Water supplies in the Platte River Basin are augmented by transbasin diversions. The Colorado-Big Thompson project diverts about 400,000 acre-ft of water annually from the Colorado River Basin to the South Platte River. Tunnels conduct the water under the Continental Divide and empty it into Big Thompson Canyon, a tributary of the South Platte River north of Denver. Water is also shifted from the North Platte to the South Platte by a diversion west of the town of North Platte, Nebraska. In sum, all the dams in the Platte River system are capable of storing more than 6 million acre-ft of water (Figure 2-6). They provide water for a distribution system of almost 90 canals on the Platte River in Nebraska that was essentially complete by 1930 (Eschner et al. 1983). Additional flows occasionally occur in

FIGURE 2-5 Kingsley Dam on North Platte River, completed in 1941, directly controls flows downstream through central Platte River. Source: Photograph by W.L. Graf, May 2003.

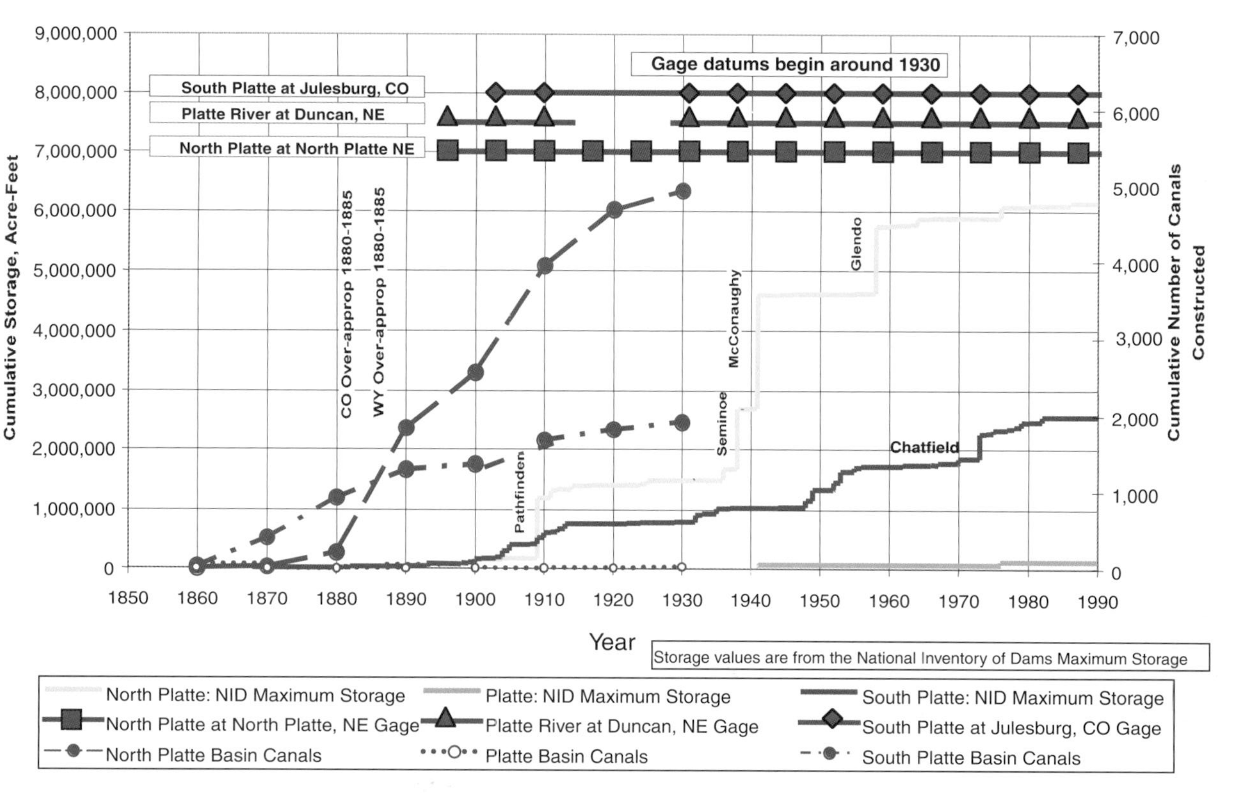

FIGURE 2-6 Cumulative storage in reservoirs, number of canals constructed, and selected stream gage periods in Platte River Basin. Cumulative storage in reservoirs increases over time in form of step functions at lower right, because as new reservoirs were added storage increased at date of closure. Extent of canals caused more gradual increase in total cumulative number of canals as shown at lower left. Canal construction was earlier than construction of large dams. Stream gage records, shown by lines at top, include most of postdam period but little of predam period. Source: Adapted from Eschner et al. 1983.

the river as a result of runoff from local rainstorms and from drainage of irrigated areas. Early gage records ("gage" is the hydrological term for "gauge") indicate that before 1930 the mean annual water yield of the Platte River at North Platte, essentially the upstream end of the portion of the river of interest in this report, was about 2.2 million acre-ft (2.7 km^3). That diminished to about 590,000 acre-ft (0.7 km^3) per year in the period 1930-1970 and was about 650,000 acre-ft (0.8 km^3) per year after 1970 (Simons and Associates 2000).

The dams, reservoirs, canals, and hydroelectric plants provide valuable benefits to the economy of the region. For example, on the north bank of the river in the Big Bend reach, the Nebraska Public Power District supplies irrigation water for 35,000 acres (14,000 ha) of productive farmland (Nebraska Public Power District 2003); the district generates enough electricity from its water-driven power stations to supply the needs of 111,000 homes. South of the river, the Central Nebraska Public Power and Irrigation District system provides additional power generation and supplies water to more than 112,000 acres (44,800 ha) of irrigated land.

Irrigated Farming

The development of storage water made water supplies for irrigation more reliable. Today, there are more than 1.1 million acres of farmland irrigated with surface water in Colorado, more than 600,000 acres in Nebraska, and nearly 300,000 acres in Wyoming. Of all this irrigated land (approximately 2 million acres), nearly 1 million acres in Colorado and Nebraska are supplied by the South Platte River, 750,000 acres by the North Platte River, and over 200,000 acres by the Platte River mainstem. Water storage and distribution systems have produced an artificial water network along the central Platte River (Figure 2-7).

In the middle and lower Platte hydrological subregion, the general area of concern in this report, agriculture consumes 1,366,400 acres-ft (1.7 km^3) of surface water annually (Nebraska Natural Resources Commission 1994); additional substantial (but unmeasured) quantities are pumped from groundwater. This subregion is the most intensely irrigated region in Nebraska (Figure 2-8). Agriculture uses 90% of the water consumed in Nebraska (Nebraska Natural Resources Commission 1994); thus, agriculture will be an integral part of any solutions related to water for threatened or endangered species. In addition to irrigation for agriculture, benefits of the water projects include flood suppression downstream and flatwater recreation opportunities on the reservoir surfaces behind the dams.

Farming in the Platte River Basin has become highly mechanized and specialized, and farms are becoming larger and fewer. Shifts in agriculture occurred with increasing availability of irrigation water, changing technology, the growing importance of economies of scale, and social changes.

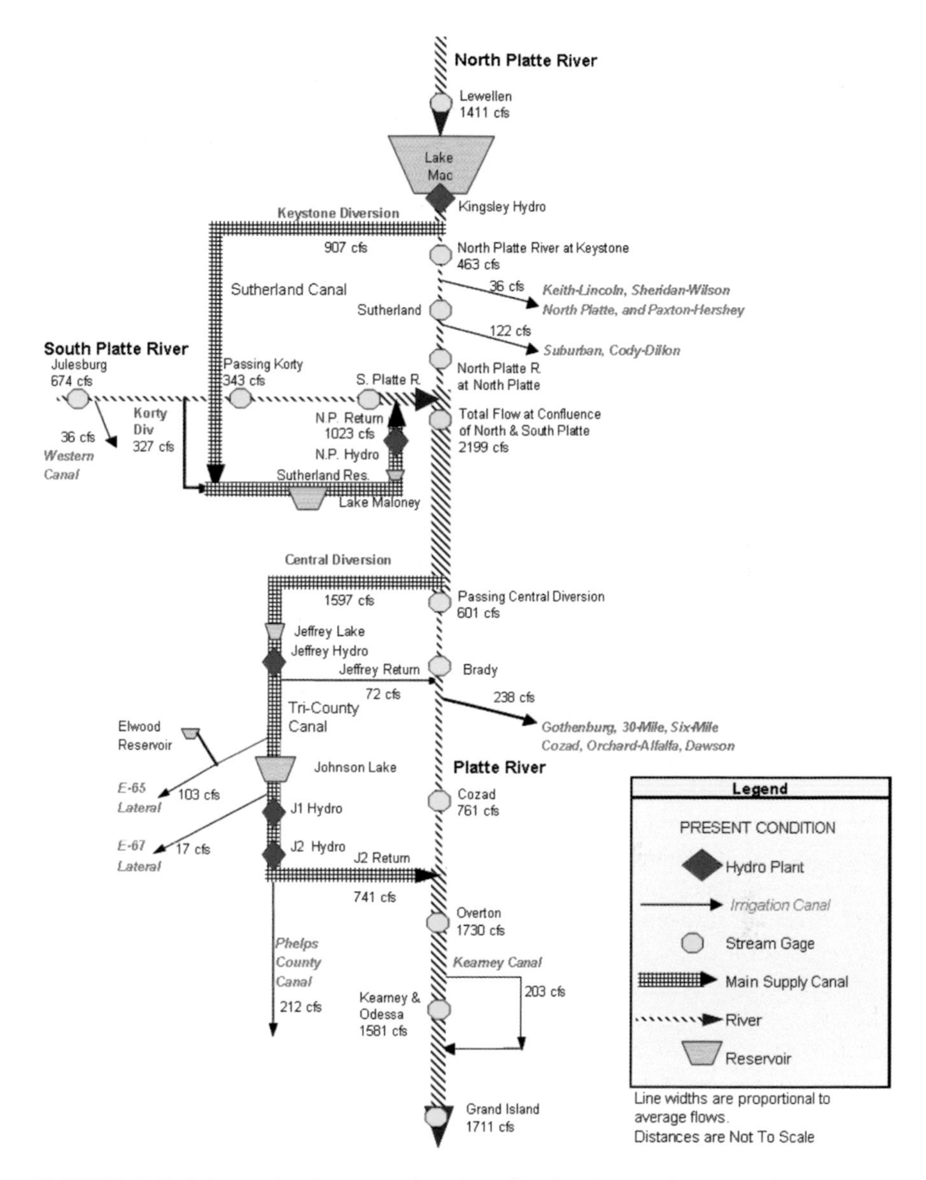

FIGURE 2-7 Schematic diagram showing distribution and connections among Platte River and various parts of its water-control infrastructure. Source: DOI 2003.

Oats, barley, and fruit and vegetable crops, formerly common, are now rarely grown. Improved efficiency in farming practices is evidenced by the number of corn acres remaining steady while production has increased by more than 700%.

Corn is the most important crop in the central Platte region of Nebraska, and the top counties in production of corn for grain (not for silage

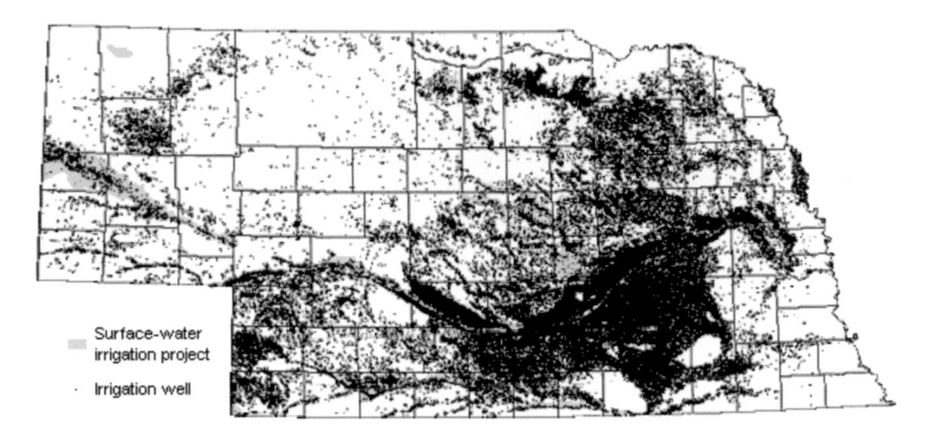

FIGURE 2-8 Irrigation water from groundwater and surface water in Nebraska (black dots are irrigation wells). Source: Winter et al. 1998.

or forage) in the state are here (Figure 2-9). Several of the most important hay-producing counties in Nebraska also are in the central Platte area. Further to the west, wheat is important.

Corn, in particular, is irrigated (Figure 2-10), although irrigation of the other crops, including sorghum and soybean, also is substantial. According to the Nebraska Agricultural Statistics Service (2004), in 2002 Nebraska farm-

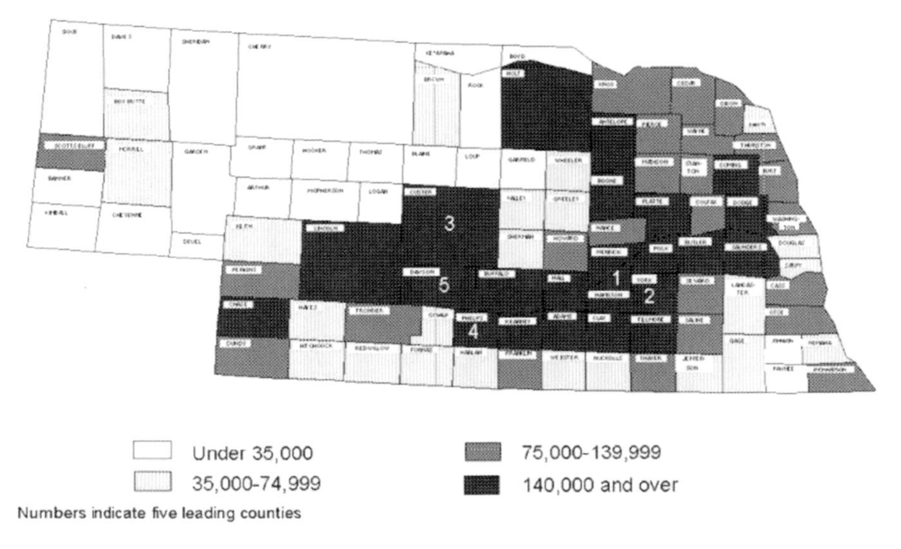

FIGURE 2-9 Distribution of Nebraska corn harvest of 1996 (a typical year), showing acres harvested for grain, 1996, and importance of agriculture and corn production along Platte River. Source: Nebraska Agricultural Statistics Service 2002.

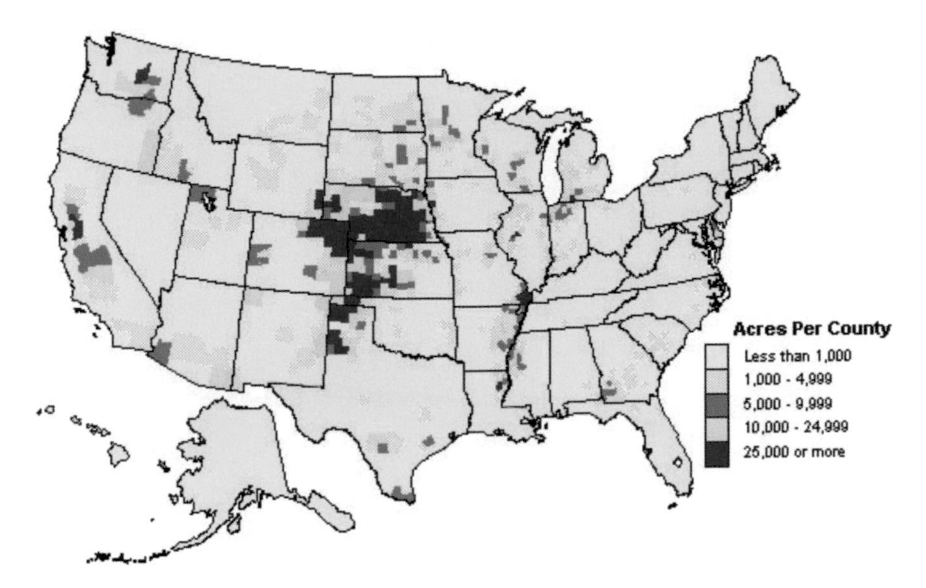

FIGURE 2-10 Distribution of irrigated corn production in 1997 (a typical year), showing irrigated corn for grain or silage by county, 1997, and prominence of Nebraska and Platte River. Source: NASS 1999.

ers planted 8,875,000 acres of corn of which 8.4 million acres was planted for grain and 475,000 acres for silage. Some 58%, about 4.9 million acres, of the corn for grain and 31% of the corn for silage were irrigated. The fraction of all acres of corn that are irrigated varies greatly in the state. It is difficult to extract the "basin acres" from the statistics, because the census is created by county and not by watershed. Counties adjacent to the Platte River have irrigation coefficients ([irrigated acreage for corn for grain]/[total acreage for corn for grain]) of 54% to 96%. About 44% of soybean crops (207,000 acres) was irrigated in Nebraska in 2002. Where soils and topography are suitable but irrigation water is unavailable, dryland crops, such as wheat, may be grown. In 2002, 8% of wheat crops (129,000 acres) were irrigated in Nebraska (Nebraska Agricultural Statistics Service 2004).

In addition to crops, agriculture in the region includes livestock—predominantly feeding operations for beef cattle. In portions of the basin with rugged topography and without adequate access to irrigation water, open livestock grazing has been the dominant land use.

Surface-water diversions for irrigation began in the central Platte area as early as the 1850s and 1860s (Eschner et al. 1983). Irrigation-ditch or canal construction proceeded during the 1890s drought years, and use of groundwater for irrigation began in the late 1800s. Groundwater use increased in the early 1900s and again after about 1950 with the development of center-pivot irrigation systems (Figure 2-11). The percentage of farmland

FIGURE 2-11 Central-pivot irrigation systems along central Platte River. Source: Photograph by W.L. Graf, May 2003.

under irrigation has continued to expand in recent years. The Census of Agriculture shows 5,807,308 acres of land in farms in the 12 counties along the central Platte[1] in 1959. Land in farms in the same area in 1997 was 5,606,895—a reduction of more than 200,000 acres. In contrast, irrigated land more than doubled between 1959 and 1997, from 914,432 acres to 2,125,781 acres. Irrigated acreage in the region now accounts for about 38% of the land in farms, compared with less than 16% in 1959. Although a breakdown of the relative quantities of surface water and groundwater used for irrigation is unavailable, it can be assumed that the growth in irrigation is mostly through increased groundwater uses, because surface water was completely allocated by about 1930. Some of the increase may also be due to increasing efficiency of water use and an ability to spread a given allocation over a larger area of land effectively.

The importance of irrigation to agriculture in much of Nebraska is illustrated by agricultural-land prices. For Nebraska as a whole, the 2001-2002 average dryland cropland value was $798 per acre (with no irrigation potential); center-pivot-irrigated cropland was valued at $1,513 per acre,

[1]Adams, Buffalo, Dawson, Gosper, Hall, Hamilton, Kearney, Lincoln, Merrick, Phelps, Platte, and Polk counties.

and gravity-irrigated cropland at $1,800 per acre (Johnson and Kuenning 2002). In the area to the north of the Platte's Big Bend, the difference is even more pronounced: gravity-irrigated cropland was valued at $1,825 per acre, in contrast with a dryland average of $680 per acre.

State and federal policy decisions affect Nebraska agricultural practices, which in turn have environmental implications. Such policies include conservation programs and decisions to help control production, prices, and environmental effects of farming. The effects of agriculture vary with the intensity of cultivation and grazing, which varies with prices received for farm and ranch products, as well as with public policy. Federal policy in the 1950s made conservation through enrollment in the Soil Bank program possible; such enrollment paid farmers to not produce (this helped to control commodity prices and soil erosion). In the 1970s, federal farm policy promoted plowing "fencerow to fencerow" (and removal of fencerows to create larger farm fields) to promote higher production for international trade, but the 1980s saw a return of a program of soil conservation and temporary land retirement with establishment of the Conservation Reserve Program. Preferential crop subsidies, indebtedness and the banking system, and international market prices all exert pressure on particular types of production.

Municipal and Industrial Water Use

With the exception of the Colorado Front Range area in the South Platte drainage (including Denver) and dispersed small towns and cities, particularly near the railroad, the overall character of the Platte River Basin is agricultural, and it has been relatively sparsely inhabited (Figure 2-12). Other than those urban areas, the watershed is generally less than 10% developed (urban and transportation land), and much of it is less than 2% (NRCS 2001). Population densities of most counties in the basin are below 20 persons/mi^2; many are below 6 persons/mi^2 (Figure 2-13).

The Denver metropolitan area, however, has grown to over 1 million people. Several of the earliest dams built in the Platte River Basin were to supply water to Denver, but the increased population and demand for water supply are becoming more challenging. Summer water-use restrictions have become the norm for the area as demands outstrip the supply. Other urban areas are likely to look to the Platte for municipal water supplies in the future.

CONFLICTS

As is the case with many western rivers, the surface waters of the Platte are completely appropriated, and conflicts around their use are legendary.

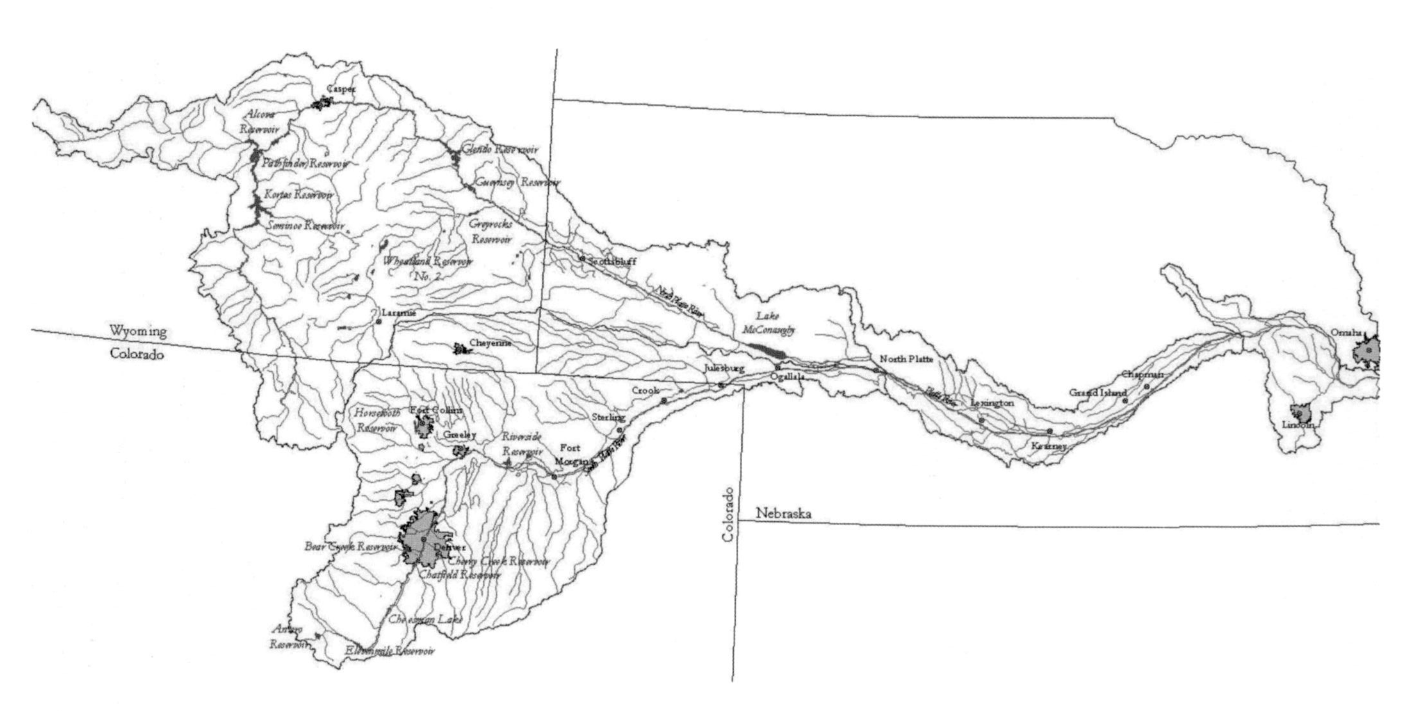

FIGURE 2-12 Urban areas in Platte River Basin. Source: Friesen et al. 2000.

FIGURE 2-13 Gothenburg, Nebraska, typical example of small cities along central and lower Platte River that are stable or growing in size and that serve rural agricultural areas. Source: Photograph by W.L. Graf, May 2003.

A major source of water along the central Platte River Valley is groundwater, but overuse has resulted in serious depletion in some areas. The development of the Platte River for irrigation, hydropower production, flood protection, and municipal water supplies has resulted in a marked shift in the natural flows in the river and in conflicts between stakeholders in the basin. The most important conflicts that have resulted from development include the following.

- *Nebraska vs Wyoming*. The principal causes of the conflict are related to the operation of existing U.S. Bureau of Reclamation reservoirs in Wyoming, the construction of new reservoirs or diversion facilities, and pumping of groundwater for irrigation. The dispute and many of its issues have been before the U.S. Supreme Court. The Platte River Cooperative Agreement (1997) is an attempt to solve some of the issues, mainly those associated with balancing the needs of water users and of threatened and endangered species in the central Platte River Basin.

• *Municipal water supply and Colorado's Front Range*. Before 1980, the Denver area relied on the development of the Two Forks project[2] to supply the next major increment of storage to meet the region's growing demand. In 1982, regional water providers came to an agreement that this source would be the most acceptable to the region and began the process of seeking permits from regulatory agencies. On June 5, 1996, the U.S. District Court for the District of Colorado upheld Environmental Protection Agency (EPA) rejection of the project. Since then, the region has explored increased conservation and the integration of existing water supplies as methods to deal with water shortages.

• *Maintenance of flows and wetlands for threatened and endangered species*. The maintenance of flows has become one of the most important issues in the basin. This dispute is focused on the habitat needs of federally listed birds (the piping plover, the interior least tern, and the whooping crane) and on the pallid sturgeon and other aquatic species. The concern has been expressed most urgently in the central Platte River Valley in Nebraska, where the need to provide adequate streamflows and habitat for those species has become the focus of conflicts between irrigation and environmental interests.

GROUNDWATER

The spatial and temporal variability that characterizes surface water in the Platte River Basin are also distinctive features of its groundwater. Stream flow and precipitation have always contributed directly to groundwater, so the amounts and locations of subsurface water have changed in response to long-term meteorological events. Irrigation influences groundwater levels, especially infiltration from field applications and leaking canals and withdrawals from pumping wells. When irrigation water is overapplied, the excess seeps past the root zone of the crop and saturates the soil beneath, ultimately connecting to the groundwater table, adding to its volume, and bringing the groundwater table closer to the surface. Despite the physical connections between surface water and groundwater, Nebraska had no legal mechanisms connecting the two until 1996, when the state initiated integrated water management.

Nebraska's groundwater supplies are extensive, and the state is the third-largest user of groundwater in the nation, behind California and Texas (both of which are much larger than Nebraska). Most of the state's domestic water supply comes from groundwater, and groundwater supports the majority of

[2]Two Forks project was a proposed $500 million concrete dam, 615 ft tall and 1,700 ft wide, to be about 1 mi downstream of confluence of South Platte River and its north fork, which would have flooded about 30 mi of river and provided 780,000 acre-ft of storage.

irrigated agriculture. The Nebraska Department of Water Resources records as of October 2002 indicated that there were nearly 93,000 registered irrigation wells and 15,000 domestic wells in the state (Nebraska Department of Environmental Quality 2002). Groundwater quality is generally good, although many wells have high nitrate concentrations—often above the EPA maximum contaminant level (MCL) of 10 mg/L—and atrazine concentrations above the MCL of 3 μg/L because of the use of agricultural fertilizers and biocides (Nebraska Department of Environmental Quality 2002).

The aquifer is part of the High Plains Aquifer that extends from South Dakota to Texas. The geology underlying the Platte River in the study area is described by Hurr (1983) as follows:

> The Platte River in South-central Nebraska is underlain by alluvial clay, silt, sand, and gravel of Quaternary age deposited in a series of broad troughs eroded into the underlying clay, silt, sand and gravel of Tertiary age, and shales and limestones of Cretaceous age. The Tertiary formation is the Ogalalla Formation. The Cretaceous rocks are the Pierre Shale and the Niobrara Formation. The Quaternary alluvium contains the principal aquifer in the area and consists of inter-fingering lenses and beds of unconsolidated clay, silt, sand and gravel. The lower one-half of this unit is predominantly clay or silt and the upper one-half is predominantly sand and gravel with some beds of clay or silt.

The thicknesses of individual layers in the aquifer vary widely over short lateral and vertical distances. Depth to the water table is small, usually less than 50 ft; near the river, it is generally 5-10 ft.

Role of Groundwater in Platte River Management

Groundwater is hydrologically connected to surface water in the study area and is likely to affect riverine conditions. Management solutions that do not include a comprehensive view of water supply in the study area are not likely to be successful over the long term. If groundwater conditions are not considered, efforts to modify the channel and provide specific water flows to shape habitat conditions may not have the desired effect. Controlling depletions from well pumping will be required in order to have an effective water management system.

Historical Perspective on Groundwater Use in Nebraska

Groundwater in Nebraska is not privately owned but can be used by overlying landowners pursuant to the reasonable-use doctrine, which allows pumping for any beneficial use without protection for senior water users. Generally, permits are not required for establishing a right to pump groundwater; landowners may use groundwater without waste, subject to sharing by competing users during periods of shortage (the correlative-rights doctrine) (Aiken 1987).

Initial development of agriculture in Nebraska was influenced by a desire to facilitate surface-water-based irrigation facilities, starting with the adoption of the state irrigation code of 1895, which was followed by the federal 1902 Reclamation Act (Aiken 1987). However, a legal decision in 1936 limited interbasin transfers (Osterman v. Cent. Neb. Pub. Power & Irrigation Dist., 131 Neb. 356, 268 N. W. 334 [1936]) and temporarily reduced the ability to import additional surface-water supplies (Aiken 1987). That decision, in combination with severe drought and the development of the turbine pump, resulted in the rapid development of huge acreages using groundwater (Aiken 1980). Increased use lowered the level of groundwater in some areas, including the central Platte River Basin (Ellis and Pederson 1986), and led irrigators eventually to resort once again to surface-water projects after the Osterman decision was reversed in 1980 (206 Neb. 535, 294 N.W. 2nd 598). Groundwater now irrigates three times as much acreage in Nebraska as surface-water sources (streams and reservoirs).

Nebraska has 23 natural resources districts, 15 of which have established groundwater-management areas over all or parts of their districts to address groundwater quality or quantity. Historically, Nebraska water law failed to address the connection between surface water and groundwater. In fact, Nebraska was the only western state that did not legally recognize tributary groundwater before 1996 (Aiken 2001). In 1996, integrated water-management authority was provided to the natural resources districts and the Nebraska Department of Natural Resources through the Nebraska Ground Water Management and Protection Act. The regions of authority are complex, and most are yet to be implemented, but the Upper Republican District has limitations on drilling new wells and limits on pumping wells because of legal concerns related to the aquifer it shares with Kansas. The North and South Platte Natural Resources Districts have established moratoria on the development of new wells that supply more than 50 gal/min in all (North Platte) or portions (South Platte) of their districts, and groundwater is now being allocated in the Pumpkin Creek Basin subarea (NPNRD 2003; SPNRD 2003). Although establishing the authority to manage surface water and groundwater is a first step, it is not clear that the natural resources districts, as currently constituted, will provide adequate management of groundwater. There are no restrictions on current (as opposed to new) groundwater pumping in the central Platte River Valley.

Current Dynamics

Groundwater flows down-gradient (from areas of high pressure to areas of low pressure) because of gravity. In the Platte River Valley, shallow groundwater and surface water often are hydrologically connected; that is, water that has flowed in the river through one stream segment may percolate into the ground through another segment, depending on substrate conditions and on whether the groundwater level is above or below the water level in the river. When the groundwater table is lower than the river, that stretch of river is a "losing" stream—the river is recharging the groundwater. If the groundwater table is higher than the river, the river will be a "gaining" stream.

A 1986 study of groundwater and surface-water conditions along the central and lower Platte (Nebraska Department of Natural Resources 1986) indicated that groundwater seepage into the river occurs from both sides of the Platte River between North Platte and Kearney. The seepage is greater now than it was under natural conditions because mounding of the water table due to canal leakage and application of water for irrigation has steepened the water table slope toward the river. Beginning at about Kearney and continuing into Merrick County, the water table slopes northeastward away from the river instead of toward it, and the river thus becomes a losing stream by providing recharge to the adjacent aquifer. In much of the same reach, the water-table slopes away from the river on the south side also and thus the river loses water by seepage in that direction too. Figure 2-14 further depicts the increase and decrease in groundwater levels in Nebraska due to groundwater and surface-water irrigation.

U.S. Geological Survey groundwater contour maps (Figure 2-15 and Figure 2-16) indicate that in 1931, the Platte River from Kearney to Grand Island generally was losing water to the aquifer, whereas groundwater contours from 1995 show that the river was gaining water from the groundwater supply. The reversal is most likely due to accumulation of a groundwater mound from deep percolation of agricultural water, primarily derived from surface water imported by irrigation districts.

The relationships among groundwater, surface water, and the elevation of the groundwater table are crucial to maintaining water levels in the river during low-flow periods, and they influence vegetation in and near the river. The portion of the flow in the river derived from groundwater is "base flow." Reductions in groundwater levels can reduce base flow, a condition that is observed elsewhere in the United States (Glennon 2002). The Platte River in the study area is unusual in that increased groundwater pumping appears to have been offset in some areas by return flows from imported surface water. Any increases in groundwater withdrawals and reductions in surface-water deliveries are likely to affect riverine habitat. Wet meadows that are important crane habitat may be supported by high groundwater levels and streamflow (Currier 1995), although alternative explanations, such as limited percolation of snowmelt and spring rains through frozen soil, have also been offered (Michael Jess, University of Nebraska, pers. comm., August 2003).

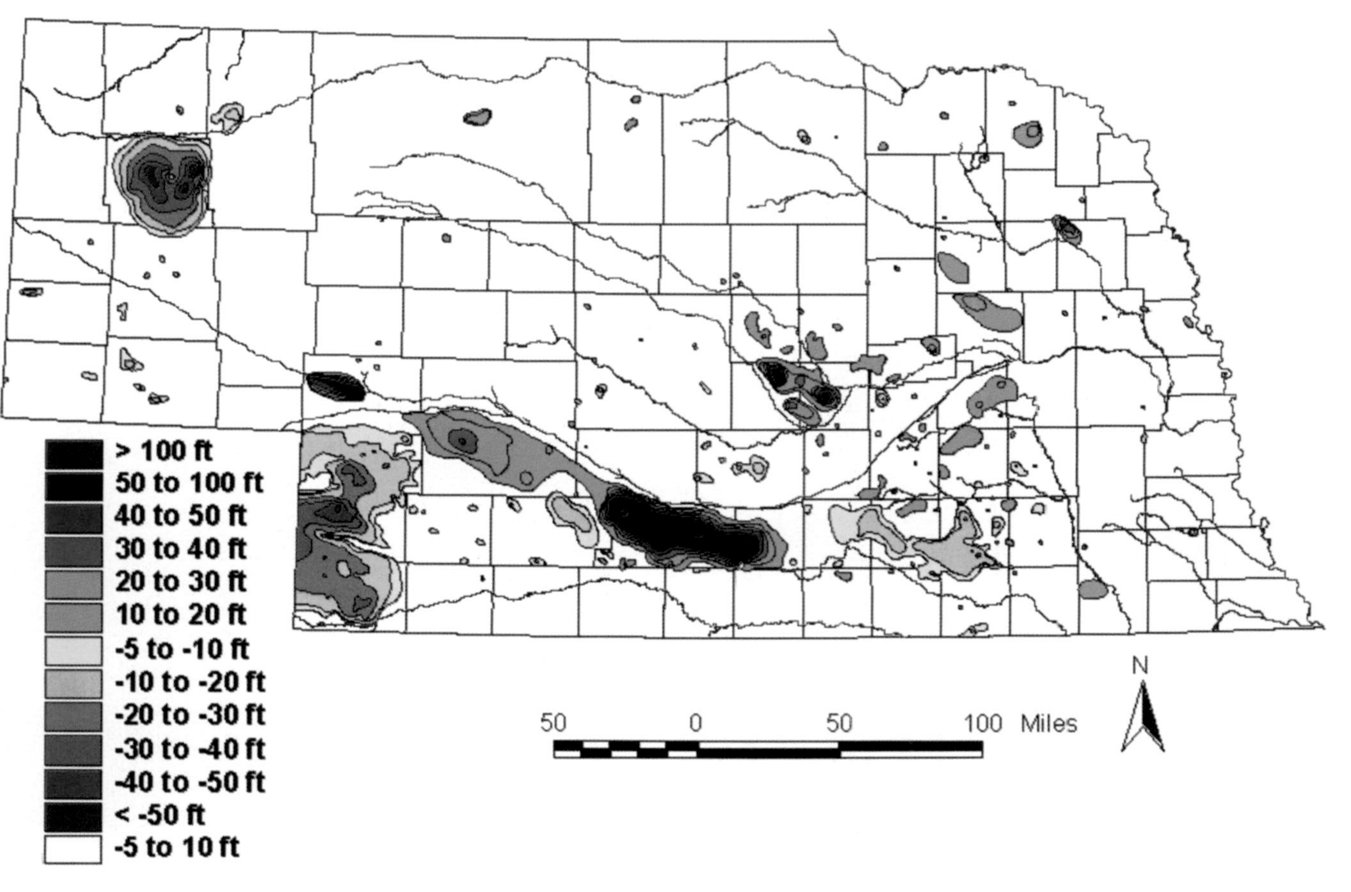

FIGURE 2-14 Fluctuations of groundwater levels, showing development of mound under region of central Platte River. Source: DOI 2003.

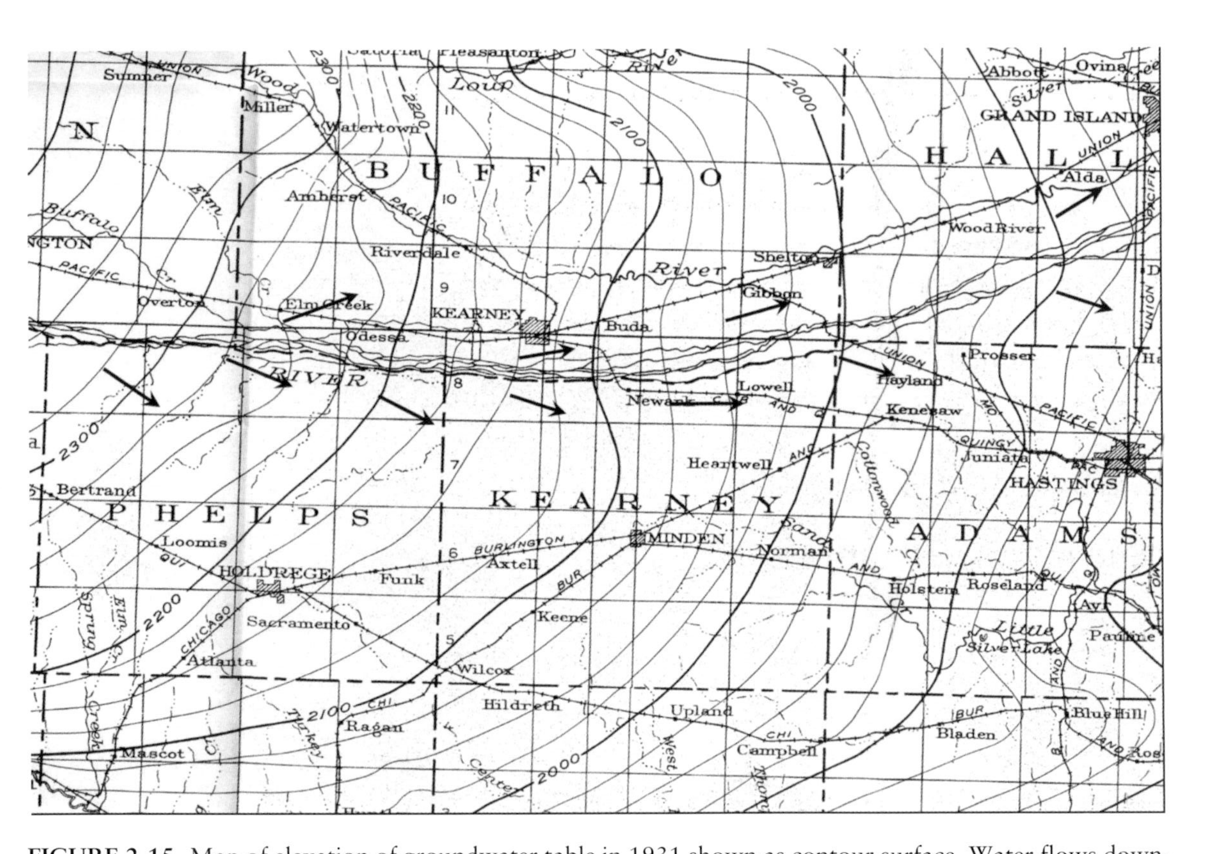

FIGURE 2-15 Map of elevation of groundwater table in 1931 shown as contour surface. Water flows downgradient and generally eastward and southeastward. Compare with Figure 2-16. Source: Adapted from Woodward 2003.

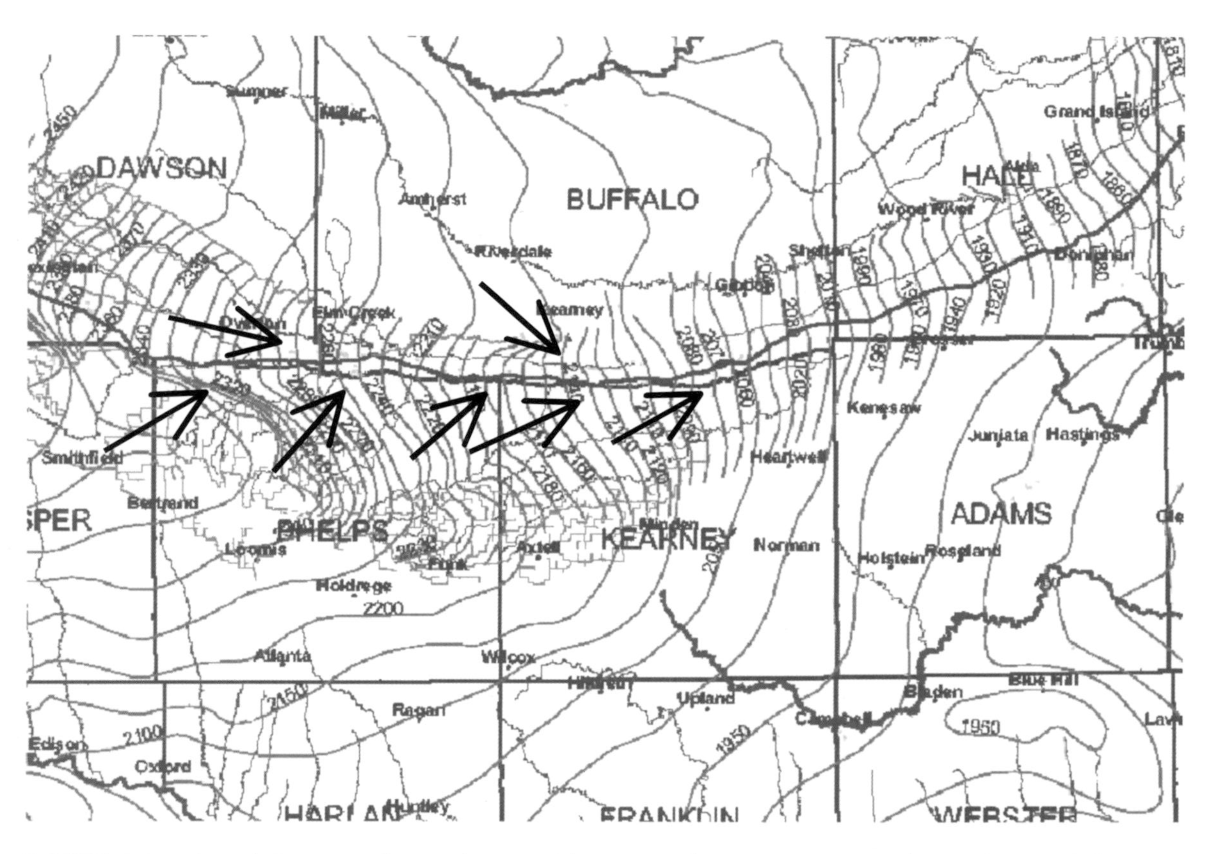

FIGURE 2-16 Map of elevation of groundwater table in 1995 shown as contour surface. Substantial elevation of surface has occurred since 1931 in western part of map. Compare with Figure 2-15. Source: Adapted from Woodward 2003.

The importance of the relationship between groundwater levels and river flows and of the impact of well pumping on river flow and habitat quality is well recognized by the drafters of the cooperative agreement (Aiken 1999):

> Under the Cooperative Agreement, all three states must develop a water depletion tracking and accounting system that will identify what water use activities are depleting Platte River flows. In Nebraska, this will also include identification of wells the pumping of which may reduce flows in the Platte over time. New surface and groundwater use begun after July 1, 1997, are subject to replacement water requirements.

The first phase of the $7 million Cooperative Hydrology Study (called COHYST), paid for by the Environmental Trust Fund as a component of the cooperative agreement, is nearing completion. Its focus is on modeling and mapping hydrologically connected groundwater in parts of 43 counties, mostly in Nebraska but extending 6 mi into Wyoming and Colorado along the North and South Platte Rivers. The area includes about 67,000 wells, mostly for irrigation; about half were installed in the 1970s (COHYST 2003). The findings of this comprehensive modeling effort will contribute to understanding of the causes and effects of changes in the groundwater table elevation near the Platte River.

Implications of Climate Variability and Change

Although groundwater levels respond more slowly to climatic conditions than do surface-water flows, long-term climate trends do affect groundwater levels. Even multiyear "snapshot" views of groundwater conditions or surface-water flows may miss multidecadal trends. Climate conditions affecting Platte River flows in the early part of the twentieth century were generally wetter than they were in midcentury, and there was another wet period from the late 1970s to the middle 1990s (Lewis 2003). Those conditions are probably related to global climate patterns that are affected by ocean temperatures and changing atmospheric circulation patterns. It is critical to understand that climate conditions since the early 1990s (when much of the habitat information the committee was asked to review was generated) are not typical of the entire twentieth century. Average annual precipitation has decreased by more than 7.5 in. in the last decade in Nebraska (NCDC 2003), temperatures have been higher than historical averages, and streamflow has been affected by serious drought. Surface-water deliveries to agriculture have also varied in response to the factors described above. Therefore, the presence or absence of groundwater mounds and quantification of depletion effects of wells need to be understood in a long-term context; but long-term data are sparse.

Potential implications of long-term climate change associated with global warming will add more uncertainty to management of the critical habitat and reaffirm the need to be cautious about conclusions related to habitat quality and water-flow requirements. Climate-model scenarios project that temperatures will continue to rise throughout the region and that the largest increases will be in the western parts of the plains (USGCRP 2000a,b). Conclusions related to increases in temperature (and therefore increases in evaporation) are relatively reliable, but implications for precipitation are less certain.

RIPARIAN VEGETATION CHANGES

The vegetation of the Platte River watershed varies sharply from west to east along a steep gradient of annual rainfall. Coniferous forest predominates in the Rocky Mountains; at lower elevations in the Great Plains, the rivers flow mainly through grassland vegetation of several types—short grass in the drier west and mixed grass and tall grass in the more humid east. In the grassland region, deciduous forest (principally *Populus* and *Salix)* is largely restricted to floodplains, such as the Platte River Valley. Changes in slope, substrate, temperature, hydrology, frequency of fire, and grazing along the Platte River from its headwaters to its mouth influence the biodiversity and ecology of individual river segments. The coarse substrate, steep slopes, and coniferous forests of headwater reaches contrast with the fine shifting substrates, gentler slopes, and riparian vegetation (currently dominated by deciduous trees and shrubs) of downstream reaches.

Early explorers indicated that the floor of the Platte River Valley was a mosaic of upland grassland away from river channels, meadow and marsh vegetation on low terraces near the river, woodland and shrubland on islands, and scattered timber on the banks. Observations during the Long Expedition (1822-1823) included the following (Long and James 1823):

> Soon after crossing the Elk-horn we entered the valley of the Platte, which presented the view of an unvaried plain, from three to eight miles in width, and extending more than one hundred miles along that river, being a vast expanse of prairie, or natural meadow, without a hill or other inequality of surface, and with scarce a tree or a shrub to be seen upon it. The woodlands, occupying the islands in the Platte, bounded it on one side; the river-hills, low and gently sloped, terminate it on the other.

Through their storage capabilities and the resulting alterations in flow regime of the Platte River, the dams and diversions have led to geomorphic and vegetation changes along the downstream corridor. At North Platte, the average annual peak flow was about 14,000 cfs (400 m^3/s) before 1930;

after that time, it was only about 3,200-3,400 cfs (90-95 m^3/s). The year 1930 is selected largely for convenience because that year saw the installation of new gaging methods and the establishment of a new gaging site near Duncan. It also predates the completion of three of the four largest dams in the Platte River Basin. Because flood flows are important in forming and changing the channel, geomorphology of the stream has undergone substantial adjustment, including a reduction in the active channel width. Initial diversions of flows reduced water discharges, exposed more river-bed surfaces, and created new and larger areas for vegetation to establish. Improved reproduction and survival of cottonwood and willow resulted in considerable woodland expansion in most reaches of the river (Figures 2-17 and 2-18); this expansion apparently had ceased by the late 1960s or earlier on some reaches (Currier 1982; Johnson 1994).

Native vegetation associated with the Platte River was broadly categorized by Currier (1982) and Johnson (1994) as grass and sedge meadows on low-lying flats bordering the outer banks of the river, open- and

FIGURE 2-17 View of Platte River near Cozad, Nebraska, showing conditions in 1866, about 20 years after extensive wood use by immigrants, soldiers, and railroad crews. Photograph probably taken when river was in flood. Source: Carbutt 1866a. Reprinted by permission of Union Pacific Historical Collection.

FIGURE 2-18 View of riparian and channel woodlands on central Platte River, showing conditions in part of river in 2003. Source: Photograph by W.L. Graf, May 2003.

closed-canopy woodland (*Populus, Salix, Fraxinus,* and *Ulmus*), shrubland (*Salix* and *Amorpha*) along channels and on small river islands, and ephemeral vegetation comprising largely annual plants (*Cyperus, Echinochloa, Xanthium,* and *Eragrostis*) that quickly germinate and fruit on newly exposed sandbars in middle to late summer (Figure 2-19). Submerged aquatic macrophytes and emergents are not common in the Platte River, but backwater areas support stands of *Typha, Phragmites, Elodea*, and *Potamogeton*.

Expanding woodland is reported as a primary cause of reduction in the suitability of the Platte River as habitat for the nesting or roosting of endangered or threatened aquatic avifauna. Sandhill and whooping cranes prefer roosting in wide river channels away from tall riparian vegetation (Krapu et al. 1984; Faanes et al. 1992). Terns and plovers prefer to nest on bare or sparsely vegetated sandbars (Ziewitz et al. 1992). Thus, central issues in the vegetation-bird connection are the extent to which vegetation in the river has changed, the causes of changes, and the degree to which vegetation limits or may limit the recovery of these avian species.

How has the vegetation of the Platte River changed historically? Investigators in the United States often use General Land Office (GLO) surveys to develop baseline (presettlement or early-settlement) conditions from which to measure the trajectory of change or to develop targets for restoration (Stearns 1949; Whitney and DeCant 2001). Maps drawn from the

FIGURE 2-19 View of highly varied habitats on portion of central Platte River, showing range of density of tree cover in some locations from dense woodland to open woodland and treeless areas that is reminiscent of reconstructions of predevelopment conditions in some reaches as suggested in Figure 2-20. Source: Photograph by W.L. Graf, May 2003.

surveyors' notes (plat maps) have been used extensively to determine the physical and ecological character of the Platte River before the construction of upstream dams and water-diversion structures. Most studies have used the bank-to-bank widths (minus any in-channel islands shown) measured on the plat maps as the baseline preregulated, open-channel (sand and water) area. For example, Williams (1978) found total (sum of all channels) bank-to-bank widths in the central Platte River to have been 1,200-2,000 m in 1865. Eschner et al. (1983) also used the GLO plat maps to determine baseline channel widths with which to compare more recent measurements from aerial photographs that date from 1938 and later.

Apparently, neither of those studies consulted the field notes on which the plat maps were based. Years after their analyses, on which many reports and legal filings were based, it was discovered that many or most of the small vegetated islands identified in the field notes as abundant were not surveyed and therefore were not drawn on the plat maps (Johnson 1994). Thus, total unvegetated channel widths measured from the plat maps were overestimated by the early studies. Moreover, the blank space between banks on the maps may have led to the impression that the presettlement Platte River was rather featureless, comprised of large expanses of sand and water, and was generally devoid of trees and other woody vegetation. They may also have led to the conclusion that any riparian vegetation in the modern river is unnatural and a product of flow regulation (Johnson and Boettcher 2000a). A partial reconstruction of predevelopment vegetation with the plat maps and field notes (Figure 2-20) was produced by Johnson and Boettcher (1999).

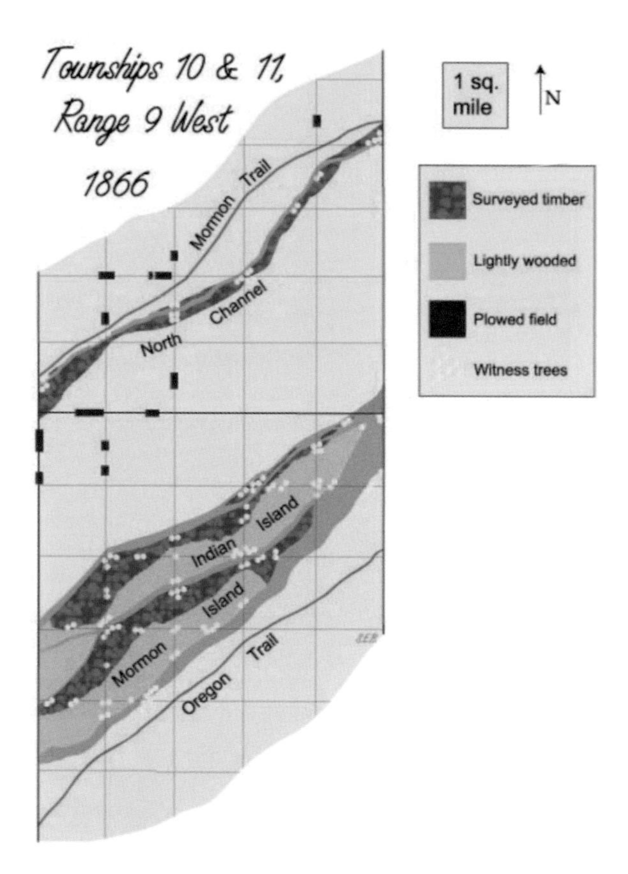

FIGURE 2-20 Reconstruction of predevelopment vegetation based on GLO plat maps and field notes (source of witness tree data) for two Platte River townships. Small unsurveyed riverine islands are not shown. Light green areas were considered as typified by grassland; two types of timbered areas are also noted, as well as some agricultural activity. Source: Johnson and Boettcher 1999. Reprinted with permission; copyright 1999, Land and Water, Inc.

The concept of a largely nontimbered river was bolstered by photographs showing a mostly treeless river near wooden bridges during early settlement (Williams 1978). However, the photographs were taken many years after Fort Kearny was built, 350,000 pioneers had passed by the Platte River on the Oregon Trail, the Union Pacific Railroad had been completed, and many farms were built and fenced. Virtually no wood was available in the vast grassland bordering the Platte River, so the high wood use associated with those human activities led to deforestation wherever there were trees, including islands within the Platte River (Johnson and Boettcher 2000a). The earliest photographs were taken too late to record the natural extent of woodland in the presettlement Platte River Basin. The most recent assessments based on the survey notes and plat maps portray the presettlement Platte as a river that had scattered trees on its outer banks but also was well studded ("immense numbers" was also used by the surveyors) with wooded islands of all sizes throughout its length. Cottonwood and willow trees dominated both the presettlement and modern floodplain vegetation of the Platte River; the exotic Russian olive and native red cedar are now abundant but were not so historically (Currier 1982; Johnson and Boettcher 2000a). The open, savanna-like nature of presettlement woodlands on large, higher islands just above floodplain level may have been the product of prairie fires, competition with grasses, and wood use and horse pasturing by American Indians. Smaller, floodplain-level islands were probably more protected from fire and had well-developed middle-story and upper-story vegetation. Current woodlands along many parts of the central Platte River have high tree density and well-developed, shrubby understories. Those conditions are different from the complex of wooded islands and large expanses of open water, sandy beaches and islands, and low-growing ephemeral vegetation shown in the earliest aerial photography from the 1930s (Box 2-1).

GENERAL SPECIES RESPONSES TO EUROPEAN SETTLEMENT

Invertebrates

The aquatic invertebrates of the Platte River in Nebraska include 18 species of unionid mollusks (Hoke 1995) and 63 taxa of insects (McBride 1995). The common taxa of insects include Ephemeroptera (*Caenis*, *Tricorythodes*, and *Heptagenia*), Plecoptera (*Isoperla*), Odonata (*Argia* and *Gomphus*), Hemiptera (Corrixidae and Gerridae), Coleoptera (Elmidae and Dytiscidae), Trichoptera (*Hydropsyche* and *Cheumatopsyche*), and Diptera (*Dicrotendipes*, *Cladotanytarsus*, *Chironomids*, and *Rheotanytarsus*). Most of the taxa belong to the collector-gatherer or collector-filterer functional feeding groups and occupy the shoreline habitats rather than the more abundant, shifting sandbar habitat. Analysis of macroinvertebrate densities

BOX 2-1
The Question of Presettlement Woodland Along the Central Platte River

What was the areal extent, composition, and structure of vegetation communities along the central Platte River before European settlement, and what controlled their dynamics? The significance of this question is that modern attempts to restore the river to greater functionality for wildlife using open riparian areas requires knowledge of its original condition. "Original" in this case refers to conditions before European settlement, the imposition of land-use changes, and the installation of river-control structures, such as dams and diversions. Present-day human activities in and near the river, and throughout its drainage area upstream, make it impossible to completely restore presettlement vegetation communities. A riverine environment less modified than the present one may be possible, however, and may be beneficial to the endangered species and to other species using the area.

The nature of the presettlement vegetation communities of the central Platte River has become better known, but some details are the subject of debate. The most recent summary discussions in the refereed journal literature are by Johnson and Boettcher (2000a,b) and Currier and Davis (2000). These and other interpretations rely on historical descriptions of the river and its vegetation, including journal entries of early travelers, soldiers, and settlers. One source of historical conditions is photography of the river, but few images are available, and their coverage does not begin until the 1860s. The earliest systematic information is from General Land Office (GLO) surveys conducted along the central Platte River in 1859-1869. The general objective of the surveys was to lay out the township and range system before organized settlement. The survey records include plat maps and surveyors' notes describing conditions along the survey lines. Two divergent views of the river emerge from the historical information: the Platte was a wooded river coursing through a prairie landscape and the Platte was a prairie river dominated by nonarboreal vegetation.

The complex braided system of the historical Platte River provided a variety of physical environments for vegetation, each with a different set of conditions for vegetation. The most commonly recognized settings include the active channel, abandoned channels, banks, low islands, and high islands.

The active channel of the river was generally without vegetation except during summer low-flow conditions, when annual plants colonized portions of the exposed bed. Although the stream was not normally more than a foot deep (except during floods), its current was swift, and the unstable sandy sediments were not a suitable substrate for vegetation. The river was so shallow that steamboats and keel boats were not used by the early fur trappers who began using the river as early as 1807, and they relied on bull boats, lightweight wooden-frame craft with stretched animal skins (Chittenden 1935). Before the construction of large storage reservoirs upstream, the active channel was much wider than the present channel, as shown by photographs and aerial photographs made before the complete effects of upstream controls became apparent.

Abandoned channels are common in other modern braided systems, and they occurred in the presettlement Platte River. When channels were abandoned by active flows, they sometimes became the location of stable cottonwood forests. A military observer during the middle 1800s described just such a case on the

Platte. Lieutenant Woodbury verbally described and mapped a heavily timbered half-mile-wide strip of trees occupying an abandoned channel near Grand Island. Such dense woodlands probably assumed a patch shape that mimicked the shape of the underlying abandoned channel, a geometry commonly observed in other modern rivers on the plains.

During the presettlement period, the outer banks of the central Platte River marking the general limits of fluvial activity were apparently the locations of cottonwood-dominated woodlands, with trees growing in isolation from each other or in limited groves. Accounts of surveyors and travelers usually described the trees on the banks as scattered, sparse, or absent in some cases from the outer, high banks of the river. Cottonwoods along the banks were often in a broken line of trees. Levi Jackman, an 1847 Mormon pioneer, penned a typical description: "cottonwood skirting the river is all the timber to be found, and very scarce at that" (Jackman 1847). This general quote probably indicates the general nature of the banks along the Platte River west of Kearney along the Oregon Trail, which joined the river at that point. The most precise information regarding vegetation along the river is from GLO surveys, cited here and elsewhere in this report, and they specify the vegetation conditions a decade or more after this example representative statement. Willow, prairie shrub, and grasses occurred along with the stream-bank cottonwoods in a community probably affected by grazing and fire before the intervention of humans several thousand years ago. Bison and deer, for example, grazed the area before the introduction of domestic livestock. The heavy use of the Oregon Trail, which followed the length of the central Platte, introduced exceptionally heavy grazing by the animals of emigrants bound for California, Oregon, and Utah. Grazing was so intense in 1849 and 1850 that travelers who previously had crossed from one side to the other in search of forage found that both sides of the river had been so heavily grazed that no forage remained (Steed 1850). The role of fire, a common feature of the prairie ecosystem, was also likely to have been a control on the structure of vegetation communities on the outermost river banks, but less so on riverine islands.

Numerous islands occurred between the high, outer banks and among the multiple channels of the river. They were of two general types: smaller, low islands rising a few inches to a few feet above the average flow of the stream, and larger, high islands that rose several feet above the flow. The small islands were apparently the locations of relatively dense woodlands with substantial understories. GLO surveyors described those islands as "innumerable," "numerous," and "scattered promiscuously." They characterized the river as being "studded with" and "filled with" islands of all sizes. The surveyors used the descriptors "brush," "brushy islands," and "covered with timber and undergrowth" for the islands. Surveyors' 1863 notes for the river just downstream of the town of Grand Island stated that "the greater part of the timber is on the small islands in the Platte River . . . there are many small islands in the river covered with timber and undergrowth" (cited by Johnson and Boettcher 2000a). Islands between the banks probably hosted relatively dense forest cover with substantial undergrowth because they were probably exposed to lower rates of fire and grazing than the prairies along the banks; present knowledge is insufficient for us to know how much lower. Grass fires can leap across wide unvegetated spaces (Wilson 1988), but the active channel of the central Platte River was a mile or more across, large enough to protect many midchannel islands. One surveyor noted that "there is no timber except that confined to the islands which is protected from the fall fires."

BOX 2-1 (continued)

Fire frequency was probably not reduced as much on the very large, high islands, such as Grand Island, which is so large that fires might have been ignited on the island itself. The island was also separated from the banks of the main river by more narrow active zones than in the case of the smaller islands. The vegetation cover of the very large islands consisted of more scattered cottonwoods, probably with more grass in the understory.

In summary, the ecological conditions of the central Platte River in presettlement times is not completely known, and the mechanisms that maintained the vegetation communities and their structure have not been completely investigated. Many details of the early vegetation communities and their structure are open to some debate. There is enough information from early accounts, however, to develop a first approximation: the wide active channels had little or no vegetation except for annual plants during low-flow periods, the banks had scattered trees and small groves with prairie-like vegetation, smaller islands were numerous and heavily wooded with dense undergrowth, and large islands had some woodland and some grassland. A successful restoration of the central Platte River ecosystem will include all those components.

in the Platte River downstream of the mouth of the Loup (Peters et al. 1989) found that rock substrates supported the highest numbers of organisms per unit area, 65,245/m^2; most were chironomids and trichopterans. Sand had the next highest density, 8,218/m^2, followed by gravel, 7,576/m^2. Invertebrate densities on silt substrates and on submerged wood totaled 6,610/m^2 and 6,572/m^2, respectively.

The invertebrate community in the Platte River before European settlement is unknown. Woody debris is now an important substrate for aquatic invertebrates, but how important it was in determining invertebrate diversity and abundance before European settlement and water development is undeterminable.

Vertebrates

It is difficult to assess the change in relative abundance or species composition of the various fish, amphibian, reptile, bird, or mammal faunas in response to changes in the hydrology or vegetation of the central Platte ecosystem since European settlement. Data to quantify accurately the comparative abundance of any species then and now do not exist.

Fishes

The present fish fauna of the Platte River includes about 100 species (in 20 families), of which 76 are native to at least a portion of the basin

(Schainost and Koneya 1999). Wide fluctuations in flows with flooding and high turbidity followed by low flows and high water temperatures impose special restrictions on the biota of the Platte River. Many fish species native to the mainstream Platte—including red shiner, sand shiner, river shiner, bigmouth shiner, western silvery minnow, plains minnow, speckled chub, flathead chub, river carpsucker, quillback, and channel catfish—are adapted to widely fluctuating conditions. Pre-1940 records also indicate that several species—such as shovelnose sturgeon, sturgeon chub, and sauger—were found in the Platte River drainage as far west as Wyoming. In addition, headwater, tributary, and spring-fed side-channel reaches support species that require clear or cool water; some, such as horny head chub, have been extirpated from the basin, but others—such as northern redbelly dace, finescale dace, plains topminnow, and Topeka shiner—are found in isolated populations in the drainage. In general, the number of native species declines in the western portion of the basin, where 30 native species have been recorded in the North Platte Basin in Wyoming and 26 in the South Platte Basin in Colorado. In addition, the proportion of nonnative fish species increases to almost 50% in the North Platte Basin in Wyoming and 41% in the South Platte Basin in Colorado (Schainost and Koneya 1999).

Herpetofauna—Amphibians and Reptiles

The herpetofauna of the Platte drainage includes salamanders (two species), frogs and toads (11), turtles (eight), lizards (11), and snakes (29). The most common representatives of these groups along the Platte River are the tiger salamander (*Ambystoma tigrinum*), the Rocky Mountain toad (*Bufo woodhousii*), the western striped chorus frog (*Pseudacris triseriata*), the painted turtle (*Chrysemys picta*), and the spiny softshell (*Trionyx spiniferus*).

Birds

A diverse assemblage of birds use the Platte River. The many life-history types include residents (such as blue jays, woodpeckers, and quail), migratory birds that are summer (breeding) residents (such as terns, plovers, warblers, orioles, and kingbirds), and migratory nonbreeding birds that use the river or floodplain only during spring and fall migration (such as cranes, ducks, geese, shorebirds, and songbirds).

The Platte River's woodlands are especially rich in species. Scharf (2003) netted 77 species on the Platte's floodplain (grassland, woodland, and channel habitats), including 19 species of summer-nesting neotropical migrants that are in decline in other parts of their ranges (Dobkin 1994; Robinson et al. 1995). Three that nest in riparian woodlands—the yellow-billed cuckoo, Bell's vireo, and western kingbird—are considered priority species for conservation in Nebraska because of their continental declines (Forsberg 1999). Colt (1997) captured 50 species of nesting birds in the Platte's woodlands, of which 31% were neotropical migrants. He also found high fledging rates and low rates of cowbird parasitism. Thus, as the Platte River's woodlands have expanded coincidentally with water development, songbird populations have increased accordingly, ironically offsetting some of the effects of the huge losses of riparian woodland caused by dam building and poor land management in much of the Great Plains (Knopf et al. 1988). Open (i.e., non-forested grassland or wetland) habitat is also important to bird species of conservation concern in Nebraska (Appendix B). Forty-one (57%) of 72 of these species occur on the central Platte and most (27 species, 66%) of the 41 species are associated with open habitat. Another eight species are characterized by using partially wooded (i.e., open woodland or savanna) habitats. Only six species in Appendix B (American woodcock, black-billed cuckoo, great-crested flycatcher, long-eared owl, ovenbird, Baltimore oriole) require more closed forest.

Large numbers of migratory waterbirds migrate through Nebraska's central Platte River Basin and adjacent Rainwater Basin area each year. Population surveys and management activities for waterfowl and cranes treat those two areas as one regional staging area because extensive interchange of birds between the two wetland systems occurs (Cox and Davis 2003). Variations in staging bird numbers from one year to another are large because migration patterns are determined by frequent, variable weather systems and by physiological needs of migrating birds. In years when spring advances slowly, larger numbers of birds will stage for longer periods in the central Platte Valley and the Rainwater Basin because conditions farther north are still frozen. In contrast, when spring develops rapidly and over a large area, migrating waterbirds may move through the Rainwater Basin area and central Platte River Basin more quickly (Cox and Davis 2003).

The Rainwater Basin area serves as important habitat for migrating waterbirds but is also susceptible to periodic drought and disease outbreaks (Chapter 5). Waterbird habitats in or near the central Platte River, in contrast, are dominated by flowing water or extensive exchange between groundwater and surface water; these habitats are not as susceptible to the occurrence of drought or disease. The central Platte River, therefore, provides important habitat for spring-staging waterbirds at times when little habitat is available elsewhere and when low temperatures temporarily freeze shallow wetlands in the adjacent Rainwater Basin area.

On the central Platte, the primary types of habitat used by staging waterbirds are open water, wetland, meadows, grassland, and cropland.

Forest communities in the central Platte River do not provide staging habitat for waterbirds. Four populations of geese, one population of sandhill crane, and several species of ducks depend on current staging habitats (Table 2-2). Over 85% of the midcontinent sandhill crane, snow goose, Ross's goose, and greater white-fronted goose populations migrate through the central Platte River region. The distribution of those waterbird populations, such as the midcontinent population of sandhill cranes, is now restricted to a narrow band of migration habitat in the central United States (Figure 2-21).

The central and lower Platte River hosts a variety of species, and restoration of the river to benefit one species should not unduly impair other species. That issue arises because clearing woodland areas to benefit whooping cranes that prefer long, open sight lines may reduce the available habitat for songbirds. The lessons from this observation are that such decisions should take into account the general distribution of the species in question, the legal standing of the species, and the ecosystem perspective. First, most of the songbirds have a general distribution in the region (Davis 2001), so the loss of some woodland in and along the river should not adversely affect the general population. However, whooping cranes are strongly connected with the river and its habitats, and there are few alternatives. Wholesale removal of woodlands without regard to the effects on other species would be irresponsible; recognition and mitigation of effects wherever possible constitutes wise management.

Second, The Nebraska Partnership for All-Bird Conservation identified 72 bird species of conservation concern in Nebraska (Appendix B). However, whooping cranes, piping plovers, and interior least terns have special legal standing in that they are federally listed as endangered or threatened species, whereas the songbirds and other bird species of concern are not. That does not mean that the interests of songbird and waterfowl populations may be safely disregarded, but special efforts must be taken on behalf of the listed species.

Finally, an ecosystem perspective would logically suggest that the central and lower Platte River should be a diverse ecosystem when viewed as a whole, with adequate accommodation for both listed and nonlisted species. The diversity ought to be manifest over the entire length of the central and lower river, whereas it might not be present in some localities and areas of restricted spatial extent (a square mile or so, for example). In the final analysis, provision should be made to care for all species, not to see the elimination of one from the total ecosystem of many square miles in extent.

Arriving at clear and achievable conservation goals in situations like these requires good science but also good policy and good process. How those elements are integrated in this report is discussed in Chapter 4.

TABLE 2-2 Proportions of Waterfowl Populations That Use Nebraska's Rainwater Basin Area and Central Platte River Valley

Species	Population Considered	Total No. Birds Population[a]	Citation	Number of Birds in Central Platte River Valley	Citation	Proportion of Population Using Central Platte
Sandhill crane	MCP	435,050	Sharp et al. 2003[b]	375,875[c]	Solberg 2002[b]	86%[b]
Mallard	MCP	7,785,800	USFWS 2003	4,097,000	Gersib et al. 1990	50%
Northern pintail	MCP	2,547,970	USFWS 2003	756,000	Gersib et al. 1990	30%
Total ducks[d]	MCP	34,527,900	USFWS 2003	23,815,000	USFWS 2003	61%
Snow goose, Ross' goose	MCP	2,490,800	USFWS 2003	2,679,300	J. Drahota, pers. comm., U.S. Fish and Wildlife Service, 2003	90%
Canada goose	GPP/WPP	651,330	USFWS 2003	558,000	J. Drahota, pers. comm., U.S. Fish and Wildlife Service, 2003	75%
Canada goose	SGPP	160,570	USFWS 2003	110,000	J. Drahota, pers. comm., U.S. Fish and Wildlife Service, 2003	25%

Canada goose	TGPP	421,900	USFWS 2003	165,000	J. Drahota, pers. comm., U.S. Fish and Wildlife Service, 2003	50%
Greater white-fronted goose	MCP	802,200	USFWS 2003	950,000	Gersib et al. 1990	90%

[a]Total number of birds considered in population is most recent 3-year average of breeding, migration, or winter survey used by Cooperative Flyway Management Plans to depict population status.
[b]These estimates of MCP sandhill crane population are indices, represent most recent 3-year average (Dave Sharp, U.S. Fish and Wildlife Service, pers. comm., 2003), and probably underestimate total proportion that migrates through Platte River Valley, because of population turnover. Telemetry and other marked-bird data for sandhill cranes suggest that as much as 99% of MCP migrates through Platte River Valley (Gary Krapu, U.S. Geological Survey, pers. comm., 2003).
[c]This estimate includes birds in North and central Platte. About 15-20% of sandhills migrating through the Platte River Basin go through the North Platte region, and remainder go through central Platte (Gary Krapu, U.S. Geological Survey, pers. comm., 2003).
[d]Includes all *Anatidae* (it includes canvasback, redhead, ring-necked duck, lesser scaup, gadwall, northern shoveler, blue-winged teal, and American widgeon); does not include scoters, eiders, mergansers, long-tailed ducks, and wood ducks.
Abbreviations: MCP, midcontinent population; GPP/WPP, Great Plains population/western prairie population; SGPP, short-grass prairie population; TGPP, tall-grass prairie population.

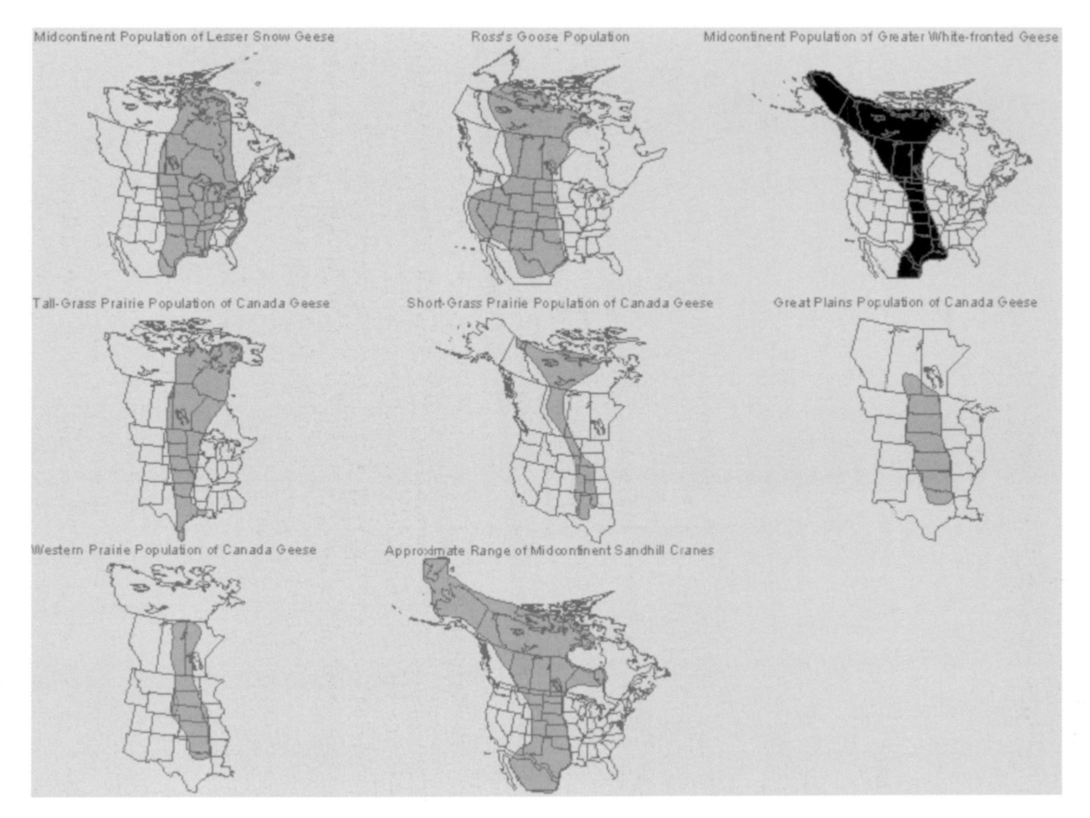

FIGURE 2-21 Generalized annual migration from southern to northern latitudes. Source: Adapted from Sharp et al. 2003.

3

Law, Science, and Management Decisions

The issues related to threatened and endangered species in the Platte River Basin are products of complicated interactions between legal frameworks and applied science. This chapter reviews the Endangered Species Act (ESA) from a legal and policy perspective, examines the practice and application of science in implementing the ESA, and concludes by placing the Platte River conflict in a broader national context.

The evolution and codification of national wildlife policy in the United States provide the context for the present issues concerning the four threatened or endangered species in the Platte River Basin. Wildlife has played an important role in America's cultural self-image. North American Indians often have regarded wildlife as sacred, and many tribes were divided into clans named for wildlife species. When European settlers founded the United States, symbols of national vitality began with the bald eagle and continued with bison and other wildlife representing the character of the nation. Fish and wildlife have continued to play important roles as cultural images, ranging from symbols of political jurisdictions, such as tribes and states, to nicknames for schools and sports teams. Thomas Jefferson's first description of the nation's physical geography began with a discussion of rivers and fishes (Jefferson 1787), and John James Audubon's works brought American birdlife to the attention of the world. Concern about the extinction of species during the nineteenth century attracted considerable attention, and at the dawn of the twentieth century Congress adopted the Lacey Act of 1900 as the first national wildlife-protection statute. The sponsor of the act, John F. Lacey, expressed the issue clearly: "[in] many of the states the native birds have been well-nigh exterminated" (Lacey 1900). In a related development, President Theodore Roosevelt established the nation's first wildlife refuge in 1903 at Pelican Island, Florida.

During the twentieth century, interest in protecting native wildlife species grew. Determined to avoid the loss of more species, as occurred with the passenger pigeon and the Carolina parakeet, and spurred by drastically declining numbers of popular, visible species—such as the brown pelican, wading birds, and some species of game ducks—state and federal policymakers established refuges and management programs. Since 1966, efforts at the federal level have included three statutes that provide the basic framework for the nation's policy regarding endangered wildlife species.

The present federal legislative authority for dealing with endangered species is the product of a progression of three major acts of Congress. The first, the Endangered Species Preservation Act of 1966, was the formal beginning of federal efforts. It was ineffective because of several minor flaws and three major shortcomings: it did not prohibit the taking of endangered species, it did not recognize all types of endangered species, and it did not provide habitat protection (Bean and Rowland 1997). The Endangered Species Conservation Act of 1969 clarified congressional intent in some ways, but it did not rectify many of the problems in the earlier act. President Richard Nixon expressed widely held concerns that existing law "simply does not provide the kind of management tools needed to act early enough to save a vanishing species" (Nixon 1972).

LEGAL AND INSTITUTIONAL BACKGROUND

Endangered Species Act

Management of lands and water in the Platte River Basin is subject to a complex web of local, state, and federal law. The federal ESA, however, is a major regulatory force that limits land and water use in the basin. Federal decisions under the ESA triggered the present committee's review and have been a primary motivating force behind development of the Platte River Cooperative Agreement.

The Endangered Species Act of 1973 remedied the shortcomings of the earlier legislative attempts, and it is the defining instrument of present policy regarding imperiled wildlife (16 U.S.C.A. § 1531-1544). Congress found that the ESA was necessary to protect the "esthetic, ecological, educational, historical, recreational, and scientific value" provided by fish, wildlife, and plants. The major purposes of the act are to provide for the conservation of endangered species and the ecosystems on which they depend. The act defines *conservation* as the use of all methods necessary for the recovery of species to the point where they can be removed from the protected list. Species are protected under the act if they are listed as "endangered" or "threatened." An endangered species is one that is in danger of extinction throughout all or a substantial part of its range, and a threatened species is one that is likely to become endangered in the foreseeable future.

The ESA is focused on individual listed entities. It therefore does not protect ecosystems themselves, but only to the extent that they are needed by listed species. The ESA may be compatible with, but it does not require, the broader protection of ecosystems or biodiversity.

Designation of Critical Habitat

The focus of this committee's review is the habitat needs of the Platte River endangered and threatened species. The ESA protects *critical habitat*, defined as the specific areas that contain physical or biological features essential to the conservation of the species and that may require special management considerations or protection (ESA § 3(5)). This report uses the term only to refer to areas that have been formally designated under the ESA.

The congressional authors of the ESA envisioned that the designation of critical habitat by the U.S. Fish and Wildlife Service (USFWS) would take place at the same time that a species was listed, but this provision of the act has not generally been carried out. USFWS believes that designation of statutory critical habitat provides little additional protection for listed species and instead has relied on other methods—including agreements with federal agencies, states, tribes, and private persons or organizations—to promote conservation of endangered species. The designation of critical habitat has often been accompanied by long legal challenges, so throughout the nation most current critical habitat designations are either by court order or by court-supervised settlements (McCue 2003). The critical habitat provisions of the ESA have long been perplexing and controversial. As enacted (in 1973), the ESA contained a regulatory provision that limited modification of critical habitat by federal actions, but it did not define the term "critical habitat" or provide a process for its designation (Patlis 2001). USFWS and the National Marine Fisheries Service jointly published their initial interpretation of the term critical habitat in 1975 (Fed. Regist. 40: 17764 [1975]). Using that interpretation, USFWS issued its first proposed determinations of critical habitat, covering six species, including the whooping crane, later in that year (Fed. Regist. 40: 58308 [1975]). In 1978, Congress added a statutory definition of and process for designating critical habitat (Patlis 2001).

Areas outside the present range of a species may be included in critical habitat but only if designation limited to the present range would be inadequate to ensure its conservation (50 CFR 424.12(e)). Because *conservation* is defined to mean progress toward recovery and delisting, critical habitat must include sufficient habitat to support a recovered population, which may be larger than the population at the time of listing or larger than a minimal viable population.

In determining critical habitat, USFWS (50 CFR § 424.12(b)) considers the species need for

- Space for individual and population growth and for normal behavior.
- Food, water, air, light, minerals, or other nutritional or physiological requirements.
- Cover or shelter.
- Sites for breeding, reproduction, and rearing of offspring.
- Habitats that are protected from disturbance or are representative of the historical geographic and ecological distributions of a species.

In addition, according to its *Endangered Species Listing Handbook* (USFWS 1994), the agency considers both species and habitat dynamics when determining critical habitat. If the species requires ephemeral habitats, for example, the designation should consider the potential location of future habitats. Unoccupied areas may be included in critical habitat if, for example, they provide landscape connectivity between occupied areas, support pollinators or organisms involved in seed dispersal, or provide areas into which the population might need to expand (Fed. Regist. 68 (151): 46715 [2003]). However, the handbook directs the service not to include areas that are unsuitable for use by the species unless they are essential to conservation of the species (Figure 3-1).

Critical habitat determination may be most difficult where more territory is occupied at the time of designation than is necessary for the survival and recovery of the species. The statute and USFWS regulations and guidance provide little indication of how the agency will identify critical habitat within the occupied range. That identification raises issues of equitable distribution of the economic and other impacts of species protection and scientific issues of what is best for the species. One recent critical habitat determination states that the agency set priorities for designation among areas that were already subject to some protection and areas with minimal habitat fragmentation (Fed. Regist. 68 (151): 46684 [2003]).

The description of critical habitat is supposed to include a list of the physical and biological features essential to the species, which are referred to as primary constituent elements (50 CFR 424.12(b)). Management of critical habitat focuses only on those features (50 CFR 17.94(c)), and critical habitat designation therefore does not limit actions in a designated geographic area that do not affect the primary constituent elements (Fed.

FIGURE 3-1 An example reach of the central Platte River with suitable habitat for whooping crane, piping plover, and least terns. This reach, near Shelton has substantial open areas with long sight lines. Source: Photograph by W.L. Graf, August 2003.

Regist. 68 (151): 46684 [2003]). Primary constituent elements are often only vaguely articulated in the critical habitat rule. A federal district court has held that identifying as primary constituent elements for the Rio Grande silvery minnow only water of "sufficient" quality and quantity did not provide an adequate standard (Middle Rio Grande Conservancy District v. Babbitt, 206 F. Supp. 2d 1156 [D. N.M. 2000]). Another federal district court has rejected a general description that the species requires a "suitable range" of temperatures or habitat patches of "sufficient size" to prevent isolation (Home Builders Association of Northern California v. USFWS, 268 F. Supp. 2d 1197 [E.D. Cal. 2003]).

Like listing decisions, critical habitat determinations must be made "on the basis of the best scientific data available." A recent General Accounting Office report concluded that critical habitat designations generally do rest on the best available science, but the available data are often narrowly limited (GAO 2003). In contrast with listing decisions, however, critical habitat designation must also take into consideration "the economic impact, and any other relevant impact," of that designation. Areas may be excluded from critical habitat if the agency concludes that the benefits of exclusion outweigh the benefits of inclusion unless it finds, on the basis of the best scientific data available at the time of determination, that exclusion will result in the extinction of the species (ESA § 4(b)(2)).

Until recently, USFWS used a "baseline approach" to economic analysis, considering only the economic impacts imposed specifically by critical habitat designation and excluding the baseline economic impacts resulting from the listing of the species. Because, as explained below, USFWS believes that critical habitat designation has virtually no effect beyond that of listing, this approach greatly simplified the economic analysis of critical habitat designation. In most cases, USFWS found that critical habitat designation would have no economic effects beyond the baseline effects of listing. In 2001, however, a federal appeals court ruled that the baseline approach was unlawful (New Mexico Cattle Growers Association v. U.S. Fish and Wildlife Service, 248 F.3d 1277 [10th Cir. 2001]). Consequently, USFWS now considers all economic impacts, including ones that are coextensive with the impacts of listing, when it designates critical habitat. In its economic analyses, USFWS attempts to forecast all costs of conducting Section 7 consultations (discussed below) with respect to the species and costs of revising or forgoing projects on the basis of the consultations.

USFWS recently began making aggressive use of the authority to exclude areas from critical habitat on the grounds that the benefits of exclusion outweigh the benefits of inclusion. This practice has not yet been tested in litigation. The agency has not articulated a general process for determining when such exclusions are appropriate; instead, it makes exclusions on an ad hoc basis. In one recent decision, it excluded entire counties in California from critical habitat for a number of vernal pool species because it determined that the costs imposed by including habitat in those counties would be disproportionately high compared with costs imposed in other areas and for other species (Fed. Regist. 68 (151): 46745 [2003]).

USFWS also has taken the position that areas can be excluded from critical habitat designation if they do not require special management. It excludes areas from critical habitat as adequately managed if a management plan or agreement that is in effect provides sufficient conservation benefit to the species, it provides adequate assurances that its conservation strategies will be implemented, and it provides sufficient assurances—for example, through monitoring and revision procedures—that the strategies will be effective (Fed. Regist. 66 (22): 8543 [2001]). A federal district court has recently held, however, that the ESA does not permit exclusion from critical habitat on the grounds that adequate management provisions are already in place (Center for Biological Diversity v. Norton, 240 F. Supp. 2d 290 [D. Ariz. 2003]). In recent critical habitat determinations, USFWS has relied on a different basis for excluding managed areas. The statute provides that any area may be excluded from critical habitat if the benefits of exclusion outweigh those of inclusion, provided that exclusion will not cause the extinction of the species (ESA § 4(b)(2)). The agency's current interpretation is that where lands are already managed for conservation of the species, the benefits of critical habitat designation are minimal and therefore easily outweighed by the potential resource costs and delays imposed by the consultation requirement. USFWS treats this basis of exclusion from critical habitat as broader than the "no special management required" criterion, such that it allows exclusion even in the absence of an approved management plan (Fed. Regist. 68 (151): 46751 [2003]).

The ESA requires designation of critical habitat "to the maximum extent prudent and determinable" at the time a species is listed (ESA § 4(a)(3)). According to agency regulations, designation is not prudent if it would increase the threat of deliberate human taking or it "would not be beneficial to the species" (50 C.F.R. 424.12(a)(1)). USFWS once took a very broad view of the "not prudent" exception, frequently declining to designate critical habitat because it would provide little incremental protection beyond that provided by listing alone. Several court rulings, however, have required the agency to make a more specific showing that the benefits of critical habitat designation are outweighed by specific threats to the species, such as the threat of increased collection activity, if it wants to invoke the "not prudent" exception (e.g., Natural Resources Defense Council v. U.S. Dept. of Interior, 113 F.3d 1121 [9th Cir. 1997], Sierra Club v. USFWS, 245 F.3d 434 [5th Cir. 2001]).

Critical habitat is not determinable when the impacts of designation cannot be analyzed or the biological needs of the species are not sufficiently well known to permit identification of an area as critical habitat (50 CFR 424.12(a)(2)). Because so little is known about many dwindling species, identifying critical habitat at the time of listing often poses a serious challenge. For that reason, commentators, including a National Research Council committee, have proposed that short-term "survival habitat" be the focus at the time of listing, leaving the detailed identification of critical habitat and the accompanying economic evaluation for the recovery-planning process (NRC 1995). The statute, however, sharply limits delay while additional information is developed. USFWS can delay designation for only 1 year on the grounds of lack of information (ESA § 4(6)(C)(ii)). At the end of that year, it must designate critical habitat on the basis of the available data.

USFWS has long believed that designation of critical habitat is "an expensive regulatory process that duplicates the protection already provided by the jeopardy standard" (Fed. Regist. 64 (113): 31871 [1999]). In recent years, USFWS has been, in its words, "inundated with citizen lawsuits" based on its failure to designate critical habitat (Fed. Regist. 64 (113): 31872 [1999]). In response to the deluge of critical habitat lawsuits, the U.S. Department of the Interior (DOI) has requested and Congress has approved a cap on spending on critical habitat determination to ensure that critical habitat work does not consume the entire ESA listing budget. During the summer of 2003, the assistant secretary of the interior for fish and wildlife and parks testified that the flood of critical habitat litigation was preventing USFWS from adding deserving species to the protected list and was delaying recovery efforts for already-listed species.

Once designated, critical habitat may be revised as new data become available, but there is no specific requirement for periodic revision. The listing agencies are required to review the status of listed species every 5 years to determine whether they should be delisted or their classification should be changed (ESA § 4(c)(2)). If review shows that critical habitat should be revised, USFWS has said that it "will take appropriate action" (Fed. Regist. 45: 13010 [1980]). However, USFWS has no funding even for new critical habitat designation. It is highly unlikely that the agency would choose to revise existing critical habitat designations in the current budget climate even if specifically asked to do so. In recent years, when USFWS has found critical habitat designation warranted in response to petitions, it has not proposed revisions. Nonetheless, there may be limits to the discretion that USFWS enjoys to delay revisions of critical habitat. In a case involving the Cape Sable seaside sparrow, a federal court has ruled that USFWS cannot indefinitely delay critical habitat revisions when, in response to a petition, it has acknowledged that revisions are necessary to adequately protect a critically endangered species. The court did not set a timetable for revisions but required USFWS to set and explain a schedule for making them (Biodiversity Legal Foundation v. Norton, 285 F. Supp. 2d 1 [D.D.C. 2003]).

Designation of Critical Habitat for Platte River Species

Several areas in the whooping crane's migratory pathway, including the main channel of the Platte River, and its immediately associated riparian habitat between junction of U.S. Highway 283 and Interstate 80 near Lexington to the interchange for Shelton and Denman, Nebraska, were designated as critical habitat for the whooping crane in 1978 (Fed. Regist. 43: 20938 [1978]) before addition of the statutory definition of critical habitat to the ESA. USFWS later proposed to designate several additional critical habitat units, including a portion of the Niobrara River and several national wildlife refuges in North Dakota and South Dakota (Fed. Regist. 43: 36588 [1978]), but that proposal was withdrawn (Fed. Regist. 44: 12382 [1979]). At the time of critical habitat designation for the whooping crane, USFWS was applying its regulatory definition, which focused on areas and elements whose loss would appreciably decrease the likelihood of the survival and recovery of the species. With respect to the Platte River unit,

USFWS noted that this area "forms the most important stopping site on the migration route of the whooping crane" (Fed. Regist. 40: 58308 [1975]):

> Historical data show that this area, sometimes called the "Big Bend" area of the Platte River, was a focal point through which the whooping cranes passed before spreading out to their wintering grounds to the south and their breeding grounds to the north. There are more old records of the presence of the species here than in any other part of the migration route, and recent confirmed records indicate continued heavy use within the last few years. Available information indicates that the combination of the Platte River channel, and adjacent wet meadows, rainwater basins, and farmlands form a unique association of habitats that is the most valuable part of the entire migration route of the species. Reduction in the quality or size of this habitat association, especially in the water level of the area, could be expected to have an adverse effect on the surviving population of the species.

When the piping plover was listed, USFWS indicated that critical habitat was not determinable. A lawsuit later resulted in a court order that required USFWS to designate critical habitat. In response to that order, USFWS issued a final rule that designated critical habitat for the northern Great Plains population of the piping plover on September 11, 2002 (Fed. Regist. 67 (176): 57638 [2002]) and that stated:

> Within the geographic area occupied by the species . . . we designate only areas currently known to be essential. Essential areas should already have the features and habitat characteristics that are necessary to conserve the species. We will not speculate about what areas might be found to be essential if better information becomes available, or what areas may become essential over time. If the information available at the time of designation does not show that an area provides essential life cycle needs of the species, then the area should not be included in the critical habitat designation.

At the same time, USFWS recognized that habitats and species are both dynamic. It stated that "critical habitat designations made on the basis of the best available information at the time of designation will not control the direction and substance of future recovery plans, habitat conservation plans, or other species conservation planning efforts if new information available to these planning efforts calls for a different outcome."

The primary constituent element identified for the critical habitat for the piping plover is "the dynamic ecological processes that create and maintain piping plover habitat." On rivers, the primary physical constituent elements were identified as "sparsely vegetated shoreline beaches, peninsulas, islands composed of sand, gravel, or shale, and their interface with the water bodies." A total of 19 units were identified as critical habitat for the piping plover, ranging across five states and covering a total of over 180,000 acres (over 74,000 ha) and 1,200 river miles. In Nebraska, the Platte River was designated as critical habitat for the piping plover from Lexington to the confluence with the Missouri River, a distance of 252 mi. The entire Loup River (68 mi) and the eastern portion of the Niobrara River (120 mi) were also designated. USFWS recognized that some of this area might not be occupied in any given year but stated that "designation is necessary because of the dynamic nature of the river. Sandbar habitats migrate up and down the rivers resulting in shifts in the location of primary constituent elements." USFWS excluded the shoreline of Lake McConaughy from its critical habitat delineation on the grounds that it was already adequately managed under plans developed by the Central Nebraska Public Power and Irrigation District. It also excluded sand pits on the grounds that they do not meet the physical and biological requirements of critical habitat.

No critical habitat has been designated for the interior least tern or pallid sturgeon. When it listed the interior least tern, USFWS concluded that designation of critical habitat was not prudent, because it would provide no demonstrable overall benefit to the tern (Fed. Regist. 50: 21790 [1985]). When it listed the pallid sturgeon, USFWS found that designation of critical habitat was not prudent and that critical habitat was not determinable (Fed. Regist. 68 (151): 46684 [2003]).

Regulatory Provisions of Endangered Species Act and Effect of Critical Habitat

Given current regulatory definitions, the designation of critical habitat has little, if any, direct regulatory impact. Nonetheless, critical habitat has frequently been a flashpoint for controversy. Objections to critical habitat designation seem to rest in part on misunderstanding of its legal significance, but may also have some justification. Apart from its narrow direct impacts, critical habitat designation may have indirect regulatory impacts. In a practical sense, it may affect the attitudes of regulators, federal agencies, and property-owners or resource-users, subtly altering the regulatory landscape. In addition, the current regulatory definitions have been rejected by one federal court and may eventually have to be revised.

Species listed as endangered or threatened under the ESA benefit from the two major regulatory provisions of the act: Section 7 and Section 9. Section 7 requires that federal agencies carry out programs for the conservation of listed species and ensure that the actions they take, fund, or authorize are not likely to jeopardize the continued existence of any listed species or to result in the destruction or adverse modification of designated critical habitat. Section 9, which applies to both federal and nonfederal actors, prohibits the taking of endangered and threatened animal species unless USFWS produces a special rule allowing limited take. Section 10 allows USFWS to authorize, by permit, acts that would otherwise be prohibited by Section 9 if the applicant submits an acceptable habitat conservation plan (HCP). A permit can be granted if the take is incidental to and not the purpose of the permitted action, the impacts will be minimized and mitigated to the greatest extent practicable, adequate funding to implement the plan is ensured, and the taking will not appreciably reduce the likelihood of survival and recovery of the species in the wild (ESA Section 10(a)(2)(B), 16 USC 1539(a)(2)(B)).

Critical habitat plays a direct role only in the operation of Section 7: federal actions must neither cause jeopardy nor destroy or adversely modify critical habitat. Section 7 is implemented through a process of formal or informal consultation. A federal agency whose actions may adversely affect a listed species must seek formal consultation with USFWS. The action agency prepares a biological assessment, detailing what it believes will be the impacts of its action. USFWS reviews the biological assessment and issues a biological opinion that the action will or will not jeopardize the continued existence of the species or destroy or adversely modify designated critical habitat (50 CFR Part 402). Jeopardy opinions must suggest any reasonable and prudent alternative (RPA) that will serve the purpose of the proposed action without causing jeopardy or adverse modification (ESA Section 7(b), 16 USC 1536(b)). The ultimate decision of whether the proposed action will cause jeopardy or impermissible effects on critical habitat remains with the action agency, which must use the best scientific data available at the time of the decision (ESA Section 7(a)(2), 16 USC 1536(a)(2)). But an agency that rejects the views of USFWS and proceeds with a project acts at its own peril. As the formal view of an agency with recognized expertise, a biological opinion carries considerable weight with a reviewing court.

Consultation can be a time-consuming and expensive process. It would be unusual, however, for critical habitat designation to increase the scope of the consultation requirement. Consultation is required for any action that may adversely affect a listed species. A biological assessment is generally required if a listed species may be present in the action area (50 CFR § 402.12), whether or not the action is within designated critical habitat. If the biological assessment results in the conclusion that the action may adversely affect the listed species or its critical habitat, formal consultation leading to a biological opinion is required (50 CFR § 402.14). Because effects on the species and its critical habitat are closely intertwined, it is hard to imagine a situation in which the critical habitat could be adversely affected without any adverse effect on the species.

Substantively, as a matter of law, the presence or absence of critical habitat currently makes little difference in the outcome of Section 7 consultations. USFWS has by regulation defined the "jeopardy" and "adverse modification" prongs of Section 7 so that they are virtually identical. An action is considered to jeopardize the continued existence of a species if it "reasonably would be expected, directly or indirectly, to reduce appreciably the likelihood of both the survival and recovery of a listed species in the wild by reducing the reproduction, numbers, or distribution of that species" (50 CFR § 402.02). An action is considered to destroy or adversely modify critical habitat if it "appreciably diminishes the value of critical habitat for both the survival and recovery of a listed species" (50 CFR § 402.02). A federal appeals court ruled in 2001 that the regulatory definition of *destruction* or *adverse modification* of critical habitat is inconsistent with the ESA, and therefore invalid, because Congress must have intended the "adverse modification" prong of Section 7 to provide protection additional to that provided by the "jeopardy" prong (Sierra Club v. USFWS, 245 F.3d 434 [5th Cir. 2001]). That decision is binding on USFWS only within the Fifth Circuit, which includes Louisiana, Mississippi, and Texas. In that circuit, USFWS has not yet formulated an official response to the decision. Outside the Fifth Circuit, USFWS continues to apply the challenged regulation while it is under review (Fed Regist. 68 (151): 46684 [2003]). If USFWS revises its regulations, critical habitat designation could have some additional substantive regulatory consequences.

In addition to its direct regulatory effects, critical habitat designation could conceivably have indirect effects that might account for some of the opposition to designation. The process of designating critical habitat requires a close examination of the ecological needs of the species. It may alert USFWS to the potential for a variety of actions that adversely affect the species and to areas that may be essential or desirable for the recovery of the species. With respect to federal actions, the consultation process should produce much the same kind of information. But those who might be affected by conservation efforts may fear any process that generates information about the conservation needs of a species. They may worry, for example, that federal agencies will choose to implement conservation efforts beyond those required to avoid jeopardy or adverse modification of critical habitat under the authority of ESA Section 7(a)(1), which directs all federal agencies to carry out programs for the conservation of listed species.

Critical habitat designation also could have some ramifications for nonfederal actions. It might make USFWS more sensitive to the possibility that specific actions could violate the take prohibition of Section 9. It could also make enforcement easier. The designation process might generate or unearth evidence or expert opinion that could be used to prove a violation of Section 9, and designation might help to persuade a jury that the area in question truly is essential to the species. Finally, by drawing attention to the importance of land or water for a listed species, critical habitat designation could affect the market price or marketability of land that is designated or that depends for irrigation on water required to maintain critical habitat stream flows.

As a practical matter, whatever its legal significance, critical habitat designation might affect how USFWS implements the ESA. Most agency decisions are made in district and field offices by personnel who might not have up-to-date training and expertise in the legal nuances of critical habitat designation or implementation. Agency personnel may believe that they have stronger regulatory grounds for prohibiting activities that could affect the species inside critical habitat than outside it. On the basis of the formal designation of critical habitat, they might also believe that they are obliged to vigorously protect those areas, which have been deemed essential to the species. Field personnel making decisions for other federal agencies and for private entities may react similarly. They might not see litigation as a desirable strategy even if they think that USFWS is overprotective of critical habitat. As a result, critical habitat may carry substantially greater practical power than close analysis of the law would suggest. It might be treated in practice as a well-defined boundary, and activities, such as water diversion or cellular-phone tower construction, might be prohibited inside but permitted outside the boundary even if the potential impacts on the species appear quite similar. That practical effect, however, seems to be substantially weaker with respect to critical habitat related to river flows than with terrestrial habitat. It is practical to draw an effective boundary across terrestrial habitat, but that is much more difficult for rivers. Designation of a single reach of a river as critical habitat can affect management of water projects just as strongly, and in precisely the same ways, as designation of the entire river.

Recovery Planning and Implementation

ESA Section 4 requires USFWS to produce and implement a recovery plan for each listed species unless the agency finds that a plan will not promote the conservation (i.e., the progress toward recovery) of the species (ESA § 4(f)). Recovery plans are frequently prepared by teams of experts drawn from various federal agencies and academe. The plans must include, to the greatest extent practicable, a description of site-specific management actions needed to conserve the species; objective, measurable criteria for species delisting; and estimates of the time and money needed to achieve the plan's recovery goals (ESA § 4(f)). USFWS tries to minimize the social and economic impacts of implementing recovery plans (Fed. Regist. 59: 34272 [1994]). Achievement of recovery goals identified in the plans may be a signal that consideration for delisting is appropriate, but it is neither necessary nor sufficient for delisting to occur. Delisting requires USFWS to determine, based solely on the best available scientific data at the time, that a species no longer meets the statutory definition of endangered or threatened.

Recovery plans have been produced for each of the listed species at issue in this report. The committee has taken the recovery goals in those plans at face value; we have not attempted evaluation of the scientific basis of the goals, which lies beyond our charge. Recovery plans provide a useful roadmap to recovery, at least under the state of knowledge at the time of their preparation. But the existence of a recovery plan does not guarantee recovery or even implementation of the steps outlined in the plan. Federal courts have been unwilling to force unwilling agencies to implement specific steps in recovery plans at specific times. Nonetheless, recovery plans often have substantial value for the species. The plans for the whooping crane and pallid sturgeon, for example, have been the basis of substantial expenditures and management activities on behalf of the species.

Endangered Species Act and Water Development in Platte River Basin

In recent years, a number of sharp conflicts have arisen over the implementation of the ESA in the context of operation of federal water projects. That those projects would be the nexus of such conflicts is not surprising. Freshwater fishes and other riparian species in the West are among the most endangered groups in the nation (e.g., Bogan et al. 1998; Ricciardi and Rasmussen 1999). Not coincidentally, most western rivers are overappropriated, with high demand for diversions from the instream flows that support riparian species. More than half of all species listed under the ESA are affected directly or indirectly by water-management projects (Losos et al. 1995), and a large proportion of water-management projects have a federal nexus. Federal water projects, which dot the arid West, are a frequent target of demands for operational changes to benefit listed species. It is generally easy to connect the projects to changes from historical hydrological patterns, which in turn may be clearly related to species declines. Furthermore, because operation of the projects is a federal action, an operating agency must consult regularly with USFWS under Section 7 (e.g., Rio Grande Silvery Minnow v. Keys, 333 F.3d 1109 [10th Cir. 2003]; Klamath Water Users Protective Association v. Patterson, 204 F.3d 1206 [9th Cir. 1999]) and make any changes necessary to avoid jeopardy. Even if a federal project is not responsible for all threats to the species, as is frequently the case, under Section 7 the operating agency must ensure that the project does not, in combination with those other threats, produce jeopardy or adverse modification of critical habitat (Pacific Coast Federation of Fishermen's Associations v. U.S. Bureau of Reclamation, No. C 02-2006 SBA [N.D. Cal., July 15, 2003]).

Many new water projects or diversions are also subject to Section 7 consultation because they have a federal nexus even if they are not carried out by a federal agency. Many require a permit from the Army Corps of Engineers under Section 404 of the Clean Water Act or a license from the Federal Energy Regulatory Commission (FERC) under the Federal Power Act. Others are funded by a federal agency or require the use of federal lands.

The Platte River Basin, like many others in the West, has been the site of bitter ESA conflicts. In the Platte Basin, those conflicts began more than 25 years ago, when USFWS issued a jeopardy opinion for the proposed Grayrocks Dam on the Laramie River in Wyoming. Litigation over Grayrocks was settled with an agreement under which Basin Electric Power Cooperative constructed the dam with less storage capacity than originally proposed and funded creation of the Platte River Whooping Crane Critical Habitat Maintenance Trust. Jeopardy opinions were issued in 1983 for the proposed Narrows Project on the North Platte River, in 1994 for diversions on National Forest land in the Front Range, and in 1997 for relicensing of the Kingsley Dam hydroelectric project. Thereafter, between 1998 and 2001, USFWS issued a series of nearly 20 jeopardy opinions for activities that would affect the central Platte (USFWS 2002a). All those jeopardy opinions concluded that proposed depletions to the Platte River would both jeopardize the continued existence of listed species and adversely modify the critical habitat in the central Platte region. Even without the formal designation of critical habitat, therefore, the opinions would have required changes in the proposed activities. In this series of biological opinions, USFWS did not flatly prohibit the proposed diversions; instead, it included RPAs calling for the diverters to contribute money to an account to be used for acquisition and maintenance of habitat along the central Platte.

In a 2002 biological opinion, USFWS announced that it had "adopted a jeopardy standard for all Section 7 consultations on Federal agency actions which result in water depletions to the Platte River" (USFWS 2002a, p. 8). The service noted that over 1,000 projects contemplated in the basin, most involving annual depletions of 25 acre-ft or less, might require Section 7 consultation in the near future. It therefore issued a single biological opinion for all such small diversions, determining that they would not jeopardize the listed species or adversely affect designated critical habitat provided that they were accompanied by conservation actions, including either replacement of the depleted water or payment of a specified amount into a habitat-mitigation fund. In the 2002 biological opinion, USFWS noted that groundwater pumping in the Platte River Basin would further degrade the ecosystem and threaten the survival and recovery of the listed species (USFWS 2002a, p. 57). Without further information, the service felt that it was not possible to predict how much groundwater depletion would occur in the action area. Because groundwater pumping generally does not have a federal nexus, it does not require consultation. However, USFWS must take the effects of groundwater depletion into consideration in Section 7 consultations on federal actions.

Role of Science in Implementing Endangered Species Act

The ESA repeatedly calls for use of the best available scientific information. Listing decisions must be based solely on the best available scientific and commercial information (ESA Section 4(b)(1)(A), 16 USC 1533(b)(1)(A)). The designation of critical habitat must be based on the best scientific data available and consider the economic and other relevant impacts of designation (ESA Section 4(b)(2), 16 USC 1533(b)(2)). Federal agencies must use the best scientific data available in fulfilling their duty to ensure that their actions do not jeopardize endangered species or destroy critical habitat (ESA Section 7(a)(2), 16 USC 1536(a)(2)).

Legislative mandates, such as those in the ESA, requiring an agency to consider the best available scientific data in their decisions have at least two aims. First, they are intended to ensure that, to the extent possible, decisions are objective and unbiased. Second, they recognize that scientific knowledge is not static and require that agencies consult the state of the science at the time of their decisions rather than relying on possibly outdated understanding. Those mandates do not, however, impose a requirement for some threshold level of certainty before agencies may act. On the contrary, in combination with the limited timeframes allowed for decisions under the ESA, they often require that an agency use its best judgment to choose a path on the basis of extraordinarily sparse information.

It is not clear how much the ESA's explicit requirement that agencies use the best scientific data adds to the background requirement that agency decisions must not be arbitrary or capricious. Under the ESA, USFWS may not ignore existing data, but its interpretation of the data and its determination of whether some data are scientifically better than others are entitled to substantial deference from the courts. USFWS has a duty to gather existing evidence, but it has no duty to undertake new studies when the existing data are ambiguous or inconclusive. Many ESA decisions, including the determination of critical habitat and the issuance of biological opinions, are subject to fairly short statutory deadlines. To meet those deadlines, USFWS often must make decisions without conclusive evidence and base its decisions on its best interpretation of the evidence available.

To implement the requirement that it consider the best available scientific data, USFWS has issued a policy detailing how it will treat information under the ESA (Fed. Regist. 59: 34272 [1994]). The policy requires, among other things, that agency scientists gather and impartially evaluate information, including information that disputes official positions; document their evaluation; and use primary and original sources when possible.

Although it repeatedly requires the use or consideration of the best available scientific data, the ESA does not require that agencies treat the data precisely as research scientists would. In particular, the burden of proof under the ESA is not necessarily the same as the burden of proof in the scientific community. Research scientists typically apply a stringent standard of proof, requiring 95% certainty before they will accept that a given event causes an observed result. That convention serves scientific purposes, but it is not necessarily appropriate for decisions under the ESA. Indeed, Section 7 of the ESA itself suggests a less stringent standard of proof: federal agencies are directed to "insure" that their activities "are not likely" to cause jeopardy or adverse modification of critical habitat. The *Endangered Species Consultation Handbook* developed by USFWS requires that the species be given the benefit of the doubt when a biological opinion must be completed in the face of substantial data gaps (USFWS and NMFS 1998). In practice, because the ESA is not highly specific about standards for listing, critical habitat determination, or jeopardy opinions and because courts are strongly inclined to defer to the agency's technical determinations, USFWS enjoys substantial discretion to determine the acceptable extent of risk and the degree of confidence that the available scientific information must provide.

Water Law in Platte River Basin

In the Platte River Basin, the ESA is implemented against a background of state water law because the suitability of the Platte River as habitat for the listed species is largely a function of operation of federal water projects upstream and the removal of water from the river for irrigation. In general, if there is a conflict between federal or state water law and the ESA, the ESA will prevail. Water law in the basin is important, though, because it has shaped the expectations of water users, and it can determine which of various possible solutions to ESA problems in the basin appear best to the parties involved.

The waters of the Platte River system arise in Colorado and Wyoming, flow into Nebraska in the North and South Platte Rivers, combine to cross most of the state in the Platte, and eventually empty into the Missouri River on Nebraska's eastern border. All three states, like most western states, use the appropriative system for allocating surface water. The natural flow of the Platte River and its tributaries was fully appropriated before 1900.

The waters of the South Platte were apportioned between Colorado and Nebraska by interstate compact in 1923. The North Platte is not subject to a compact; instead, it has been the subject of prolonged litigation. Its waters were equitably apportioned among Wyoming, Colorado, and Nebraska by the U.S. Supreme Court in 1945 (Nebraska v. Wyoming, 325 U.S. 589 [1945]). The Court left open the possibility of modification in light of changed circumstances. In an action begun in 1986, Nebraska sought modification to require (among other things) that Wyoming and Colorado supply additional water for habitat protection. The United States, because of U.S. Bureau of Reclamation (USBR) projects in the basin, was also involved in this suit. After many years and two interim Supreme Court opinions, many aspects of the dispute were finally settled in 2001. The settlement capped irrigation withdrawals from the North Platte system in Wyoming, including both surface-water diversion and pumping of hydrologically connected wells; called for Wyoming to transfer its habitat-mitigation lands in the central Platte to USFWS once storage has been increased in Pathfinder Dam; allocated the available waters of the USBR North Platte Project between the states in low-water years; and created an interstate committee to monitor compliance with its terms. The settlement did not resolve the issue of flows for wildlife or habitat, however, because the parties agreed that those issues were better addressed through the cooperative agreement process, discussed below. The Supreme Court approved the settlement and issued a new decree incorporating its terms, again subject to modification if conditions change (Nebraska v. Wyoming, 534 U.S. 40 [2001]).

Nebraska allows appropriation of instream-flow rights (Nebraska Revised Statutes §§ 46-2,107–46-2,119 [1984]) but only to the extent necessary to maintain the instream uses for which appropriation is requested, such as protection of habitat for aquatic species. Because recognition of instream-flow appropriations is a recent phenomenon, instream-flow rights are junior to existing surface-water appropriations.

Nebraska does not require appropriative permits for development of groundwater. As a default matter, landowners have the right to remove groundwater. Whereas the state Department of Natural Resources is responsible for surface-water allocation, local natural-resource districts have the authority to regulate groundwater pumping (Nebraska Revised Statutes 46-656.25). Most natural-resource districts have not implemented any groundwater pumping restrictions. Many observers expect that full implementation of the cooperative agreement will likely lead to regulation of groundwater pumping in Nebraska (Aiken 1999; Sax 2000).

Nebraska has a state statute that protects endangered species: the Nebraska Endangered Species Conservation Act (NESCA), adopted in 1975 and closely patterned after the federal ESA (Aiken 1999). The NESCA provides that federally listed species that occur in Nebraska shall be protected under Nebraska law (Nebraska Revised Statutes 37-806 [1998]). Therefore, all the federally listed species in the Platte River Basin are also protected by state law. The NESCA, however, allows the Nebraska Game and Parks Commission (NGPC) to identify critical habitat independently of any federal designation (Nebraska Revised Statutes 37-807 [1998]).

The NESCA requires the Department of Water Resources to consult with the NGPC before approving applications to appropriate water. Applications must be denied if the department concludes that granting them will jeopardize the continued existence of a listed species or destroy or modify critical habitat (Nebraska Revised Statutes 37-807 [1998]). In the middle 1990s, the department denied an application by the Central Platte Natural Resources District to construct a diversion dam at Prairie Bend, in part because the area was in the critical habitat designated by USFWS and the department concluded that the dam would alter the habitat in a way that would be detrimental to the whooping crane. The state Supreme Court upheld the determination (Central Platte Natural Resources Dist. v. City of Fremont, 250 Neb. 252 [1996]), and a concurring judge noted that the state act allows the department to "err on the side of caution" (id. at 268-69, White, C.J., concurring). Aiken (1999) details a number of instances in which the state agency has denied applications for diversions from the Platte River on the basis of impacts on endangered species or their habitat.

Platte River Cooperative Agreement

The series of jeopardy opinions on Platte River diversions, in particular the drawn-out relicensing proceedings for Kingsley Dam, drove the basin states and DOI to enter into a cooperative agreement in 1997. The agreement is overseen by a Governance Committee consisting of one representative of each state, two federal representatives (one from USFWS and one from USBR), two representatives of environmental interests, and three representatives of water users (one from the North Platte, one from the South Platte, and one from downstream of Lake McConaughy). The purpose of the agreement is to implement elements of the recovery plans for the central Platte species in a manner that will prevent future jeopardy opinions for existing and new water-development activities.

The parties have committed to increasing flows by 130,000-150,000 acre-ft per year at Grand Island and protecting or restoring 10,000 acres of habitat in the central Platte region by 2010 (Cooperative Agreement, Attachment 3). The agreement commits USFWS to recommend specific RPAs for new activities with a federal nexus that will deplete the Platte River. For depletions of 25 acre-ft or less per year, USFWS will recommend replacement of the water or payment of a mitigation fee determined by a formula that it developed in 1996. For depletions of more than 25 acre-ft, USFWS will recommend replacement within the state of the diversion, outside the irrigation season, at a time of shortage for the species. All consultations for such activities, however, remain subject to reinitiation if a full program is not implemented. The terms of the cooperative agreement have been incorporated in the 1997 biological opinion on renewal of the FERC license for Kingsley Dam and all later biological opinions.

The cooperative agreement initially envisioned that within 3 years the Governance Committee would develop a long-term program and the program would go through the National Environmental Policy Act review process. It set specific milestones to be accomplished within the 3 years, such as development of a water-accounting system for depletions, including those from new groundwater wells, and identification of the impacts of activities undertaken under the cooperative agreement on the listed species. At the end of the 3 years, the program was far from completion. The parties agreed to extend the cooperative agreement to June 30, 2003. The program is still not complete, and it is unclear to the committee how many of the milestones have been met. The agreement has led to the creation of an Environmental Water Account encompassing 10% of the water stored in Lake McConaughy and to the postponement of a proposed diversion project in Wyoming (Sax 2000).

The intent of the agreement is to develop a long-term program that will provide certainty with respect to future water development and will facilitate recovery of the listed species. The 130,000-150,000 acre-ft annually promised in the initial phases of the agreement are likely to prove far less than what is ultimately needed to recover the species, but the federal agencies viewed implementation of the program as sufficiently valuable to justify limiting demands for additional water during the first increment period (Sax 2000). The full program is supposed to identify and provide for the amount of added flow that will be needed by the species in the long term. If the program never materializes, USFWS will be forced to return to project-by-project Section 7 consultation, and it will probably impose more onerous conditions as RPAs.

SCIENCE AND UNCERTAINTY

This report was solicited in large part in response to controversy attending the use of available information by USFWS in its designation of critical habitat for listed species found in the central Platte River and its recommendations for instream flows. We have been asked to assess the underlying scientific basis of agency decisions that have substantial economic and social implications. At issue is the quality of the data used in decisions and whether those data have been appropriately interpreted and applied. More specifically, we have been tasked to examine the scientific foundations of a number of conclusions that relate the habitat needs of species to river operations. Use of the term *science* in this regard requires clarification. It is also important to remember that science need not be the decisive factor in environmental-management decisions. Those decisions are sometimes based on costs, values or political preferences, experience, assumptions, or any combination of those. Management of threatened and endangered species in the Platte must rest to some extent on evaluation of scientific data. The federal ESA requires that many decisions—including listing, designation of critical habitat, and evaluation of jeopardy—take into account the best available scientific evidence. That does not mean, however, that science does or even can prescribe all the management choices for those species.

Data Collection and Evaluation

Science is a process, not an outcome or a product. At its most formal, the process has two major steps: gathering of data designed to differentiate among alternative explanations of phenomena of concern; and communication of the methods used, data gathered, and analysis and interpretation of the data for review by the relevant community of peers. Through an iterative process of data-gathering, communication, and re-evaluation, scientists can generate increasingly robust explanations of how physical, chemical, and biological systems function; how species behave and interact in those systems; and how manipulation of environmental conditions is likely to affect species and ecological communities. By making their results and interpretation public, scientists allow others to verify and build on their work. Because science is a process, scientific knowledge and the solidity of scientific conclusions are expected to advance with time and additional work. Scientific conclusions are always contingent, subject to modification or re-evaluation as more data are gathered and more hypotheses are tested.

Data derived from experimentation in controlled systems, as in laboratories, can provide the strongest evidence in favor of or against a hypothesis because potential confounding factors can be eliminated. But such controlled experimentation is often impractical or impossible in large natural systems. Therefore, scientists necessarily and legitimately rely on other kinds of data to understand such systems. Only very rarely do data derived from controlled experiments clearly resolve issues in species conservation.

Models can be useful tools in conservation planning, especially when data gaps make it difficult or impossible to answer management questions directly. Models are constructs of a system and are developed by application of basic principles to evaluate the effects of change in specified factors on the modeled system. Models may be simple or complex. They should clearly describe the input variables, algorithms, and calibration procedures that they use. Modeling algorithms can use statistical or deterministic approaches or can provide continuous simulations or estimates of steady-state conditions. The critical issues in model evaluation are whether scientific principles are appropriately used in model construction and whether the model is sufficiently transparent for all assumptions and manipulations to be apparent. Where feasible, model outputs should be evaluated through comparison with field measurements.

Models are increasingly used in conservation planning for at-risk species. Population viability analysis (PVA), which has informed listing decisions, recovery plans, and habitat conservation plans under the ESA (Shaffer et al. 2002), uses demographic models to predict the likelihood of population persistence over a specified period under various circumstances, for example, where greater or smaller amounts of habitat are available, when qualities of habitats vary, under different magnitudes of mortality from diverse sources, or with varied reproductive success and offspring survival. PVA typifies both the value and limitations of using models to support policy and management decisions. Although it allows for explicit projections of population fates under real scenarios that imperiled species face, it is greatly constrained by the quality and quantity of available data, difficulties in estimation of parameters, and the usual lack of model validation. The committee references population viability model outputs in its assessments of whooping cranes, piping plovers, and interior least terns in Chapters 5 and 6, acknowledging both the reasonable inferences that can be drawn from the modeling efforts and assumptions that constrain application of the results. The PVA developed by the committee was constrained by the short study period. It did not include systematic sensitivity analyses and did not base stochastic processes and environmental variation on data from the Platte River region. A more thorough representation of environmental variation in the Platte River could be developed from regional records of climate, hydrology, disturbance events, and other stochastic environmental factors. Where records on the Platte River basin itself are not adequate, longer records on adjacent basins could be correlated with records on the Platte to develop a defensible assessment of environmental variation and stochastic processes. In addition, a sensitivity analysis could demonstrate the effects of wide ranges of environmental variation on the outcomes of PVAs.

To support management decisions—such as whether to list a species, where to designate critical habitat, or whether a proposed action will cross the jeopardy threshold—available data must be synthesized and applied. That requires the use of professional judgment. Professional judgment combines available data with professional knowledge of the systems or species in question on the basis of direct experience and knowledge of relevant scientific literature. Ideally, managers using their best professional judgment reject unreliable data, document their interpretations so that others can understand and evaluate the decision process, identify pertinent uncertainties, and clearly articulate the conclusions they draw. That decision-making process is not the classic process of "doing science." Instead, it is "using science." But it shares the underlying scientific commitment to comparing assumptions about the natural world with the available empirical data. Managers often must make their best guesses on the basis of sparse data. Their guesses are limited by the data available; to be consistent with professional practice, their judgments must be consistent with existing data and, at the next decision point, subject to reconsideration in light of any new evidence that becomes available.

Uncertainty and Decision Making

Uncertainty is typically high in the context of ESA decisions because available data are incomplete and environmental systems are inherently variable. Nonetheless, the ESA frequently demands that decisions be made quickly. Decision making in the face of uncertainty is familiar in a variety of other contexts, and some general methods for dealing with uncertainty have been developed.

Uncertainty can, in some cases, be well characterized and taken into account through mathematical formulas. Possible outcomes and their probabilities may be accurately understood. For example, before rolling a fair die, one cannot say for certain what the outcome will be, but the possible outcomes (a 1, 2, and so on) and their probabilities are not in doubt; there is a 1/6 probability of rolling any chosen number. Uncertainty of that type cannot be resolved by collecting additional information; but because the probabilities of the known events are understood, decision making and management are not unduly complicated. Because the behavior of environmental systems is far less regular than the behavior of a die, that type of uncertainty is not typical of environmental systems.

In other situations, the possible outcomes may be understood but their probabilities not precisely predictable. Weather forecasting is an example. Climate records that can be used to understand the variability of weather and the possible weather events may exist, but tomorrow's weather, or even the probabilities of different outcomes, cannot be objectively determined. To arrive at weather predictions, forecasters might use extensive scientific data. Multiple scientifically valid models may be available to interpret the data. In some cases, results of different models might point in different directions. Forecasters combine the raw data and modeling results with their own professional experience and use their scientific and professional judgment to arrive at weather forecasts. Not surprisingly, different weather forecasters may arrive at different forecasts on the basis of the same information. In circumstances of uncertainty, appropriate use of the best available scientific data will not necessarily produce a single, universally agreed on result. Thus, the existence of disagreement does not establish that the best available data have not been used or that they have not been interpreted according to applicable professional standards.

Adaptive Management

It may be possible to reduce uncertainty about the functioning of natural systems, and thereby to improve management decisions, through adaptive management. The concept of adaptive management encompasses a spectrum of practices designed to maximize learning and the feedback of new information into management decisions (Box 3-1). It can be done passively through the systematic collection and review of information about the effects of management actions or actively by designation of management actions as experiments. Whether and to what extent adaptive management should be used depends on the ability to learn, on the value of information that might be generated, and on whether latitude and resources are sufficient to allow alternative approaches to be implemented. Adaptive management and other forms of learning may reduce uncertainty; but in the context of conservation of endangered or threatened species, it will rarely be possible to eliminate uncertainty.

The committee believes that adaptive management—at least in the sense of careful data collection, evaluation, and periodic reassessment of management choices—could be useful for the management of Platte River species. Specific recommendations for monitoring and data-gathering are provided in Chapter 8. That information-gathering will serve little purpose if what is learned is not put to use in future management decisions.

BOX 3-1
Adaptive Management

Adaptive management for water resources is an approach that operates with a feedback mechanism. First, scientific information and explanations provide recommended behavior patterns for managers who desire to benefit given species in an ecosystem. For the Platte River, the recommendations might be for instream flows and variations in them to cause changes in the river channel to benefit threatened and endangered species. Management actions, such as the release of prescribed flows of water from a reservoir, are undertaken with particular ecosystem goals. In the case of the Platte River, releases might artificially simulate floods that occurred before dams were constructed on the river and might include some high flows to sweep away seedlings on bars and beaches, which would result in open sight lines and stopover locations for whooping cranes. Next, adaptive management requires monitoring of the system to determine whether the desired outcome has been achieved. For the Platte River, observations on the river channel might confirm the expected changes and increases in suitable habitat for whooping cranes, or they might reveal unexpected outcomes. Finally, adaptive management uses the observations from monitoring to fine-tune the original management plans to improve the outcomes. The committee endorses the concept of adaptive management as presented in NRC (2004 a,b).

Scientific Validity

Scientific validity is a contextual determination. It may mean that a reasonable scientist in the field would accept a conclusion as justified in light of the evidence available. In some circumstances, specific conventions have been developed that require a specific level of certainty before a conclusion is accepted (e.g., a 95% confidence level). Those conventions rest on value judgments about the relative costs of false-positive and false-negative errors. The costs may be very different between the management context and the context of abstract scientific research. Moreover, because science is not fixed in time, neither are judgments regarding scientific validity.

The degree of scientific certainty needed to justify a particular management decision is a social decision that requires evaluation of the relative costs of different types of errors. The committee has not attempted to make those decisions, which we regard as outside our charge. Instead, we have tried to evaluate the strength of the scientific evidence supporting current management decisions (Box 3-2).

Data and Uncertainty in Implementation of Endangered Species Act

Scientific research and the knowledge it produces play important roles in the administration of the ESA. Answering the question of whether a species is endangered or threatened requires the use of scientific observation. Ecological analysis leads to an understanding of habitat needs. The development of recovery plans requires an understanding of why populations have declined so much that a species is threatened or endangered. Such research usually requires not only biological approaches to understand the dynamics of the species itself but also approaches that include earth, water, and atmospheric sciences to understand the changing physical basis of its habitat. USFWS and action agencies are required to use the best available scientific information in conducting consultations and in determining jeopardy opinions.

In the context of ESA implementation, available data typically are sparse, and what data do exist are usually derived not from laboratory experiments but from less-structured methods. Listing decisions, critical habitat designations, and biological opinions are usually informed by data from simple species surveys, counts of individuals at circumscribed locations through time, and observations of resource use, behavior, and reproductive success under varying circumstances. In a few cases, more complex and comprehensive studies that provide additional data on species behavior or biology, such as survival of offspring to maturity and long-distance dispersal, are available.

In addition to being sparse, available data often have been collected from diverse sources that can have varying reliability. Sources may range from observations by laypersons through systematic inventories undertaken by trained professionals to systematic development of dynamic process

BOX 3-2
Criteria for Assessing the Degree of Scientific Support for Decisions

The charges to the Committee on Endangered and Threatened Species in the Platte River Basin generally required the committee to assess the degree of scientific support for decisions reached by Department of the Interior agencies regarding species and river processes. In determining whether and to what extent decisions are supported by existing science, the committee considered:

- The extent of data available.
- Whether the available data had been generated according to standard scientific methods that included, where feasible, empirical testing.
- Whether those methods were sufficiently documented to allow others to repeat them and whether and to what extent they had been replicated.
- Whether either the data or the methods used had been published in documents made freely available to other researchers and the public to facilitate criticism or correction and whether they had been formally peer-reviewed.
- Whether the data were consistent with accepted understanding of how the systems function and whether they were explained by a coherent theory or model of the system.
- Whether the decisions were publicly explained with clear reference to supporting data, models, and theories so that the rationale for the decisions was apparent and open to challenge by stakeholders.

No one of the above criteria is decisive, but taken together they provide a good sense of the extent to which any conclusion or decision is supported by science. Because some of the decisions in question were made many years ago, the committee felt that it was important to ask whether they were supported by the existing science at the time they were made. For that purpose, the committee asked, in addition to the questions above, whether the decision makers had access to and made use of state-of-the-art knowledge at the time of the decision.

The committee was also asked to assess the scientific validity of the methods used to develop instream-flow recommendations. The criteria applied in answering that question were similar to those above but focused more directly on methods:

- Whether the methods used were in wide use or generally accepted in the relevant field.
- Whether they had a sound theoretical basis or were supported by a generally accepted understanding of the system.
- Whether they were sufficiently documented to facilitate replication.
- Whether sources of potential error in the methods have been or can be identified and the extent of potential error estimated.
- Whether the methods have been formally peer-reviewed or published in documents made freely available to other researchers and the public.

Again, no one criterion is decisive. Considering all those factors, the committee made a judgment as to the validity of the methods that it was asked to evaluate.

studies carried out by scientists. Data-collection efforts typically are not commissioned or handled directly by resource managers; therefore, they often do not directly address key management questions. It can be expensive, time-consuming, difficult, or even impossible to collect directly the key data needed to address management questions.

The most popular method of organizing available data to assess the likelihood of species persistence is the aforementioned PVA, a modeling tool that uses information on genetic, demographic, and environmental sources of variability. The output of PVA is not an explicit prediction of time to population extinction but an estimate of the probability that a population of a given size will persist for some specified period. As in other fields of science, some uncertainty is inherent in PVA.

Scientific uncertainty pervades species-conservation issues, including those related to the central Platte River. Regan et al. (2002) term such uncertainty "epistemic." Sources of epistemic uncertainty include survey shortcomings, measurement errors, variation and inherent randomness of the natural system and species responses, and the application of subjective judgment in the analysis and interpretation of available information. For the Platte River species, epistemic uncertainty is pervasive. The contribution of Platte River stopover resources to the fitness of individual whooping cranes can only be surmised through indirect data and subjective model assumptions; assessing the likelihood of regional persistence of Platte River piping plovers and interior least terns requires many assumptions about local structure and dispersal and about interactions among birds across larger landscapes; and information on the pallid sturgeon is virtually nonexistent beyond locations of capture, from which only the most basic inferences regarding habitat requirements and use can be made.

An additional challenge for the effective use of science in ESA regulatory decisions is that these decisions typically incorporate value judgments. The judgments have not been made either by Congress or by USFWS at a level specific enough to inform management decisions. Neither the ESA nor USFWS's implementing regulations, for example, define *threatened*, *endangered*, or *jeopardize the continued existence of* in quantitative terms by identifying acceptable threshold levels of extinction risk. Nor do the statute and regulations specify how the agency should choose among occupied habitats, not all of which may be required to maintain a viable population, identify critical habitat, or determine what magnitudes of economic or other costs may justify exclusion of specific areas from critical habitat.

Those aspects of ESA implementation decisions are not scientific in the sense that they could, even in theory, be decided solely through evaluation of empirical, objectively gathered data. They require social or political value judgments that are inevitably subjective. The committee believes that these judgments should be made transparent; that is, USFWS should clearly explain in a decision document both its evaluation of the scientific data and its use of nonscientific factors to reach a final decision.

WATER MANAGEMENT: CONNECTING LAW AND SCIENCE

The Platte River serves a large and growing urban population as a potable-water supply, provides water for irrigation of millions of acres of cropland, and provides habitat for a wide array of species, including some that are threatened or endangered. Competition for use of the Platte River's waters has fueled serious debate within the basin states of Colorado, Wyoming, and Nebraska and beyond. To understand the details of the conflicts that have emerged in this watershed, it is useful to look at the forces that have created similar conflicts throughout the United States, and they are set forth here. The next section discusses some of the common elements of the conflicts and relates them to the Platte River controversy.

Overappropriation of Water

Many watersheds in the west are "overappropriated"; that is, more water has been legally allocated to users than can be physically provided in all years. There are many reasons for overappropriation, from lack of good technical information on which to base allocations to lack of political willingness to limit allocations (Meyers 1966). Perhaps the most famous example of overappropriation is the Colorado River, whose waters were allocated between upstream and downstream states by using data from one of the wettest periods in the last century. As a result, the Colorado River Compact allocates some 10% more water than is available in a typical water year.

Although the natural flow of the Platte River was fully appropriated before 1900, new uses of groundwater and surface water continue to be allowed. Surface-water and groundwater uses are necessarily connected; use of surface water beyond some point depletes hydrologically connected groundwater, and vice versa.

Changing Values and Public Perceptions

The values that the American public expects water management to serve have evolved over the last 50 years. Expanded focus on protecting endangered species, providing for recreational uses, and providing water supplies for Indian settlements have introduced major new pressures on river systems that were already overappropriated. Protecting these recently recognized values in the context of highly altered hydrological regimes increases possibilities for conflict.

In addition, public perceptions are often shaped by beliefs about natural systems that may not be based on adequate, defensible, or any empirical data. Entrenched ideological commitments complicate efforts to find compromise solutions or resolve conflicts collaboratively.

Institutional Inflexibility

The ability to solve problems in a changing physical and social context is affected by the ability of water-management institutions to respond to changes. Water-management institutions generally focus on protecting the status quo, including the rights of existing water users, and may not adapt quickly to changing circumstances.

Irreversible Decisions and Unacceptable Consequences

Many decisions in water management are effectively irreversible, and this limits the usefulness of adaptive management. For example, construction of a major dam is, as a practical matter, ecologically irreversible. The dam is unlikely to be removed in the short run; and although it might be removed at some distant time, it is likely to leave enduring effects on the landscape (Figure 3-2).

FIGURE 3-2 Keystone Diversion on the North Platte River, upstream from the central Platte River, represents an example of an effectively irreversible feature of the watershed that has enduring effects on flows. Source: Photograph by W.L. Graf, May 2003.

The Platte River system is already highly modified by the construction of dams, canals, and diversions that have important economic consequences. A large constituency has come to rely on the continued operation of those artifacts, but their environmental effects have only recently been recognized. It is not surprising that new water-management proposals have generated considerable concern among water and power interests.

Dispersed Costs and Concentrated Benefits

Federal water-planning guidelines typically forbid funding of new projects unless their benefits will exceed their costs. However, the costs of federal projects are spread among taxpayers nationwide, whereas the benefits typically are geographically concentrated. The concentration of benefits makes it possible to bring concerted political pressure to bear; such pressure has led in many cases to the development of water projects whose benefits do not exceed their costs. Historically, conflicts over the allocation of benefits and costs have been exacerbated by funding rules that provided full funding for some purposes (such as flood control) and less than full funding for others (such as municipal water supply).

Lack of symmetry between the costs and benefits of water projects can also contribute to pressures to maximize appropriations and diversions at the expense of environmental values (Farber and Frickey 1987). A nationwide constituency benefits, in a diffuse way, from the protection of endangered or threatened species and the other environmental values of river systems. In contrast, a focused, clearly identified community of water users benefits from diversion. Water users therefore typically find it easier than environmental interests to organize and exert political pressure on decision makers.

Evaluation of Cumulative Environmental Impacts

A difficult challenge encountered by water-resources planners and managers is how to evaluate cumulative environmental impacts (see NRC 1986, 2003). A classic example is the regulation of multiple discharges of waste into a single river. It is possible that none of the individual waste streams will threaten environmental values, but together they may produce unacceptably poor water quality. It can be difficult both to understand what total impact will pass acceptable threshold levels and to divide responsibility for avoiding unacceptable levels equitably. In the Platte, cumulative impacts include those of upstream activities (including withdrawals) in two other states and those of withdrawal of hydrologically connected groundwater.

Insufficient or Poor Data

As suggested previously, a primary source of conflict in water management can be the lack of information to support informed decisions. Although state and federal agencies historically collected an enormous amount of basic information concerning water resources across the United States, data-gathering efforts have recently been curtailed by funding concerns. Water-quality data are generally less available than water-quantity data. In general, data gaps leave a large number of uncertainties. As a result, decisions must be made under circumstances in which their environmental impacts, both favorable and unfavorable, are inapparent or difficult to predict.

Categories of Water-Resources Conflicts

The numerous water-resources conflicts across the United States can be categorized in a variety of ways, but one convenient approach is to define them in terms of water use. Conflicts frequently occur between upstream and downstream uses, instream and out-of-stream uses, groundwater and surface-water uses, and present and future uses. That categorization can help to create a framework for conflict resolution. The Platte River Basin today is experiencing all four types of conflict.

Perhaps the most common source of water conflicts is associated with upstream vs downstream uses. In many circumstances, water is diverted from a stream and only a portion of it is returned, leaving downstream users with less than the natural flow. Much of water law in the West, in particular the prior-appropriation doctrine, is devoted to providing a framework for resolving such conflicts. They are particularly intractable in larger watersheds, where the upstream and downstream uses may be separated by hundreds of miles and by state or international borders.

As water demands and water uses change, the character of upstream vs downstream water uses may change dramatically, increasing the potential for conflict. Disagreements between upstream and downstream states can be resolved by negotiation of interstate compacts or, if that fails, by litigation. The Platte has been the subject of both. The basin states successfully negotiated a compact for allocation of the waters of the South Platte, but the North Platte has been the subject of protracted litigation and is allocated largely by a court decree.

Conflicts between instream and out-of-stream uses in the West have increased dramatically since the 1970s, as water demands for municipal and agricultural uses have expanded and stresses on aquatic systems have been recognized. In some states, instream-flow rights have been established to protect water flows for biological communities. It is important to note that until the 1980s, many states did not specifically recognize the public values of such flows, and maintaining flows for habitat was not guaranteed.

Quantifying ecological instream-flow requirements is often challenging in that anticipating the response of biological communities to a variety of flow regimes is difficult and estimating the effects of cumulative stressors on the biological communities is a young science. Two typical approaches are taken to establishing instream flows in rivers: establishing water rights based on a percentage of the natural flow on a monthly or weekly basis and providing minimal flows on the basis of the biological needs of a specific species. Instream flows have now been established for numerous rivers, often after one or more species have been identified as threatened or endangered or as part of a larger environmental licensing process (such as for hydropower production or municipal water supplies). Other users of water may view the establishment of such instream flows as direct threats to their access, noting that such requirements erode their ability to obtain water for out-of-stream uses. The flows that have been suggested by USFWS in the central Platte are examples of instream flows to protect endangered species, and there has been litigation over attempts by the NGPC to obtain instream-flow rights to support the central Platte's endangered and threatened species.

Conflicts are increasingly emerging between surface-water and groundwater use. The relationship between surface water and groundwater was often ignored in early allocation of water rights. In some states, including Nebraska, surface water is regulated and groundwater remains largely unregulated. As described in the groundwater section of Chapter 2, the depletion of groundwater and its effects on surface flows have created substantial problems throughout the Southwest and elsewhere. Although the relationship between surface water and groundwater has been qualitatively understood for decades, legal and institutional constraints have impeded conjunctive management of these resources. The complicated hydrological relationship of surface-water and groundwater flows makes their regulation difficult, and often appropriate management can be achieved only after a detailed understanding of their interactions is obtained. The current efforts on the Platte to develop a comprehensive groundwater–surface-water model mark a good first step in developing a strong understanding of the hydrology of the river.

At the center of many water conflicts is the general notion of sustainability, which is related to the balance of present needs with future needs. Although many definitions of *sustainability* exist, a commonly used one from the Bruntland Report (Bruntland 1987) is "meeting economic, environmental and social needs of the present without compromising the ability of future generations to meet their own needs." That definition is especially appropriate for watershed planning because it goes beyond the simple allocation of today's water to include the needs of people and the broader biological community for many generations into the future. The concept of sustainability is a broader management goal than resolving the individual categories of conflict mentioned above. However, a sustainable system must have a mechanism to address those conflicts.

Sustainability does not imply managing a watershed in a static fashion, that is, ignoring changes in water demands, water use, water availability, and the economic, environmental, and social systems that rely on water. It is difficult to imagine a sustainable system that is not "adaptive." Adaptive management of watersheds—alteration of management of a watershed based on the system state and the current inputs and desired outputs of the system—may be required much more in the future as pressure on water supplies continues to increase.

LESSONS FOR THE PLATTE FROM OTHER WATER CONFLICTS

Every water-resources conflict can appear unique to those involved in it. Every watershed does have distinctive characteristics, but many of the causes of water conflicts and approaches to their solution are similar across a large number of basins. The components and categories of water-resources conflicts noted above are found again and again in other water-resources conflicts. Three lessons from water conflicts in the Colorado, Cedar, Klamath, and Snake Rivers are offered for those dealing with science and decision making for the Platte River.

The first general lesson is that uncertainty is always associated with the data available for decision making associated with water resources. Lack of complete data cannot be an excuse for making no decision; complete data will never be available. Adaptive management when incomplete data are available is an important hedge against mistakes derived from uncertainty, because in adaptive management, monitoring and measurement provide a constant stream of new data to evaluate the outcomes of decisions. If resource changes are not as expected, the new data can support revised decisions.

The second general lesson is that the level of uncertainty in the Platte River Basin is at least similar to that in other basins in the western United States. In some cases, the quality of data is better than in other circumstances, so although decisions are required in the face of incomplete data and some uncertainty, our scientific understanding of the Platte River and its resources provides support for decisions to a degree similar to the support for decision makers in other basins.

Finally, conflict resolution in water resources requires a comprehensive view of the entire watershed. The comprehensive view must take into account the interests of all the stakeholders and economic sectors even if they are far removed from the site of intensive management. The comprehensive view of the watershed also dictates that downstream restoration and management are inescapably connected to management of the land and water resources in the upper watershed.

SUMMARY AND CONCLUSIONS

The issues faced in the Platte River Basin are similar to water-related conflicts in other parts of the United States. Protection of federal and state listed species in the central Platte River Basin has had, and will continue to have, a profound effect on management of the waters of the Platte River and many other rivers in the country. Critical habitat designation, which was the focus of the call for this study, however, has not been an important driving force in management decisions. Every biological opinion that has identified the need to restrict or modify federal actions in the Platte River Basin has found both that the proposed actions would jeopardize the continued existence of listed species and that they would adversely modify or destroy designated critical habitat. In other words, no biological opinion has said that the effects on critical habitat were unacceptable but there would be no jeopardy, and no biological opinion has ever relied solely on the effects on critical habitat to mandate changes in a project. USFWS's views on the stream-flow and habitat needs of the protected species have been an important factor in management decisions, but the formal designation of critical habitat has not.

The best available scientific data are often incomplete or inconclusive, particularly in the context of endangered-species management, because information about the biological needs of species is difficult to gather and systematic data-collection efforts often are of recent origin. Management judgments may accord with professional standards and be considered scientifically valid even if they are made in the absence of conclusive data and even if they turn out to be wrong when judged against later-accumulated data. Science is a process of incremental learning, but managers do not have the luxury of waiting until conclusive evidence is accumulated; they are required to act on the basis of their best understanding of the system at any given time.

Many decisions made under the ESA, including decisions on critical habitat designation, cannot be fully determined with scientific data even if the data are complete. Those decisions require judgments about social values, including the relative importance of competing values at stake. Identification of endangered or threatened species requires choices about the level of acceptable risk. Designation of critical habitat requires some balancing of species' needs against economic impacts—again, a determination of acceptable risk. In addition, when a species at the time of listing occupies a broader range than appears to be required to maintain a minimal viable population, critical habitat designation must include decisions about which areas to protect and which to allow to be destroyed. Those choices include scientific considerations, to the extent that some areas appear more important than others to the species, but also require balancing of equities with respect to competing human uses.

The uncertainty facing Platte River managers is high but not unusual for river systems in the United States. Furthermore, uncertainty about the Platte River is an element of all management decisions, not just decisions that restrict diversions. Just as there is uncertainty about stream-flow requirements of the listed species in the central Platte, there is uncertainty about the efficacy of the tradeoffs authorized by USFWS in its recent biological opinions authorizing new diversions in return for habitat-restoration efforts or payment of mitigation fees into a habitat acquisition and restoration fund.

Adaptive management is one approach for identifying and taking steps to close key information gaps. Adaptive management is being implemented, at least in small ways, in some other river systems. Platte River management might benefit from more-systematic efforts to accumulate data about the effects of flow levels on the physical and biological system and to incorporate the data into future management decisions.

4

Scientific Data for the Platte River Ecosystem

Background discussions in Chapter 2 and the analyses of environmental law and science in Chapter 3 lead to an assessment of our understanding of the ecological foundations for threatened and endangered species in the Platte River Basin. The agencies of the U.S. Department of the Interior (DOI) have evaluated research into the habitat requirements of the whooping crane, piping plover, interior least tern, and pallid sturgeon for survival and recovery and have based their recommendations for instream flows and river management on the relevant studies. Critics question the studies.

The charge to the committee regarding the Platte River ecosystem and its management generally takes the form of assessing the "scientific validity" of DOI decisions. For the purposes of this report, a management decision and the conclusions that led to that decision have scientific validity if the scientific knowledge that existed when the decision was made is identifiable and verifiable and was the product of research methods and techniques that were generally accepted by the scientific community at the time of the decision. The data and the information that resulted from processing them must be identifiable and archived so that they are recoverable by subsequent investigators if needed. The conclusions drawn and the theories and models used to understand and explain the conclusions must be verifiable, that is, subsequent investigators must be able to replicate the research and arrive at the same conclusions. The methods and techniques must be similar to those used by other workers in similar applications and be commonly found in the scientific literature or discourse of professional meetings. The committee assessed the science used by DOI agencies as it existed when the agencies made their decisions. Since those decisions were made, there have been advances in science and engineering that may improve management.

This chapter evaluates the types, relevance, and quality of science used by DOI to understand and manage the individual endangered species and the ecosystem associated with the river, and it assesses the validity of the science for policy decisions. The scientific basis of listing a species and designating critical habitat is discussed in Chapters 5, 6, and 7. This chapter begins with a consideration of how science is connected with the goals of restoration of the Platte River for the benefit of threatened and endangered species. It then describes the basic connections that sustain the Platte River ecosystem (including its hydrology and geomorphology) and the habitats important for its threatened and endangered species. Next, it addresses specifically the validity of the science underlying DOI decisions related to instream flows and ecosystem connections that managers use to preserve and enhance habitat for threatened and endangered species. The chapter concludes with some special scientific considerations for decision makers that have not yet been fully explored.

SCIENCE AND MANAGEMENT TARGETS FOR SPECIES

Management of the Platte River for the benefit of threatened and endangered species entails a preliminary decision that deeply involves science and the state of our knowledge. It is likely to require restoration of the physical system of the river, its hydrology and geomorphology, to create habitats useful for sustaining the species. *Restoration* in this sense implies managed and designed changes to alter the existing river to some other target condition. The target of restoration in most applications is the presettlement condition because those arrangements supported in relative abundance the species that are now endangered or threatened. It is rarely possible to completely attain such a restoration because of human effects such as land-use changes in the watershed and water-control infrastructure, but as a general objective the presettlement conditions represent the endpoint of a spectrum of possibilities. There are two fundamental approaches to restoration: first, through knowledge of the presettlement conditions, and second, through knowledge of the present connections among physical systems, habitats, and species. In the case of the Platte River, restoration of the river to its prehuman or pre-European-settlement condition is faced with three issues: first, knowledge about prehistoric systems is sparse; second, they were always changing; and third, it is not possible to reconstitute them. First, knowledge about the prehuman ecosystem of the Platte River is highly limited by the lack of direct observations. Proxy measures of the environmental conditions are sparse because of the general dryland nature of the northern Great Plains. Although geologic evidence can inform us about the long-term general environments and their adjustments on a scale of tens of thousands of years, the evidence is not detailed enough to reconstruct specific environmental conditions along reaches of the river a few kilometers long. These issues commonly face restoration efforts, and they are not unique to the Platte River case.

We have considerable knowledge about the nature of the Platte River system's general characteristics for the period between the arrival of early settlers and the twentieth century. By the time of official Government Land Office (GLO) surveys, the most exacting early assessments of landforms and vegetation that were conducted in the 1860s, substantial alteration of the vegetation by Europeans had already occurred. By the 1890s, the water system was also subject to human-induced adjustments. As a result, there are three distinct time periods of our knowledge base: the presettlement period for which we have little direct information during which the now listed species were abundant; the post-settlement period of the nineteenth century for which we have some knowledge during which the species were under considerable pressure, and the twentieth century period for which we have the most information when the species populations declined to very low levels.

Restoration of the central and lower Platte River ecosystems to their presettlement conditions is not possible, even if the prehistoric target conditions could be specified. The central river and lower river are at the downstream end of a far-flung drainage basin and river system, and they exhibit characteristics that are determined by processes throughout the watershed. Land use and land cover are now substantially different from the prehistoric conditions, and the watershed hosts several large dams and many smaller control structures. Those features change runoff and stream flows and could not sustain a prehistoric reconstruction in the central and lower Platte River.

A second approach to defining restoration goals (in addition to knowledge about undisturbed ancient conditions) is to establish the sorts of conditions that we know from research in present environments favor the threatened and endangered birds and fish but are also consistent with our knowledge of presettlement conditions. We can then create an environment that contains those conditions (Box 4-1). This approach has the advantage of working from an observable premise: the connections among river flows, geomorphology, vegetation, and wildlife. Those connections are complex and are not completely understood; but given our partial knowledge about them, restoration for species is possible. As this normative approach to restoration proceeds, corrections and adjustments, particularly in the flow regimes of the river, can provide for experimentation through adaptive management.

BOX 4-1
Restoration for the Future Platte River

The primary route for managing the Platte River to benefit threatened and endangered species is to "restore" the river flows. The basic hydrological question, however, is to restore the flows to what target condition? One choice might be entirely "natural" flow regimes, such as ones that might have existed before the arrival of humans or at least before the imposition of numerous water-control structures that occurred after European settlement.

It is not possible to return the flow regime of the Platte River to either of those "natural" models, for three reasons: knowledge, limitations of the present system, and social considerations. First, we lack detailed knowledge about the true nature of prehuman flow conditions, and we have only minimal understanding of flows before the construction of dams and diversion works. We can compensate somewhat for this lack of direct knowledge by application of theory. For example, we know how modern braided rivers work, and we know the prehistoric Platte River was a braided stream, so we can make some useful generalizations about the presettlement river. Second, the present hydrological system includes widespread changes in hydrology throughout the watershed. For example, widespread intensive grazing of many areas may or may not mimic grazing intensity and patterns of the original native animals that once roamed the region. Mountain watersheds now include storage reservoirs that did not exist in presettlement periods. Therefore, even if adjustments could be made for some "natural" target in the central and lower Platte River, those adjustments are not likely to be easily coupled with changed runoff and storage conditions upstream of those reaches. Finally, the water-control infrastructure of the river is in place as a result of public decision processes that sought to create the present hydrology, characterized by suppressed flood peaks and a dependable water supply even in dry months. A complete return to the presettlement flows would mean the abandonment of social and economic goals that have driven substantial investment in the system—an unlikely scenario.

A more likely alternative for the target of flow restoration is the blending of objectives, whereby some flow characteristics benefit key wildlife species and attempt to mimic presettlement conditions to the extent possible. Other hydrological characteristics also remain in place to serve additional (such as social) needs. Such a compromise arrangement represents a more "normalized" flow regime that mimics natural rhythms, magnitudes, and durations but within constraints that recognize the changed nature of the basin and other competing economic demands on the water resource. In that way, restoration of the river flows is restoration part of the way toward the "natural" objective.

Any restoration and management program will have to go forward with incomplete knowledge, but decision makers can use the best available scientific knowledge in guiding their choices. Researchers and decision-makers can provide education for a public that may expect higher levels of scientific certainty than are possible. Ecosystem research is not as controlled or exacting as a bench science, so its input to public decisions is accompanied by more uncertainty than is the case for laboratory sciences.

BASIC CONNECTIONS IN THE PLATTE RIVER ECOSYSTEM

Issues defining the current conflicts in the central Platte River are related to seven interconnected elements: threatened and endangered species, other species (not regulated by the Endangered Species Act), flow of water through the river, sediment transported by that flowing water, groundwater, agriculture (and other human systems and uses), and riparian vegetation. Habitat is controlled by those elements, and they are intricately linked by connections that ultimately respond to stream flow and other ecological processes (such as fire, grazing, and human use). Water flowing through the river constitutes mass and energy, and changes in stream discharge trigger changes in sediment, morphology, and vegetation. The mass of water, for example, supplies sustenance for vegetation, and the energy that water expends as it descends through a channel contributes to erosion, transport, and redeposition of sediment. The energy expended by water flowing downgradient through the river performs geomorphologic work. It rearranges sediment, building such landforms as islands and bars attached to the banks and erodes previously existing forms. The significance of those features is that they are the foundations of vegetation and species habitats.

Sediment, in turn, influences the morphologic characteristics of the channel. As outlined in Chapter 2, the central and lower Platte River have a braided channel, but the variety of forms is shaped by interactions between the volume of sediment to be transported and the water, with its accompanying energy, that is available for the work of transport. When the available energy is less than that required to transport the available sediment, deposition occurs, bars develop, islands expand, and beaches encroach into the channel area. When the available energy is sufficient to entrain and transport the available sediment that is temporarily stored in bars, islands, and beaches, erosion results, and these features become smaller.

Riparian forests are influenced by abiotic ecological processes, such as the dynamics of water, sediment, and morphology of the river. Surfaces created by deposition are potential seedbeds for young plants. In the absence of fire and grazing, if flows do not sweep away or bury new plants and their substrate, the vegetation is likely to become relatively permanent and be in the form of a cottonwood and willow woodland. The resulting vegetation then has a feedback effect on the physical processes: the trees introduce hydraulic roughness to the channel and its nearby surfaces and thus induce additional deposition, which leads to higher floodplain surfaces that further influence vegetation. Under low natural flows, a broad, braided channel (with many subchannels) may be converted to a narrower system of fewer channels. That chain of events also can be reversed by higher natural flows, or clearing of vegetation, which can expose surfaces to energetic flows that might entrain sediment and erode the surface.

Those hydrological connections are particularly relevant for threatened and endangered species. DOI's perspective on the connections is outlined by the U.S. Fish and Wildlife Service (USFWS, unpublished material, June 16, 2000). Whooping cranes prefer roost sites that include shallow water bars that are surrounded by deeper channels and that have long sight lines (unvegetated areas)—a set of conditions common along the Platte River of a century ago. Piping plovers and interior least terns prefer a habitat that has unvegetated areas with sandy surfaces exposed by receding river flows during the breeding season of late spring. If nests are to produce fledglings, they cannot be inundated by pulse flows. Pallid sturgeon appear to prefer streams with sandy bottoms for foraging and a series of annual flows that have natural fluctuations, with high spring flows and lower flows in late summer (USFWS, unpublished material, June 16, 2000).

The causal chain of adjustments in the Platte River ecosystem begins with alterations of ecological processes, such as hydrology, and perhaps fire, grazing, and invasive species. Flows respond to two primary influences: climate and human regulation. Climatic influences through variation in the rainfall and snowpack that supplies river discharge change on time scales of a decade or more. Because much of the water flowing through the central and lower Platte River originates as precipitation over the Rocky Mountains, climatic changes in the mountainous areas are more relevant to flow in the central and lower Platte River than are climatic changes in Nebraska.

Human influences on Platte River hydrology are long term and short term. Large storage dams in the upper reaches of the Platte River Basin through the 1900s changed river discharges by lowering flood peaks and by releasing more water during summer months than was released during the summers before the dams were built. Other major changes include the installation of diversion works on the river system that divert some or all of the flow during portions of the year. Water often returns to the channel from return canals or groundwater seepage. The short-term effects of human regulation of the river include management of water delivery from storage sites to downstream water-users, adjustments in flow to generate hydropower during periods of high demand for electricity, and scheduling of return flows. Groundwater, which is influenced by pumping and seepage from fields and unlined canals, also contributes to stream flow.

Human-induced changes in the controlling flows of water in the Platte River Basin are larger and more important than climate-induced changes in controlling the Platte River ecosystem and habitats for endangered species. The storage reservoirs in the Platte River Basin store about 6 million acre-ft of water, and they have reduced flood peaks by 80% or more in the central Platte River (Figure 2-6). Those changes are at least half an order of magnitude greater than any changes envisioned for river systems as a result of climatic change from global warming (Arnell 1996). Simpson (2003) observed a similarly important role for human activities in changing channel conditions on the South Platte River in Colorado. Although storage reservoirs, diversions, power houses, and return flows have apparently caused changes in habitats along the river, these same engineered features offer an opportunity for ecosystem restoration. If the water-control infrastructure can be operated in such a manner as to partly mimic the natural flows that once dominated the river and if other ecological processes can be re-established, it is likely that habitats will respond by reverting to conditions that are more like those preferred by the listed species (Murphy and Randle 2001).

Given the magnitude of engineered changes and the economic value derived from them (such as flood control and water supply for agricultural or urban uses), it is not possible to completely restore predevelopment flows and the habitats they created. Bison will never again mass along the shores of the Platte. Even so, it is possible to partially recreate the presettlement conditions, given appropriate knowledge about the connections among the various subsystems that make up the Platte River ecosystem. In geographic locations where more extensive restoration is not possible, some management actions (such as wetland creation) may be necessary to complement restoration activities. In general, because habitat requirements of individual endangered species that use the central Platte are linked to more natural functions of the entire ecosystem, it is possible to pursue ecosystem restoration goals even when the presettlement conditions cannot be fully attained. In contrast with ecosystem restoration, simply combining habitat-management goals for individual species quickly leads to conflicting management options that may be mutually exclusive, especially when other species are affected. Management of songbird habitat as opposed to crane habitat is one contentious example that was illustrated in presentations to the committee. Compared with individual-species management, ecosystem-restoration approaches tend to be more stable, will include more (but not all) species successfully, and provide a larger geographic and temporal scale on which work can be accomplished. Restoration of conditions more similar to presettlement conditions than the present arrangements will be likely to benefit a wide range of species, including valuable waterfowl. Such restoration is also likely to result in more stable conditions and populations.

If the hydrological regime is adjusted to become more like the original, natural regime with the influence of dams, native species are likely to be favored because they established communities in a river regime without dams. Often, regimes that are highly unnatural eliminate the advantages of native species, and invasive species are more successful.

DOI research and investigations by others supported two types of management decisions targeted to maintain habitat and enhance benefits to the listed species. First, DOI identified a series of operating rules for the flow of water in the river; these rules became the agency's recommendations for instream flows. Recommended instream flows are important because they represent DOI's vision of the kind of river hydrology that most directly affects habitat of individual species, particularly the movement of sediment, the forming of the channel, and periodic high or low flows. According to the DOI view, instream flows can benefit the threatened and endangered species by hydrologically creating or maintaining preferred habitat conditions. Second, DOI identified connections among water, sediment, morphology, and vegetation that predict how the Platte system may respond to recommended instream flows. Besides mechanical removal of vegetation and some wetland manipulations, DOI did not appear to consider restoring any ecological processes.

Available Data on the Platte River Ecosystem

One purpose of this report is to evaluate the scientific validity of DOI's conclusions about what the instream flows should be and how they influence other aspects of the Platte River ecosystem. Part of that evaluation entails a review of the available data that DOI used to reach its conclusions. The following paragraphs assess the data available to DOI for its instream-flow and ecological research. Assessments of the data pertaining to the threatened and endangered species are presented in later chapters, where each species is considered in detail.

The most important data used as input for explanations and predictions of Platte River habitat needs of the listed species are related to water, sediment, channel morphology, and riparian vegetation. The water-discharge data on the Platte River are from a set of gaging stations that have varied lengths of record (Figure 4-1). The gages with the longest records in the central and lower Platte River are on the mainstem Platte River above the Loup River confluence and near Duncan, Nebraska, where measurements began in 1890. Both records are discontinuous, so their value for analysis and modeling is diminished. The longest continuous record (1925+) for the central and lower Platte River is from gages on the river near Overton, Duncan (Figure 4-2), and Ashland, Nebraska. Data from a nearby location at Lexington extends the Overton gage record back an additional decade. As the twentieth century progressed, the U.S. Geological Survey (USGS) installed additional gages on the river; by 1955, there were six recording sites in the central and lower Platte River, three on local tributaries, and six more in the upstream reaches of the North and South Platte Rivers.

Examination of the stream gage data for the central and lower Platte River shows that before 1925 measurements were sporadic and accom-

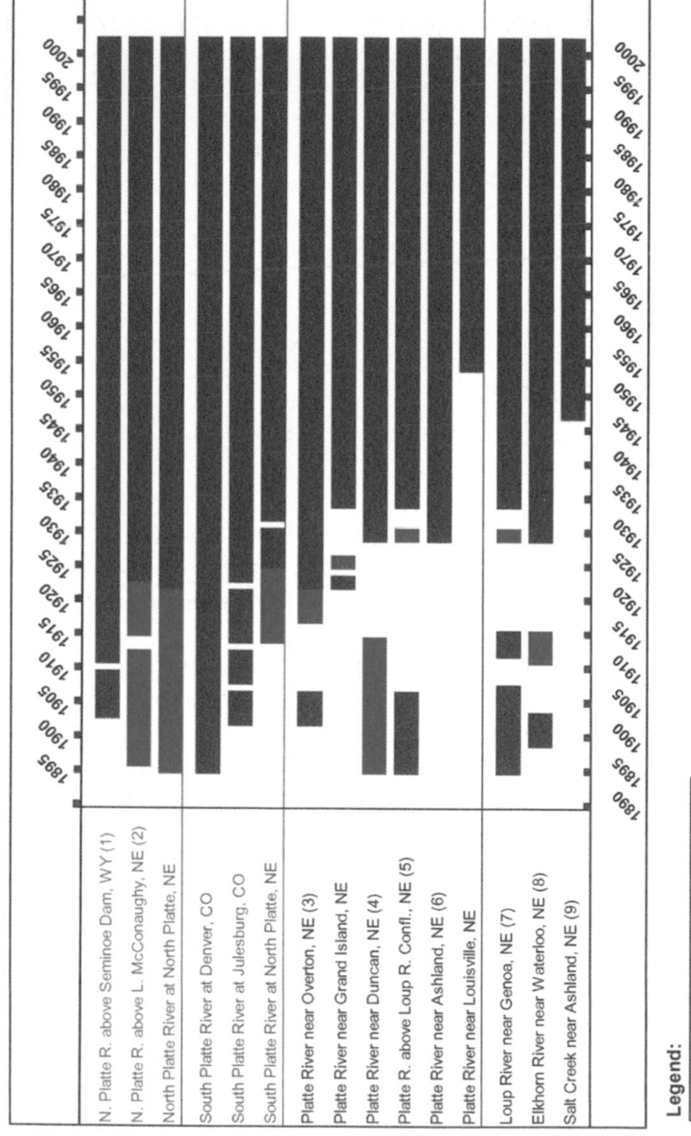
N. Platte R. above Seminoe Dam, WY (1)
N. Platte R. above L. McConaughy, NE (2)
North Platte River at North Platte, NE
South Platte River at Denver, CO
South Platte River at Julesburg, CO
South Platte River at North Platte, NE
Platte River near Overton, NE (3)
Platte River near Grand Island, NE
Platte River near Duncan, NE (4)
Platte R. above Loup R. Confl., NE (5)
Platte River near Ashland, NE (6)
Platte River near Louisville, NE
Loup River near Genoa, NE (7)
Elkhorn River near Waterloo, NE (8)
Salt Creek near Ashland, NE (9)
1890
1895
1900
1905
1910
1915
1920
1925
1930
1935
1940
1945
1950
1955
1960
1965
1970
1975
1980
1985
1990
1995
2000
Legend:
Missing record
Non-continuous and/or estimated
Continuous gage record

plished with techniques that would be considered unreliable today. In addition, the Platte River is a difficult stream to gage because its channel banks and bed are unstable; this complicates the calculations necessary to produce a stream-discharge measurement. That situation comes about because early calculations involved only two variables: channel width and depth of flow. Such an approach was incapable of taking into account the rapid changes in channel geometry that are common in sand-bed braided rivers. Methods became more sophisticated and discharge measurements more reliable after about 1925. The post-1925 stream-gage record is useful for supporting explanations of channel dynamics and interactions with sediment and vegetation, and the record documents short-term changes of a few years in duration. DOI analysts can specify with confidence the process connections between stream-flow and sediment transport, riparian vegetation, and inundation of various surfaces in and near the channel. Long-term changes in discharge are still difficult to specify, however. In searches of the stream-gage record for cyclic changes in discharge, for example, a common rule of thumb for hydrological analysis is to require a record that has 2-4 times the length of the suspected cycle; in this way, the cycle is repeated often enough in the record to be confidently identified. Hydrological adjustments to climatic changes occurring in cycles of several decades to a century, therefore, are not yet identifiable in the stream-flow record for the Platte River. The additional complications of variations in exchanges between surface water and groundwater are also difficult to discern in the stream-flow record.

The long-term gaging data from the North Platte, South Platte, and Platte Rivers show that river discharge has changed (Murphy and Randle

---

FIGURE 4-1 Periods of stream-flow record from gaging stations in Platte River Basin. (1) Includes four sites: Saratoga, WY; upstream of Seminoe Dam; upstream of Pathfinder Dam (historical); and downstream of current Pathfinder Dam site (historical). (2) Includes data from records for Camp Clarke (1896-1900), Mitchell (1901; 1907-1911), Scottsbluff (1912), Oshkosh (1929-1930), and Bridgeport (1902-1906; 1915-1928; 1931-1998). (3) Includes data from historical record for Platte River near Lexington, NE, before 1925. (4) Includes data from records for Platte River near Central City (1922-1927) and Columbus (1895-1914; June 15-Oct. 31, 1928). (5) Sum of Platte River at Columbus, Loup River near Genoa, and Loup River Power Canal. (6) Consists partly (Oct. 1, 1960-July 2, 1988) of record synthesized by combining Platte River at Louisville with Salt Creek near Ashland and Platte River at North Bend with Elkhorn River at Waterloo. (7) Includes data from record for Loup River at Columbus, NE, before Oct. 11, 1978, with Loup River Power Canal diversion added beginning Jan. 1, 1937. (8) Includes data from historical record for Elkhorn River at Arlington, NE. (9) Consists partly (Sept. 30, 1969-Dec. 31, 1998) of record synthesized by regression based on Salt Creek at Greenwood, NE. Source: Adapted from Stroup et al. 2001.

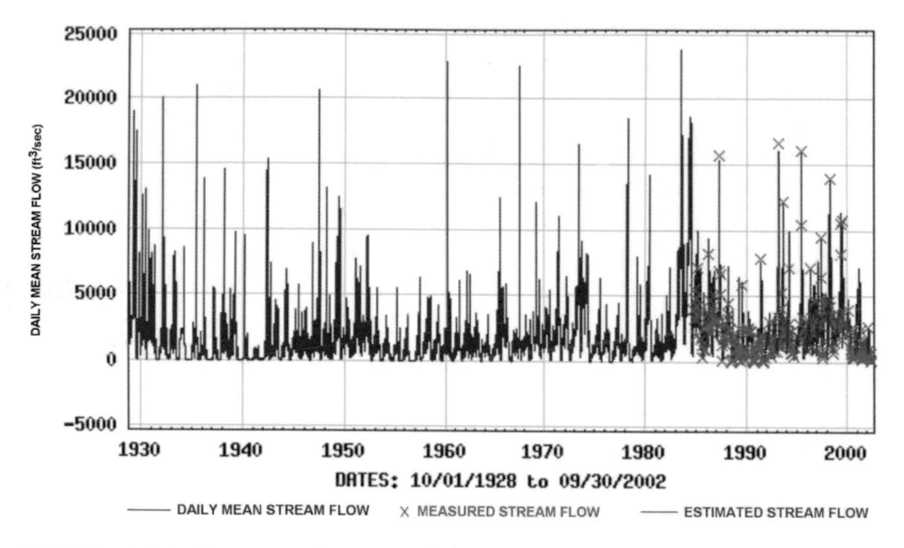

FIGURE 4-2 Daily stream-flow record for Platte River near Duncan, Nebraska, showing length of record and variability of flow. Source: USGS 2003.

2003). The flow record can be divided into four periods, separated by the gaging record and installation of engineering structures in the system: 1895-1909, 1910-1935, 1936-1969, and 1970-1999. There was a steady decline in the mean daily discharge of the rivers at a variety of measurement points from 1895 to 1969. Since 1969, however, flows have increased (Table 4-1), probably because of transmountain diversions into the river from outside the basin and return flows from high groundwater. The increased annual discharges through the Platte River in the later parts of the record have not

TABLE 4-1 Annual Mean Platte River Flows, cubic feet per second

Gaging Station	1895-1909	1910-1935	1936-1969	1970-1999
North Platte River at North Platte, NE	3,190	2,750	646	862
South Platte River at North Platte, NE	582	492	322	619
Platte River at North Platte, NE	3,780	3,240	968	1,480
Platte River near Cozad, NE	3,550	3,020	461	981
Platte River near Overton, NE	3,660	3,160	1,140	2,100
Platte River near Grand Island, NE	3,580	2,950	1,080	2,110

Source: Randle and Samad 2003.

had appreciable effects on the system, because the system adjusts its geomorphology only during high-discharge (flood) events. The increased annual discharges are passing through the system as longer-duration, relatively low-conveyance flows.

The magnitude of the annual flood has followed a course of change similar to the changes in annual discharges, but the changes in annual floods have been smaller (Table 4-2). As a result, annual floods during the period 1970-1999 are not nearly as large as those in the system before 1909. In the extreme case, at the gage site on the Platte River at Cozad, the magnitude of the recent annual flood peaks is only 15% of the magnitude observed before 1909. The large upstream dams were built to exert this flood control and to save the water from the flood peaks in reservoirs for later, slower releases for irrigation and urban water supply. The significance of the earlier and larger annual floods is that they performed most of the geomorphic work of changing the channel configuration, maintaining a wide active channel, adding and deleting islands and bars, and sweeping seedlings from exposed sand accumulations. Those functions have not been performed by the recent and smaller floods.

To investigate the flow movement and distribution in the entire basin, a central Platte OPSTUDY hydrological model has been developed as a management tool to incorporate all hydraulic structures, water-rights demands, and so on. The basic function of this mass-balanced hydrological model is to simulate all the flow inputs and outputs at various locations from the inflows of the North and South Platte River down to the flow at the lower Platte River. Reservoir flow storage, flow diversions, irrigation demands, and groundwater inputs and outputs are included in the model.

The earliest descriptions of the vegetation and morphology of the channel appear in journals and accounts by explorers beginning in the early 1800s. Qualitative assessments described a river with scattered trees on its

TABLE 4-2 Annual Flood Peaks in Platte River Basin (Flows with Return Interval of 1.5 Years), cubic feet per second

Gaging Station	1895-1909	1910-1935	1936-1969	1970-1999
North Platte River at North Platte, NE	16,300	8,150	2,160	2,380
South Platte River at North Platte, NE	2,330	1,430	712	1,420
Platte River near Cozad, NE	17,600	9,140	1,980	2,590
Platte River near Overton, NE	19,400	9,000	3,490	4,750
Platte River near Grand Island, NE	17,300	10,100	4,500	6,010

Source: Randle and Samad 2003.

banks and immense numbers of wooded islands all coursing through an open prairie landscape (Johnson and Boettcher 2000a). Explorers occasionally reported quantitative estimates of channel dimensions. Channel depths were estimated at 2-4 ft, and bank-to-bank widths were up to 2 mi in some places (Mattes 1969). Starting in the late 1850s, the GLO conducted surveys to implement the Township and Range Survey system that is the basis of all modern property descriptions in the Midwest. The surveyors' notes and plat maps derived from them included channel characteristics, descriptions of large islands, general vegetation, size and species of witness trees, soils, geology, human uses of the land, and depths of flow in the river. The GLO surveys generally provide the first systematic record of the predevelopment environment of the central and lower Platte River. With the construction of the Union Pacific Railroad, photographers began recording images of the river (Figure 4-3).

The advent of modern surveys of river cross sections was associated with bridge construction from the late 1800s on. The data are probably accurate, but they were limited to crossing sites. Bridge cross sections do not necessarily reflect the kinds of channel dynamics that occur in more-natural reaches inasmuch as they tend to be narrower and more stable than other cross sections. The first complete coverage of the geomorphology of the river resulted from the USGS topographic mapping program, which produced maps for Nebraska and the river based on conditions in 1896-1902. Private maps followed, showing parts of the river at various dates

FIGURE 4-3 Photograph of central Platte River near 100th Meridian in vicinity of Cozad, Nebraska, in 1866. River has wide active channel mostly devoid of vegetation except for few trees, apparently cottonwood, on island. Source: Carbutt 1866b. Reprinted by permission of Union Pacific Historical Collection.

and stages. Channel structure and vegetation were systematically photographed from the air for the first time in 1938. Later aerial photography and other remote-sensing imagery, including infrared imagery, provide views of the system at roughly decade intervals to the present.

Cross-sectional surveys of the river as input data for hydraulic modeling date from about 1980. Eschner (1983) surveyed six cross sections across the central Platte River during 1979-1980 and related changes in channel geometry to changes in discharge. Thereafter, DOI researchers surveyed various cross sections at various times, and other state agencies and individual researchers surveyed additional cross sections. There is no central repository of cross-sectional data; for many river locations, only one set of cross-sectional data exists. For reaches with multiple surveys, DOI investigators have produced quantitative information about channel change. Researchers documented these repeat survey sites (Randle and Murphy 2003), and they now provide a benchmark for monitoring and input for modeling.

Sediment-discharge measurements are not available for the central and lower Platte River, so investigators have used model calculations to estimate sediment transport through the system. Input for sediment-transport models includes daily water discharge and estimated (or assumed) amounts of sediment discharge. Such fluvial geomorphology and hydrological engineering models are numerous (Simons and Senturk 1992). The initial model output is in the form of mathematical curves that generalize the connection between water and sediment discharge. Further calculations use the sediment rating curves and gaged water discharges to produce the estimates of total sediment transport. The best understanding of sediment discharge through the central Platte River derives from DOI work summarized by Randle and Samad (2003), who used three sediment-transport functions to make connections between water and sediment discharge, models by Kirchner (1983) and Simons and Associates (2000), and a model constructed by Randle and Samad. Although the exact numbers for results varied from one model to another, sediment trends were highly similar in all models. High correlation among the models was somewhat expected in that the major variable in the calculations was water discharge, and the same values were used in all the calculations.

The results of sediment-transport and discharge studies in the Platte River show that the roles of the North and South Platte Rivers as sources of sediment for the Platte River in Nebraska have changed. From 1895 to 1935, the North Platte supplied much more sand to the Platte than did the South Platte (Murphy and Randle 2003); but from 1936 to 1999, the South Platte supplied at least as much as the North Platte. Patterns of sediment transport through the Platte River also changed. From 1895 to 1935, volume and rate of sediment transport were consistent along the entire length of the central Platte; but after 1936, the loads from Cozad were lower than the amounts transported past the town of North Platte. The decrease was probably the product of sediment deposition in the 70-mi reach from North Platte to Cozad, which is also the location of the Johnson-2 (also known as J-2) return flow (Figure 4-4). Also after 1936, the amount of sediment transported for reaches downstream of Cozad and the Johnson-2 return flow were higher than the amount at Cozad. The increase was probably the result of erosion in upstream reaches.

The central Platte River has two distinct segments from the perspective of sediment: the segment upstream of Cozad and the Johnson-2 return, which is accumulating sediment (aggrading), and the segment downstream of Cozad and the Johnson-2 return, which is losing sediment (degrading). That circumstance is likely to lead to different geomorphic forms in the two reaches. As might be expected, the distribution of sediment transport over time and through space is explained mostly by changes in stream flow, largely as a result of controls by dams, diversions, and return flows. All hydrological and geomorphic processes operate amid cyclic weather patterns typical of western extremes and more gradual, longer-term climatic changes.

Information on riparian vegetation has many of the same sources as geomorphic data. Early descriptive accounts from explorers and settlers provide qualitative information and were followed by more quantitative data from the GLO survey, aerial photography, and finally satellite imagery. Johnson and Boettcher (2000a) reviewed the journals and diaries of explorers and surveyors from the early 1800s to the 1870s. They concluded that scattered timber was found along the outer banks, which were susceptible to prairie fires, while heavier woodland occurred on numerous islands of all sizes in the channel. Surveyors recorded the species and size of hundreds of living witness trees but also found the stumps of many trees, indicating extensive deforestation in progress by settlers, soldiers, and railroad crews. Little quantitative information is available on the Platte's woodland cover between the 1870s and the first available aerial photographs in the late 1930s. Johnson and Boettcher (2000a) suggested that cutting of trees by prairie farmers continued for decades after settlement and contributed to the open appearance of the river in the early photographs. Currier and Davis (2000) questioned whether such clearing took place, but U.S. Surveyor Amherst Barber noted in 1899: "Bearing trees were not used at aforesaid corners for the reason that the whole of the river bed and islands is constantly searched by farmers from the prairies, who cut and carry off all poles large enough to cut, and the small sapling I found, of 2 in. or so, would soon be removed."

The foregoing discussion about woodland occurs in the following important context. Woodland probably has always existed along the central and lower Platte River during the past two centuries, but it has occurred in

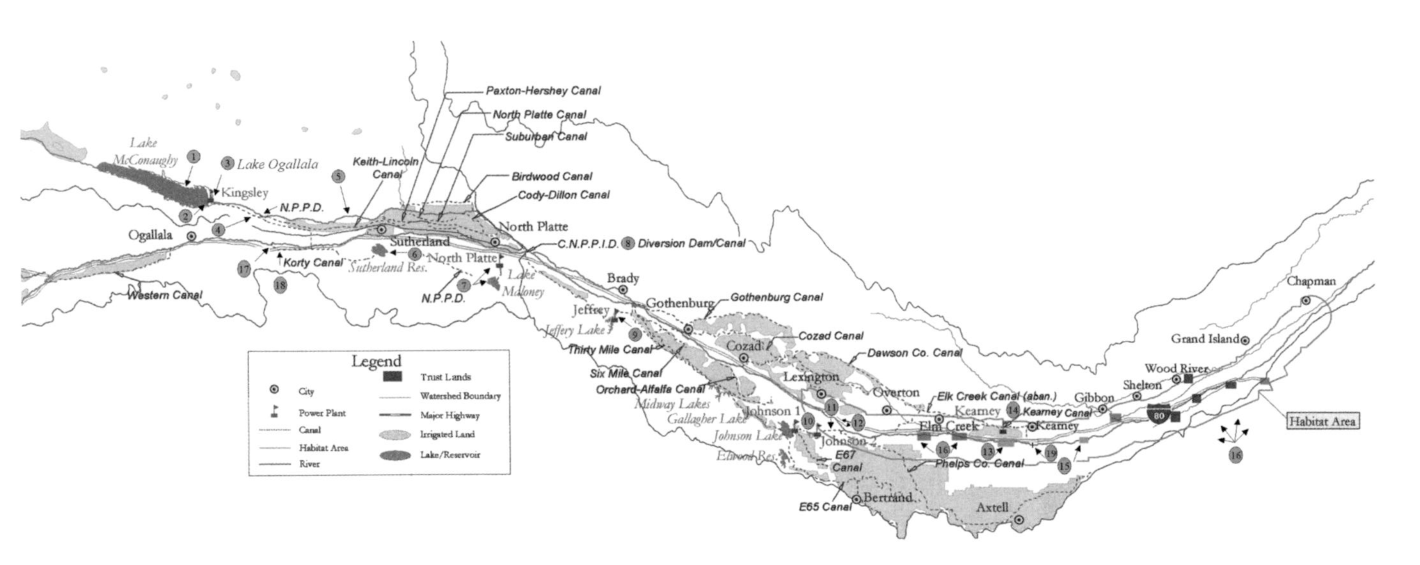

FIGURE 4-4 Central Platte River with associated water control infrastructure and places of interest mentioned in this report. (1) Lake McConaughy. (2) Kingsley hydroelectric plant. (3) Lake Ogallala. (4) Keystone/Sutherland Canal. (5) River channel Kingsley to North Platte. (6) Sutherland Reservoir/Gerald Gentleman Station. (7) Lake Maloney/North Platte hydroelectric plants. (8) Central diversion. (9) Jeffrey return. (10) Johnson Lake/Johnson-1, Johnson-2 hydroelectric plants. (11) Johnson-2 return. (12) encroached reach (river channel below Central diversion to Johnson-2 return). (13) Cottonwood Ranch. (14) Kearney Canal. (15) Audubon Rowe Sanctuary. (16) Platte River Trust lands. (17) Korty diversion. (18) Korty Canal. (19) Kearney bridge. Source: Adapted from USFWS, unpublished, June 5, 2000.

different amounts. By the mid- to late 1800s, it may have been at a minimum, but at the very least it occupied mid-channel islands. In the twentieth century, woodland had expanded and covered more areas than in the late 1800s. In all cases, however, there has been a gradient of environments ranging from quite open situations to closed woodland. More specifically, woodland cover apparently was minimal during the late 1800s and early 1900s. Comparison of the woodland distribution in the riparian zone from 1938 (when aerial photography of the region began) to present times shows extensive woodland expansion along many reaches of the central Platte River (e.g., Williams 1978; Eschner 1983; Johnson 1997). Those long-term records indicate that woodland cover expanded and reached a maximum in many reaches of the river beginning in the 1960s. Since the 1980s, managers have cleared substantial areas of woodland for habitat-management purposes, but other change has been minor.

The riparian vegetation associated with the Platte River has changed dramatically in the last 150 years. The extensive island woodland was cleared for timber by the late 1800s. Woodland expanded sharply in the first half of the 1900s in connection with the construction of dams and diversions that caused an increase in the recruitment and a decrease in the mortality of riparian woody plants. By the 1960s, equilibrium among flows, sediments, land form, and woodland appeared in most reaches of the river (Johnson 2000).

Scientific Data for Ecosystem Connections

Various datasets on the corridor of the central and lower Platte River support systematic understanding of connections among the major ecosystem components: water discharge, sediment movement, morphology of the river, and riparian vegetation. (Other ecosystem components, including grazing and invasive species, have not been analyzed.) Collectively, those characteristics determine the nature of habitat available for wildlife in general and of suitable habitat for threatened and endangered species in particular. DOI agencies accomplished three tasks in the process of listing the four protected species and in designating critical habitat for the whooping crane and piping plover. First, DOI developed an understanding of the species habitat needs and then an understanding of the river system dynamics. Second, DOI determined the magnitudes, frequency, duration, and timing of flows in the river to benefit the listed species. Finally, DOI recommended a set of instream flows to benefit the listed species. This section addresses river ecosystem and habitat interconnections, the next section reviews instream flow issues, and a third section addresses the habitat suitability issue.

Data outlined earlier in this chapter paint a general picture of environmental change in the corridor of the central and lower Platte River. The relative abundance of habitat conditions favorable to the listed species seems to have declined over the last century. The alteration of flows by human controls has substantially changed the behavior of the river. Human controls have reduced flood peaks, eliminated many pulse flows, and decreased the total amount of water flowing through the system. The decline of discharges since 1895 has reversed in recent years, but not to the point of recovery to predevelopment levels.

The water-control infrastructure and related water-flow changes have larger effects throughout the river's subsystems. Flow diversions effectively dewater the river in some reaches, resulting in the deposition of sediment, particularly upstream of the Johnson-2 return (Figure 4-5). Flow in the lower reaches is sufficient to move sediment, so erosion proceeds there because sediment influx from upper reaches is interrupted (Figure 4-6). Reductions in flow events, particularly those about bankfull (a water discharge that fills the channel cross section but does not spill over onto the floodplain), and changes in the sediment transport regime are forces that play themselves out on a geomorphologic stage. The characteristic forms that once dominated the river—a braided configuration laced with intertwined channels, bars, islands, and beaches of great complexity—have become simplified. Features that once would have been unvegetated and mobile have become vegetated and stabilized in the central Platte River (Figures 4-7 and 4-8).

Extensive reports document generalizations about the dynamics of water, sediment, morphology, and riparian vegetation. Summary explanations, supported by numerous references, of the physical changes in water, sediment, and morphology of the river appear in Nadler and Schumm (1981), Simons and Simons (1994), and Murphy and Randle (2003). The changes

FIGURE 4-5 Johnson-2 diversion on central Platte River, a feature that influences local distribution of flows in river and provides valuable irrigation water for agriculture. Source: Photograph by W.L. Graf, May 2003.

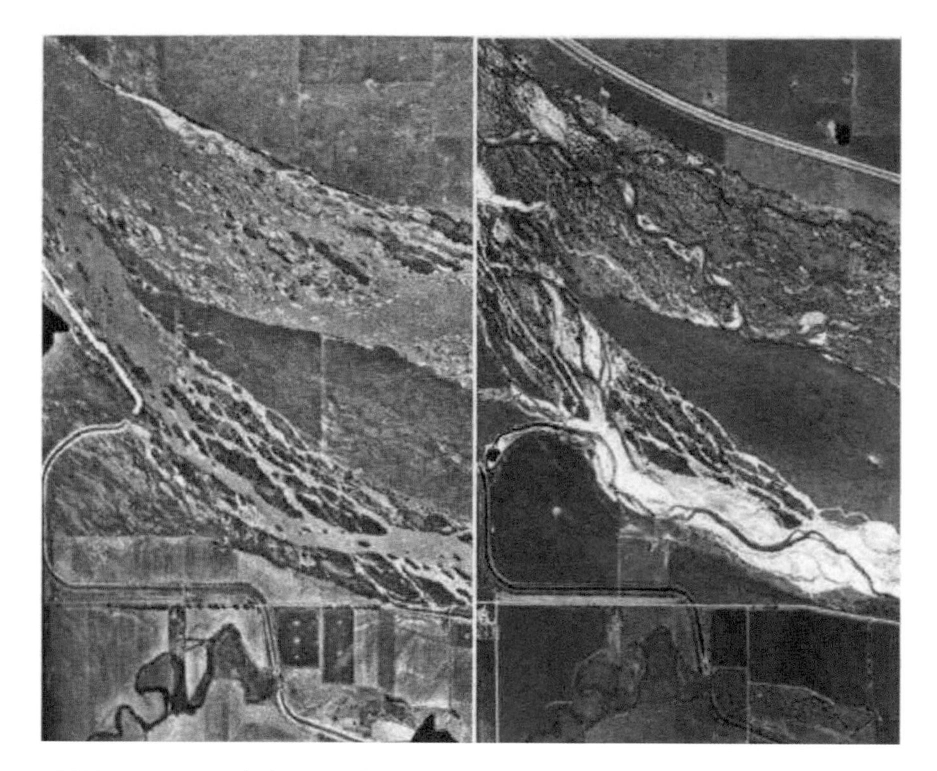

FIGURE 4-6 Aerial photographs from 1938 (left) and 1998 (right) of Platte River at Johnson-2 irrigation return site (for location, see site 11 south of Lexington on Figure 4-4). Irrigation return drain is on left side of each image. In 1938 view, channel above return is open and sandy with wooded islands; in 1998 view, channel is covered with vegetation except for relatively narrow active channel and clearer area downstream of return point. Channel is bifurcated in this reach. Source: USBR 1998.

associated with riparian vegetation and connected physical changes have been the focus of some debate, but clear understanding has emerged (Johnson 1994, 1997; Johnson and Boettcher 2000a; Currier and Davis 2000). Currier (1995) has explained further connections with groundwater. Many studies of the channel history of the Platte River have been conducted. Most time-series graphs start with channel widths measured on GLO plat maps, which are followed by estimates made from aerial photographs, first available in the late 1930s. Standard methods have not been used or developed to measure width or area, so comparisons among studies are difficult; studies vary widely in reaches sampled, dates of photography, sample sizes (number of transects or sample points), maximal vegetation allowable to qualify as "open" or "unvegetated" channel, and use of error analysis.

Eschner's (1983) oft-cited graphs show a steep decline in channel width between the time of the plat maps and the first aerial photographs (Figure 4-9). The decline would be less steep if nonsurveyed vegetated islands could

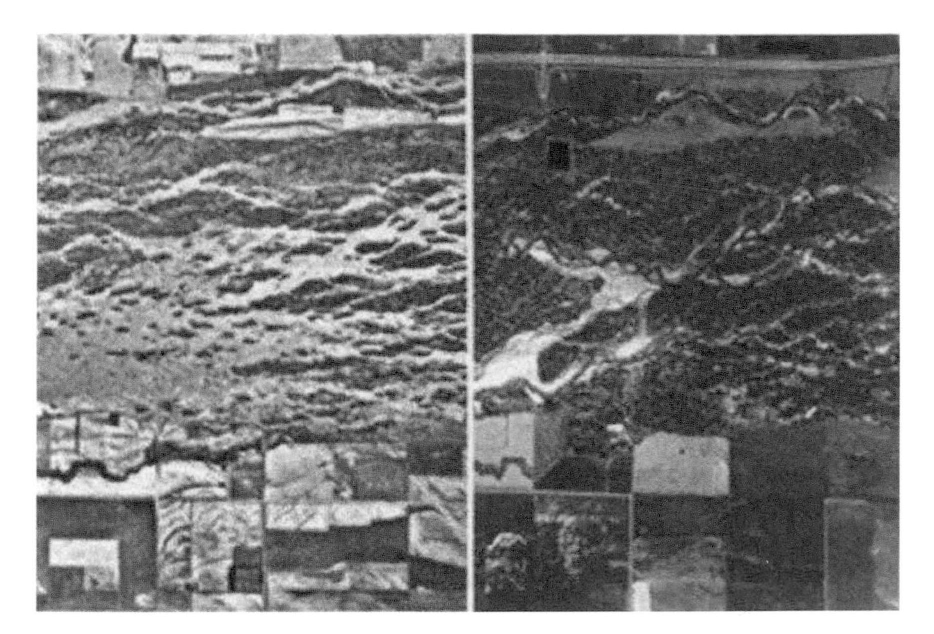

FIGURE 4-7 Aerial photographs from 1938 (left) and 1998 (right) of Platte River at western edge of Cottonwood Ranch (for location, see site 13 south of Kearney on Figure 4-4). During 60-year period between two dates, woodland has expanded to cover most of river area. Source: USBR 1998.

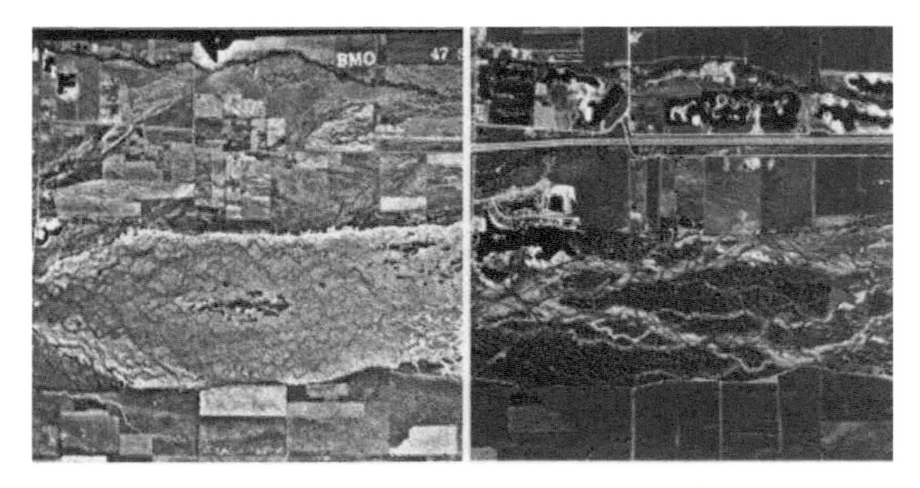

FIGURE 4-8 Aerial photographs from 1938 (left) and 1998 (right) of Platte River immediately downstream of Kearney bridge crossing (see Figure 4-4 for location). In 1938, abutments of bridge and approach berms for road constricted channel, leading to narrowing at western edge of 1938 image. At that time, channel was devoid of vegetation except for forested island in center of image. In 1998, bridge and its approaches were removed, and channel was largely vegetated except for narrow, active portions. Source: USBR 1998.

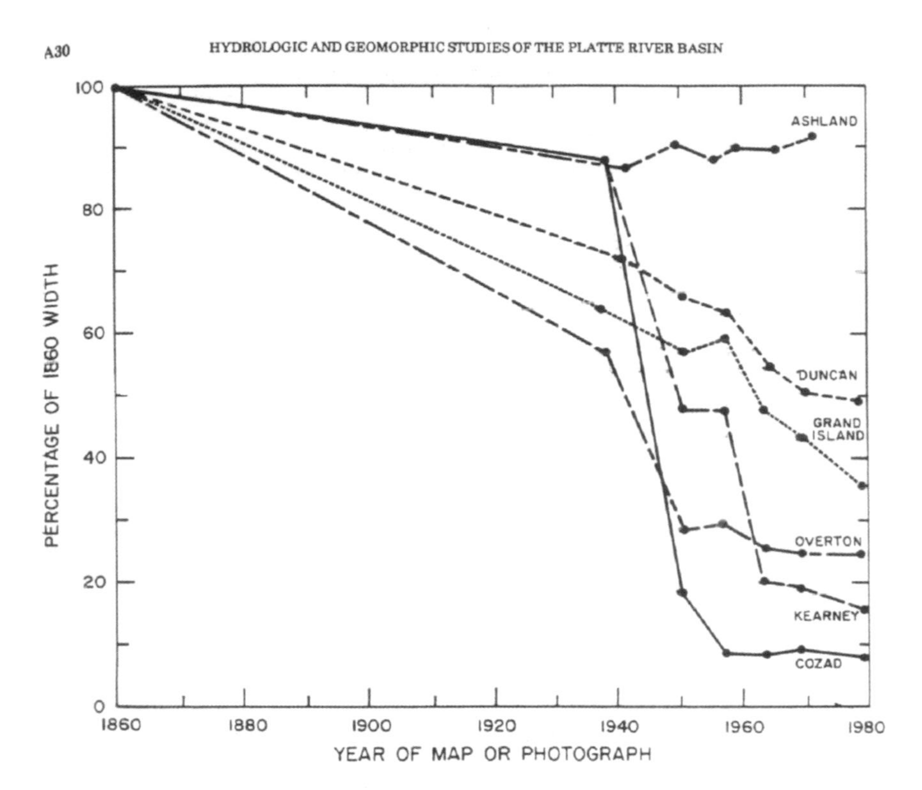

FIGURE 4-9 Changes in channel width at various cross sections of Platte River as interpreted by T. R. Eschner on basis of GLO plat maps (1860s) and aerial photographs (1938 and later). Greatest amount of narrowing has occurred in upstream or western cross sections, with decreasing amounts of change in downstream or eastward direction. Source: Eschner 1983.

have been accounted for in the width calculations on the plat maps. Johnson's channel-area estimates (1994, 1997) did not begin with the plat maps, because of uncertainties in accounting for the missing islands. Rather, they began in 1938 when aerial photographs first became available; that is, any woodland expansion before 1938 was not estimated. Several quantitative patterns since 1938 are evident (Figure 4-10). Channel-area decline has been more extreme in upstream reaches of the Platte River; this suggests downstream attenuation in the effects of flow regulation. Channel-area decline occurred earlier in upstream than in downstream reaches. Channel area in many reaches stabilized or began to increase in the 1960s, so equilibrium or steady-state conditions probably began nearly 4 decades ago. Declines in channel width and area were caused by vegetation colonizing and surviving in the active channel. The vegetation varies considerably in

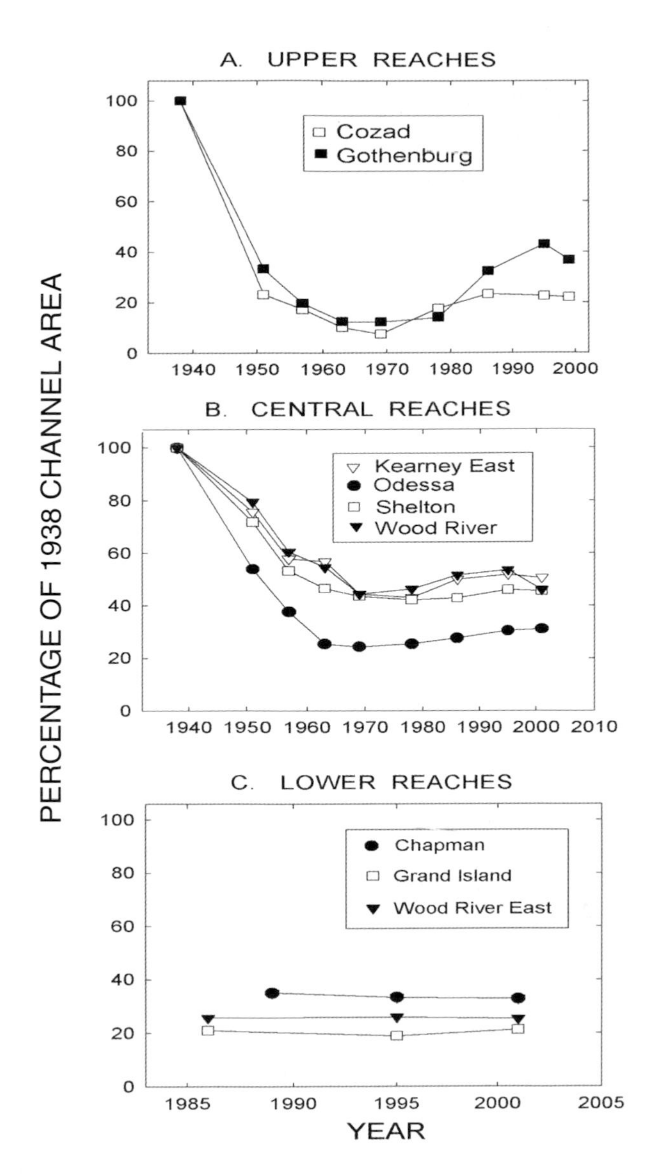

FIGURE 4-10 Changes in channel width at various cross sections of the central Platte River as interpreted by C. Johnson on basis of aerial photography. In later part of record, after 1960s, decreases in width remained stable or reversed. Source: Johnson 1997. Reprinted with permission; copyright 1997, John Wiley & Sons Limited.

composition and in structure and is highly dynamic in time and space (Currier 1982; Johnson 1994). The initial composition of vegetation on new sandbars depends on the season of drawdown and exposure. Generally, exposure early in the growing season (May and June) allows for colonization by perennial, pioneer woody plants (cottonwood, *Populus;* and willow, *Salix*) and many annual plants (such as *Echinochloa, Eragrostis, Xanthium,* and *Cyperus*), whereas first exposure late in the growing season (August) produces domination almost exclusively by annuals. Sandbars with only annuals convert back to unvegetated surfaces the next year. Sandbars with young perennial vegetation (especially if cottonwood and willow seedlings are present) will continue to develop more-complex and more-diverse vegetation unless it is killed by erosion or sedimentation. Woody vegetation on sandbars that has survived 3-5 years is relatively resistant to both processes.

What caused the channel narrowing and woodland expansion? Several single-factor explanations have been offered, including a decline in peak flows (USFWS 1981a) and increases in summer low flows (Nadler and Schumm 1981). The most recent published analyses (Johnson 1994, 1997, 2000) support the following multifactor explanation. Reduced stream flow in late spring (the period when reservoirs are filled) during the peak seed-dispersal period for pioneer cottonwood and willow trees best correlates with the historical rates of woodland expansion. Lower flows during this period after regulation enabled more extensive reproduction of trees on the exposed sandbars than in the predevelopment period. Reduced flows at other periods tended to increase tree-seedling survival and thus contributed to the expansion of cottonwood and willow woodland into formerly active, unvegetated portions of the channel.

Survival of tree seedlings also may have been increased by higher late-summer (irrigation-return) flows in the regulated river; however, the GLO notes indicate that the Platte River rarely went dry, contrary to popular belief based on some settlers' diaries (see Edmonds 2001 regarding pitfalls of written records). Additional causes of modification of local conditions in some areas are grazing and invasive species.

Three invasive riparian plant species are of most concern to Platte River managers: a Eurasian exotic, purple loosestrife (*Lythrum salicaria*) (Figure 4-11) and invasive nonnative variants of two grasses—common reed (*Phragmites australis*) and reed canary grass (*Phalaris arundinacea*). *Lythrum* and *Phragmites* are known to propagate and spread from root and stem fragments, so the practice of mowing and disking infested Platte River sandbars may be causing proliferation.

The relationship between open channel and wooded areas is dynamic. Essentially, the modern Platte River is a microcosm of its earlier, predevelopment form but operating under modified ecological processes.

FIGURE 4-11 Purple loosestrife, an introduced, nonnative species that aggressively occupies some niches along Platte River. Source: Photograph by W.L. Graf, August 2004.

When stream flow was reduced, the river was less able to rework or otherwise influence all its formerly active floodplain. About half the floodplain of the central Platte River reverted to woodland that became relatively stable in area and in location; however, bank erosion along active channels in these wooded areas creates new areas of open channel incrementally (Figure 4-12). The remaining relatively unwooded area of the floodplain that comprises open channel and young vegetated sandbars is, conversely, highly dynamic. Vegetation comes and goes, depending on flow conditions. For example, from 1986 to 1995, considerable areas of former channel became vegetated, and former vegetation became channel (Figure 4-13). Although there was considerable turnover and relocation of channel and vegetated categories, net channel area was generally constant. Channel area increases during periods of high stream flow and decreases during periods of low stream flow.

The present conditions have resulted from increased width and depth in the remaining active channel as other minor channels were deactivated by the growth of new woodland. Instead of being dispersed among several

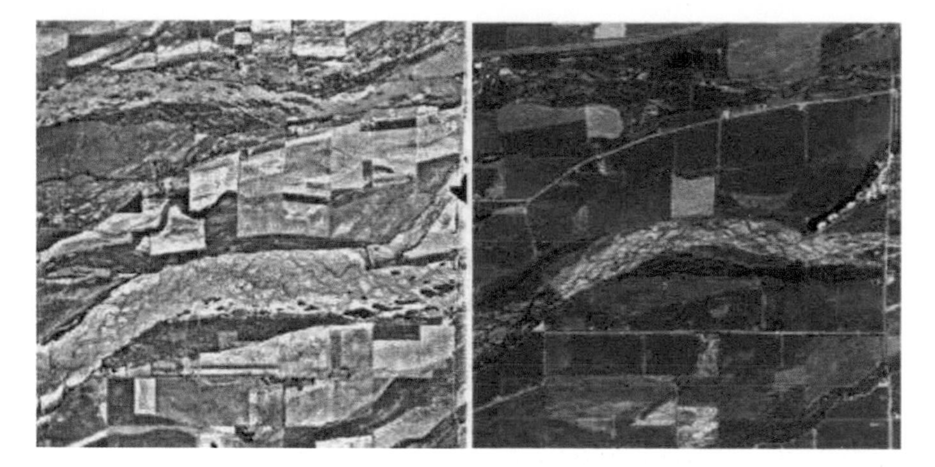

FIGURE 4-12 Aerial photographs from 1938 (left) and 1998 (right) of Platte River at Audubon Rowe Sanctuary (for location, see site 15 southeast of Kearney on Figure 4-4). Channel in this reach is in two parts separated by large island. In 1938, both parts were open and mostly without vegetation except for small islands that were wooded. In 1998, active channels are more narrow, but southern channel remains braided. Source: USBR 1998.

subchannels, a single channel carries most of the flow in some reaches (Figure 4-14). That process has reduced the recruitment and increased the mortality of tree-seedling populations. Mortality of tree seedlings is extremely high in the modern river. Johnson (2000) found that on about 90% of the new sandbars all seedlings died by the end of their first year of life. The flow now fits the remaining channel better, and summer stream-flow pulses from thunderstorms and unused water deliveries during wet weather and moving ice in winter have become increasingly effective in killing young tree seedlings by erosion and sedimentation. Summer desiccation is another mortality factor, but it occurs at a much lower frequency than mortality due to flow and ice.

Habitat implications of the newly established woodland cover are complex. In reaches where the total distribution of woodland has no net change in coverage, the remaining open area may no longer have sufficient size or channel structure to support species that require open and wide channels. Increased woodland cover has caused increases in other species, such as deer, turkeys, and songbirds.

Some reaches may not have reached equilibrium or may have disequilibrated. Currier (1997) reported that channel narrowing and vegetation expansion had occurred between the middle 1980s and the middle 1990s in several reaches not examined by Johnson (1994). Johnson (1997) and Currier (1997) both measured rapid vegetation expansion in a rela-

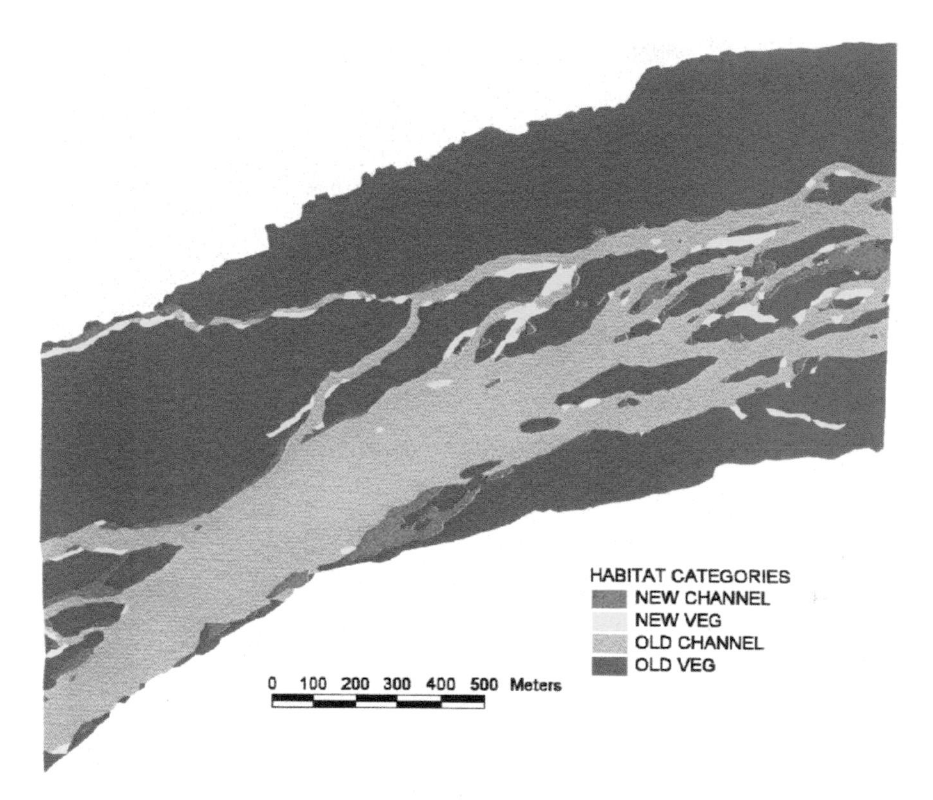

FIGURE 4-13 Map of areas of habitat change in cover from 1986 to 1995 for reach of Platte River near Shelton, Nebraska. Source: Johnson 1997. Reprinted with permission; copyright 1997, John Wiley & Sons Limited.

tively wide channel near Grand Island during the same period but interpreted the cause differently. Johnson (1997) speculated that release of sediment after the clearing and leveling of islands upstream of Grand Island had aggraded the riverbed, altering the flow splits among channels, dewatering the main channel, and causing the vegetation expansion. Currier (1997) speculated that the cause of the narrowing at Grand Island was alteration of stream flow by upstream water development. Resolution of the equilibrium issue was difficult because the authors used different methods, reach boundaries, and aerial photographs.

Most of the Platte's woodlands now have a mature canopy of cottonwood and willow, with an understory of elm, ash, and cedar. As a result of systemic controls that include natural and human influences, there is little regeneration of cottonwood and willow, and the mean age of the existing woodland is increasing in the absence of ingrowth. Cottonwood and willow will decline in importance over time, with corresponding declines in biodiversity, unless measures are taken to promote their regeneration and establishment. There is also

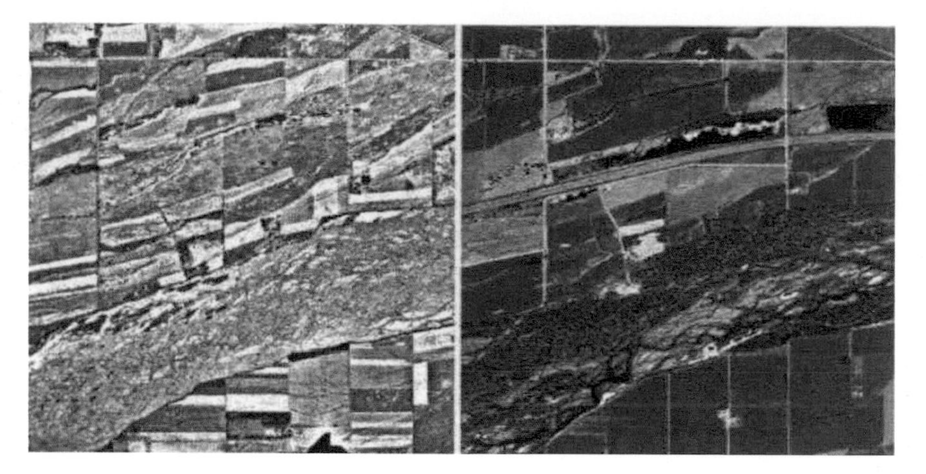

FIGURE 4-14 Aerial photographs from 1938 (left) and 1998 (right) of Platte River west of railroad bridge near Gibbon (see Figure 4-4 for location). Between 1938 and 1998, channel converted from braided open channel to modified version with braided and single-thread components side by side, with woodland covering portion of 1938 active channel area on north side. Source: USBR 1998.

some variety of densities, with some open areas and some areas with gradational densities between open and closed woodland.

Despite varied interpretations, studies of woodland processes in the central Platte River generally agree on the following. First, woodland was an important and permanent part of the presettlement Platte, but its aerial coverage has increased substantially in the modern, regulated river. Second, woodland expanded in many areas from the 1930s to the 1960s. Third, a relatively stable relationship between the areas of woodland and open-channel area has developed in most reaches since the 1960s, with stability of composition and structure. Fourth, some reaches continue to exhibit channel narrowing, but the cause of the narrowing in these reaches has not been completely explained. Fifth, the different conclusions reached by investigators on these issues have been difficult to resolve because of differences in study areas and methods.

A MODEL FOR RIVER AND HABITAT CHANGE

General explanations provide the basis of a more exacting analysis of the interactions among water, sediment, form, and vegetation. USFWS used the Physical Habitat Simulation System (PHABSIM) to make connections among the various components of the ecosystem of the river. PHABSIM is a collection of interactive programs that accept as input such hydrological measures as water depth, velocity of flow, substrate, and cover (Figure 4-15). The program

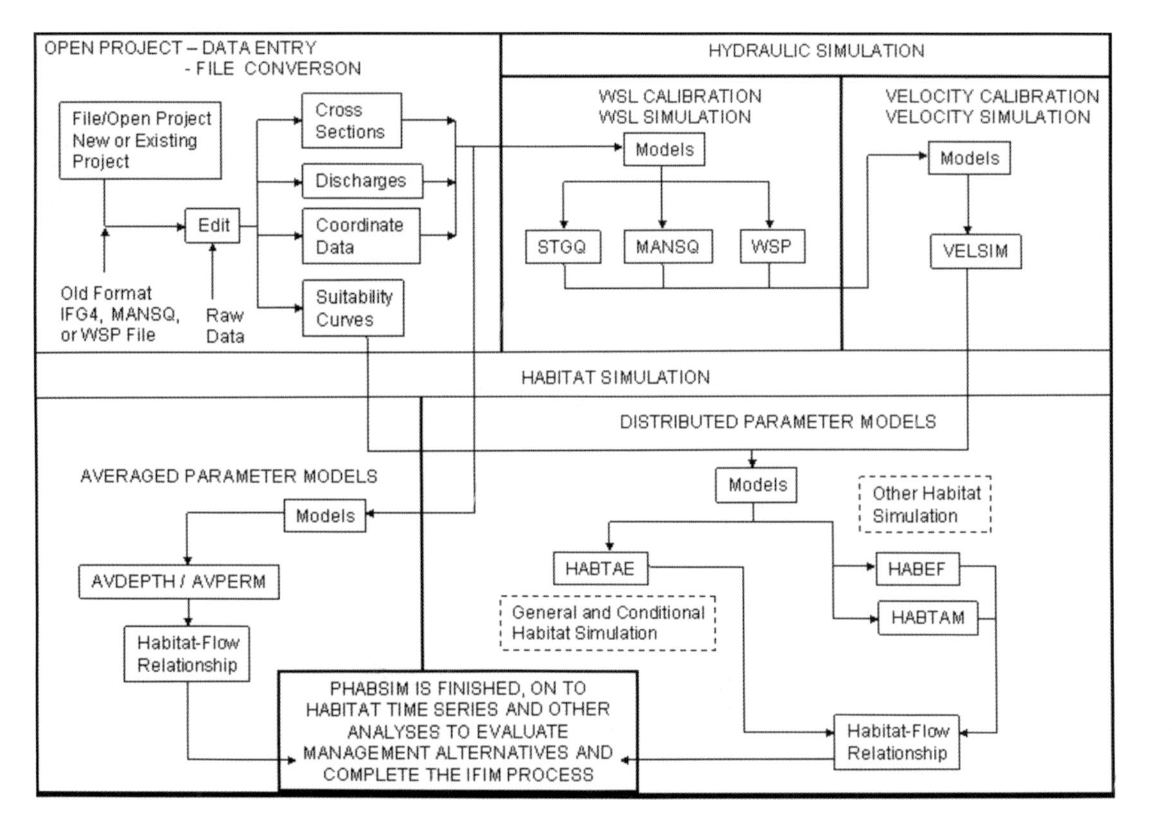

FIGURE 4-15 Schematic diagram of basic operations of Physical Habitat Simulation System (PHABSIM) used by DOI agencies to specify connections among habitat characteristics, habitat preferences, and river discharges. Abbreviations: WSL, Water Surface Elevation. IFG4, STGQ, MANSQ, WSP, VELSIM, AVDEPTH/AVPERM, HABTAE, HABEF, HABTAM, are models and programs. Further information and definitions available at http://www.fort.usgs.gov/products/Publications/15000/appendix.html. Source: Waddle 2001.

takes into account coincidental data about the amount of suitable habitat for listed species in and along the river at a variety of discharges, and it builds a series of mathematical relationships between the magnitude of flow and the amount of usable habitat per unit length of the stream. The general mathematical relationship can be expressed graphically (Figure 4-16). There may be many such curves for each species of interest, with each curve representing the amount of suitable habitat available during different life stages of the species.

PHABSIM assesses the habitat value of a reach of stream over a range of discharge for a target species or life stage (Bovee et al. 1998). The model has seen service in a wide variety of environments and for many species and has even been used to assess recreation potential and aesthetics (Gillilan and Brown 1997). In most applications of PHABSIM, patches of habitat are characterized at different rates of discharge in terms of three "microhabitat" variables: depth, water velocity, and substrate size. Substrate size in each patch is estimated from field data, and depth and velocity are estimated by a spatially explicit hydraulic model; both one- and two-dimensional models have been used. At each discharge, the habitat value of each patch is evaluated independently for each variable according to "suitability curves" that range from 0 to 1, and the product of these values and the area of the patch

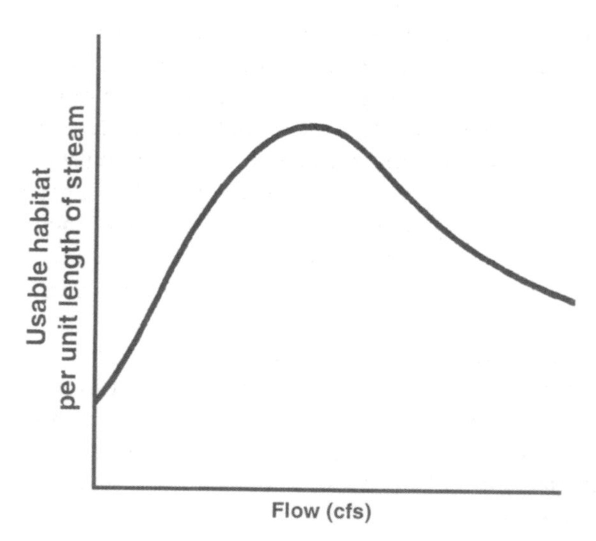

FIGURE 4-16 Schematic example output of PHABSIM. Graph represents amount of habitat for particular species that is available at each flow discharge of river. System generates one such graph for each life stage of species of interest. At some flow discharge, amount of available habitat is at maximum peak of curve. Source: Adapted from Gillilan and Brown 1997.

yields an estimate of weighted usable area (WUA). These are summed over patches, producing a curve of WUA over discharge for the reach.

Although PHABSIM is widely used, it has been challenged on the basis of both its approach and its application (Castleberry et al. 1996). It has some limitations. The one-dimensional hydraulic models use dubious methods to estimate the variation in velocity across the channel, and the typical spatial scale of the two-dimensional models does not match that of the suitability curves (Kondolf et al. 2000). The accuracy of the velocity estimates is seldom tested, and reported tests show weaknesses in results (Kondolf et al. 2000). The biological meaning of WUA is unclear: Bovee et al. (1998, p. 71) say only that it is a "microhabitat metric." However, the modeling implicitly assumes that habitat preferences do not vary with physical and biological conditions in the stream and that density is a valid indicator of quality; both these assumptions are dubious (Bult et al. 1999; Holm et al. 2001; van Horne 1983). Applications of PHABSIM often do not report confidence intervals for estimates of WUA (Castleberry et al. 1996). Those reservations aside, PHABSIM has been the standard for habitat assessment along rivers. The U.S. Bureau of Reclamation (USBR) is developing an alternative sediment-vegetation model, SEDVEG, to evaluate the interactions among hydrology, river hydraulics, sediment transport, and vegetation for the Platte River, but it has not been widely used.

The application of PHABSIM to the Platte River beginning with investigations in the late 1980s and the early 1990s produces a valid starting point for assessing the complex web of connections among water flows, sediment processes, geomorphology, and habitats. There remains considerable uncertainty regarding exact specifications of particular flow discharges to accomplish habitat restoration because of the complexity of the channel system and the lack of substantial empirical studies to connect the fluvial processes with the resulting habitat. In many cases, judgments had to be made about the most likely approaches to successful restoration of the river processes. Uncertainty also enters the analysis because it is unclear that if habitat is reconstructed, it will serve the needs of the threatened and endangered species. However, as discussed in Chapter 3, science does not offer perfect certainty. Rather, it offers reasonable approaches based on observable experiences. The application of PHABSIM and later improved strategies, such as the establishment of normative flows, are steps in a process of improving understanding of the habitat needs for the species rather than sharply defined end points.

Instream-Flow Recommendations

Once analysts define the connections among various measures of flows and habitat for species of interest with the PHABSIM model, the calculation of flow recommendations with a second family of models collectively known as Instream Flow Incremental Methodology (IFIM) is possible. IFIM includes the label "incremental" because the system of equations and data can simulate the habitat implications of incremental changes in flow. IFIM uses computer software to evaluate microhabitat and macrohabitat characteristics that occur at various discharge levels and can project the resulting outcomes of flow scenarios through time in the form of "habitat time series." In the parlance of IFIM, *microhabitat* refers to small spaces or areas up to a few meters in extent. *Mesohabitat* refers to small areas of the stream and includes depth and velocity of stream flow, substrate characteristics, and vegetation cover, often associated with a geomorphic unit of the stream, such as a pool, run, or riffle. *Macrohabitat* refers to the longitudinal aspects of the stream important to species, and includes water quality (particularly temperature in many applications), channel morphology, and overall water discharge. Macrohabitat includes a longitudinal portion of stream within which physical or chemical conditions influence the suitability of the entire stream segment for an aquatic organism. The scale of analysis is sometimes kilometers. *Total habitat* in IFIM models refers to the entire wetted perimeter of the channel along a given length of stream that may extend several kilometers. Geographically explicit predictions can define the expected areas and locations of useful habitat for the species in question because such habitat is a function of discharge. For a complete discussion of IFIM and its relationship to PHABSIM, consult Waddle (2001) or the website version.

IFIM relies on a complex system of hydrological models that operate with computer software (Figure 4-17). Because IFIM requires as input cross-sectional survey data and discharge records and calculates the amount of available habitat for many reaches of the river and for several species (in the Platte River application), its full development required several years to complete. USGS maintains the software and assists in its application by USFWS and other DOI agencies. The software is in the public domain, and it is used by university researchers and private consulting firms for contract work.

During the period when USFWS was determining its instream-flow recommendations for the central and lower Platte River, the agency used IFIM and applied it to the three threatened or endangered avian species along the river. When the agency completed its calculations and made instream-flow recommendations, PHABSIM and IFIM were the state-of-the-art technology for instream-flow purposes and were used for many rivers in the United States, Europe, Australia, and New Zealand (Gillilan and Brown 1997). The software development began in the 1970s; by the 1990s, it had become "the most sophisticated and comprehensive method of quantifying instream flow needs" (Stalnaker et al. 1995). Research by USBR, USFWS, and USGS personnel—including the survey of additional

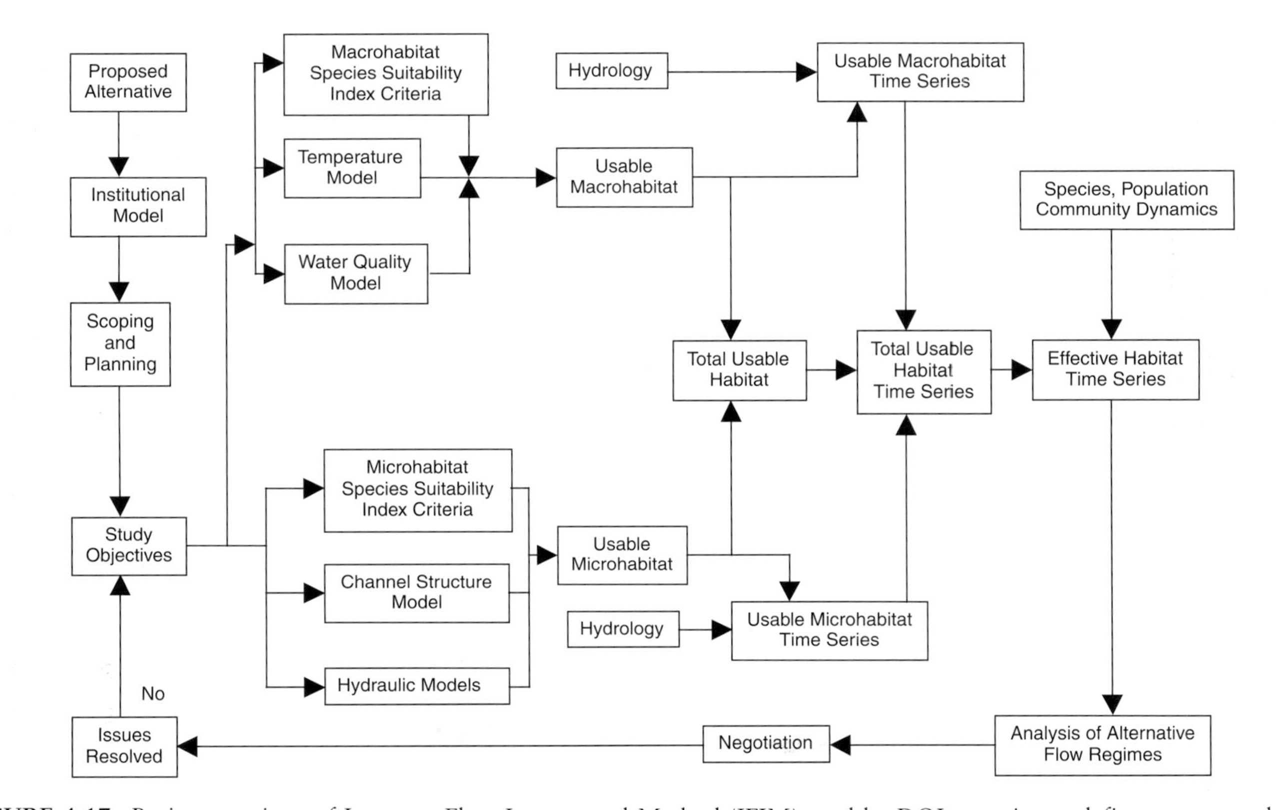

FIGURE 4-17 Basic operations of Instream Flow Incremental Method (IFIM) used by DOI agencies to define recommended flow magnitudes, frequency, duration, and timing. Source: Waddle 2001.

cross sections, additional measurements of habitat relationships, and modifications of the programs that make up IFIM—resulted in continuous improvement of the models through the 1990s. Continued improvement of the models makes a "rerun" with original data possible.

Application of IFIM models to the Platte River by DOI agencies produced a series of instream-flow recommendations. A 1990 workshop brought together interested researchers to discuss the problem of establishing instream-flow recommendations, partially stimulated by relicensing requests to the Federal Energy Regulatory Commission for power projects along the Platte River owned by the Nebraska Public Power District and the Central Nebraska Power and Irrigation District (M.M. Zallen, DOI, unpublished material, August 11, 1994). By 1994, DOI agencies had used IFIM to generate their recommendations, and after some revisions the agencies recommended three types of discharges: species flows, annual pulse flows, and peak flows. Later modifications in the recommended target flows resulted from work in 1996, 1999, and 2001. The relevant documents describing the rationale for these flows and for specific recommended values are Bowman (1994), Bowman and Carlson (1994), Altenhofen (J. Altenhofen, Fort Collins, CO, unpublished memo, March 4, 1996), Boyle Engineering Corporation (1999), and Murphy et al. (2001).

At the time of the National Research Council review, the most recent summary of DOI recommended target flows was that of USFWS (2002b). Tables 4-3, 4-4, and 4-5 outline the agency recommendations. Instream-flow targets for general purposes or specific species represent discharge conditions that are intended to result in favorable habitat for the threatened and endangered avian species along the central Platte River and for pallid sturgeon in the lower Platte River (Table 4-3). The recommendations specifically recognize the variability among wet years (the wettest 33% of years on record), dry years (the driest 25% of years on record), and normal years (all others) that result from short-term weather extremes. The general recommendations seek to ensure that water flows in the channel in sufficient quantities to support the species in question but not enough to harm

TABLE 4-3 Species Instream-Flow Recommendations of USFWS for Central Platte River, Nebraska

Period	Wet Year Discharge, cfs	Normal Year Discharge, cfs	Dry Year Discharge, cfs
Jan. 1-Jan. 31	1,000	1,000	600
Feb. 1-Mar. 22	1,800	1,800	1,200
Mar. 23-May 10	2,400	2,400	1,700
Oct. 1-Nov. 15	2,400	1,800	1,300
Nov. 16-Dec. 31	1,000	1,000	600

Source: USFWS 2002b.

TABLE 4-4 Annual Pulse-Flow Recommendations of USFWS for Central Platte River, Nebraska

Recurrence Interval	Recommended Flow, cfs	Notes
3 of every 4 years (75%)	2,100-3,600 (Feb.- Mar.) 3,000 (May-June) 3,400 (May-June)	30-day duration for Feb.-Mar. 7- to 30-day duration for May-June 10-year running mean of 30-consecutive-day exceedance
Every year (100%)	2,000-2,500 (Feb.-Mar.)	30-day duration for Feb.-Mar.

Source: USFWS 2002b.

other listed species. The values are derived from the observation that many habitat values are at their maximum when flows are 2,000-2,500 cfs and that below that value suitable habitat areas diminish in size and become restricted in their distribution. The final values in Table 4-4 represent a compromise between habitat needs and a management objective to store as much water as possible. For specific discussions of the needs of each species, see Chapters 5, 6, and 7.

Pulse flows (Table 4-4) maintain the general functionality of the river by supporting the movement of sediment, the woodland and open-channel balance, and the shaping of the channel by physical processes. The movement of sediment through dryland rivers in pulses is a common observation (Graf 1988), so in returning the Platte River to a condition that is more similar to predam conditions, pulse flows must occur with their natural timing during spring or early summer. Pulse flows must reflect natural duration to be effective and provide some variation in flow conditions—a characteristic that is a departure from artificially controlled flows that remain similar from month to month with little change (Richter et al. 1996, 1997).

TABLE 4-5 Peak-Flow Recommendations of USFWS for Central Platte River, Nebraska

Recurrence Interval	Recommended Flow, cfs	Notes
1 of every 5 years (20%)	16,000 Feb.-June	5-day duration At least 50% of flow between May 20 and June 20 May-June preferred
2 of every 5 years (40%)	12,000 Feb.-June	Feb.-June for channel maintenance
10-year running average of 5-consecutive-day exceedance	8,300-10,800 Feb.-June	5-day duration

Source: USFWS 2002b.

All natural dryland rivers are subject to occasional large floods that accomplish considerable geomorphic work by maintaining or changing the channel configuration (Rhoads 1994; Schumm 1977). From a biological perspective, the forms and processes maintained by these infrequent peak flows are the substrate for suitable habitat. USFWS recommended that peak flows be large-discharge events lasting about 5 days (Table 4-5). Their exact timing by year is less important than their occurrence about once during each 5-year period. From a management standpoint, they therefore might be created through controlled releases only during relatively wet periods. The recommended once-every-5-year pulse flows represent flood discharges that are of a general magnitude that is similar to the pre-1909 annual flood flows (the period before construction of large dams upstream). The magnitude of the 10-year running average for recommended pulse flows is about 50-60% of the magnitude of pre-1909 annual flood flows. Therefore, in magnitude they are similar to predam flows, but in the restored river these high flows would be less frequent than the predam flows (compare Tables 4-3 and 4-5).

The literature in geomorphology, hydrology, and ecology supports the basic importance of maintaining species, pulse, and peak flows for river restoration and species protection as applied by USFWS and USBR. In geomorphology, literature defining the importance of general flow maintenance, pulse flows, and peak flows for channel formation and maintenance includes Schumm (1977), Graf (1988), Richards (1982), Leopold (1994, 1997), Brookes and Shields (1996), Knighton (1998), and Petts and Calow (1996a,b). The supporting hydrological literature includes Chow (1964), Shen (1971), Simons and Senturk (1992), and Chang (1998). The supporting ecological literature includes Petts and Calow (1996c), Clark and Harvey (2002), Franklin (1993), Millington and Pye (1994), and NRC (2002a).

The recommended pulse flows are smaller than the floods observed before the 1941 closure of Kingsley Dam, when river processes and flood magnitudes were more natural than they are now. Historical aerial photography, including the paired examples in this report, show that channel shrinkage has occurred since 1941, so small pulse flows are appropriate to the shrunken channel system. The pulse flows are likely to be contained within the banks of the existing river and are therefore unlikely to threaten property damage.

The proposed instream flows that resulted from the DOI agencies' analysis and that are summarized in Tables 4-3, 4-4, and 4-5 appear to the committee to be in the correct magnitude and timing to achieve the desired results of using river processes to foster habitat for the threatened and endangered species. The flows represent reasonable calculated estimates of the magnitude needed to readjust the channel geomorphology, channel deposits, and channel-side landforms to suit the needs of the species better than the flow regime that results from no management for habitat benefit. The timing of pulse flows reasonably mimics the timing of such events before the installation of large numbers of dams upstream, particularly before the closure of Kingsley Dam. Annual high flows during May are likely to restrict the success of seedlings and thus preserve open spaces needed by the species. Occasional pulse flows will mimic flood events that once partly controlled the river's geomorphology and vegetation. Specific management questions—such as whether the periods between peak flows are appropriate for maintaining useful habitat system and whether the proposed system of flows is sustainable over a period of many years—cannot be resolved at this time. Monitoring of the system response to the instream flows will be required to inform managers about the efficacy of their management strategies and will suggest possible adjustments—a working example of adaptive management.

Habitat Suitability

Habitat is the physical space within which an organism lives and the biotic and abiotic resources in that space. Habitat is a species-specific or population-specific concept. It is the sum of the resources that are needed by an organism, including those used for forage, shelter, reproduction, and dispersal. The quality of habitat for any given species varies from location to location, from season to season, and from year to year. Organisms are adapted to the features of the environment that enhance their survival. Analysts determine the degree of habitat suitability for a species by assessing the presence and abundance of organisms in relation to one or more physical characteristics of the environment. The suitability of any habitat can therefore be determined by relating species and abundance to habitat features on the assumption that greater density reflects greater habitat suitability. Habitat suitability guidelines have a history in USFWS more than 25 years long and have been formalized in habitat evaluation procedures and their applications in habitat suitability index modeling (USFWS 1980, 1981a,b). Habitat suitability relationships are based on expert opinion and data-collection campaigns that related species to particular environmental conditions. Frequently, common characteristics—such as river depth, velocity, and bedload for fish or vegetation height, density, and composition for birds—are key factors in the process. Relative abundance connected to those environmental variables provides a guide to identifying habitat suitability. PHABSIM is especially beneficial in the process of sorting out habitat suitability because it uses the same ecosystem elements as do the habitat characteristics.

Over the last 2 decades, USFWS developed habitat suitability guidelines for the four listed species in the central Platte River. The guidelines define the physical characteristics of habitats deemed useful to each species for stopover, resting, breeding, or feeding. The guidelines aid in connecting knowledge about species needs with management strategies designed to enhance or create habitat to support them. USFWS based habitat suitability guidelines on the best available information, even in the cases of the whooping crane and pallid sturgeon, for which available data were particularly sparse. USFWS's specifications for suitable habitat relied on data collected by DOI personnel, research by them and other professionals, and wide-ranging expert opinion. Chapters 5, 6, and 7 provide complete reviews of the background literature on each of the species. In reaching its conclusions, USFWS relied on 54 pieces of published literature (including many standard reference books and articles in peer-reviewed journals) that form a convincing basis of habitat suitability findings. The agency incorporated several measurable physical-habitat elements—including depth of water, vegetation density, size of sandbars, and river-level variability (water-level fluctuation)—as physical measurable dimensions that are relevant to the listed species. The resulting habitat suitability guidelines offer an estimate of the degree to which particular physical environments can serve as viable habitat for a species. The guidelines can be broadly defined or sensitive, depending on how much is known about the selected species and the degree to which the physical environment can be codified or quantified.

Habitat suitability guidelines for the listed species on the central and lower Platte River emerge from processes outlined by USFWS (unpublished material, June 16, 2000). USFWS defined specific optimal conditions for the whooping crane, piping plover, interior least tern, and pallid sturgeon beginning with the hydrological behavior of the river as related to habitats for the species. For example, in the case of cranes, USFWS identified common characteristics of known roosting sites, such as a wide channel, lack of forest vegetation, sandy substrate, unobstructed long views, and shallow water nearby. For the piping plover and interior least tern, USFWS deduced common characteristics among river-channel habitat, nesting sites, and feeding locations for each species. For the pallid sturgeon, USFWS identified favorable river conditions, including the presence of sandy bottoms, islands or bars, and sediment-rich waters. USFWS extended the earlier study by exploring restoration measures that could re-establish suitable habitat conditions where they had been lost.

Additional Scientific Data Issues for Decision Makers

In addition to the issues of interconnections among ecosystem components, instream flows, and habitat suitability, several issues remain to be addressed by DOI researchers: how species other than the listed ones that use the Platte River corridor are affected by management, alternative approaches to measuring and managing the effects of water-control infrastructure on river flows and other ecosystem components, the physical implications of present restoration efforts, the other ecological characteristics of the system that need to be re-established, and evaluation of water-quality and climate-change issues.

THE PLATTE RIVER AS AN ECOSYSTEM

The Platte River ecosystem is one of the most diverse in the Great Plains, with an especially rich assemblage of vertebrates and higher plants. For example, 58 species of fish live in the central Platte River (Chadwick et al. 1997), and more than 300 species of vascular plants grow on the Platte's floodplain (Currier 1982). About 50 species of birds nest in the Platte's floodplain woodlands, nearly half of which are neotropical migrants, a group of birds in decline in other parts of their ranges (Robinson et al. 1995). Perhaps as many as several hundred bird species, including large numbers of waterbirds, use the Platte's channel and woodland communities twice each year during transcontinental migration (Chapter 2). This bountiful biodiversity is maintained by substantial habitat heterogeneity on the Platte's floodplain, adjacent wet meadows, and nearby agricultural fields. Heterogeneity in the river system is now controlled largely by the hydrological regime.

The elements of USFWS's target flows are based on perceived needs of the listed species. For example, a pulse flow is recommended to scour vegetation to provide more suitable sandbar habitat for terns, plovers, and cranes; higher spring and fall flows are proposed to favor whooping crane use during migration; and higher summer flow minimums are proposed to reduce the mortality of forage fish used by terns. However, developing a single set of target flows for one river to favor four very different species (one large migratory bird, two small nesting birds, and one large fish) that have divergent threats, different ecological optima, and different space-related and time-related uses of habitat comes with costs to other species (Clark and Harvey 2002). In addition, specific flow prescriptions that benefit one listed species may actually cause harm or preclude benefits to other listed species. For example, using more water from reservoir storage in fall for whooping cranes may reduce the opportunity to provide adequate spring pulse flows to build higher sandbars for tern and plover nesting.

Criticism of species-focused approaches to develop flow targets for rivers has stimulated the development of alternatives. Most of them emphasize a holistic or ecosystem-level approach that involves returning rivers closer to their predevelopment hydrograph in an effort to regain natural stream-flow variability. Shifting the current hydrograph toward the "natural flow regime" should simultaneously benefit large numbers of native riverine species adapted to preregulation conditions, including those currently protected. This approach should have a lower probability of unintended losses or shifts in biodiversity than reshaping the hydrograph on the basis of the expected response of a small number of rare species.

A related approach to systemic river restoration is that of the "normative river" (Stanford et al. 1996). Normative habitat conditions are those established from what is possible in a natural-cultural context, as opposed to striving for pristine conditions. The primary goal is to regain as much as possible of the former structure of the hydrograph (peaks, pulses, base flows, and timing) given system constraints (storage capacity, water rights, and property damage). The difference between a normative-river approach and that currently proposed by USFWS for the Platte River is that the latter may produce a disarticulated target hydrograph that differs in fundamental ways from predevelopment river flows. USFWS recommended peak-flow target for the Platte, for example, is scheduled earlier than when most predevelopment peaks occurred, possibly to prevent flooding of interior least tern and piping plover nests. The primary advantage of the development of a normative hydrograph is that the ecosystem-based approach emphasizes the rebuilding of key physical processes and high biodiversity of the preregulated river.

The Yakima River Basin Enhancement Project is working toward the establishment of normative flows for salmon (NRC 2002b); this project is a good example of a basinwide riparian restoration project that couples basic research with clear management objectives and stakeholder participation, and it may serve as an alternative approach to that recommended by USFWS for the Platte River. Another river-restoration project after which the Platte could be modeled is that of the Colorado River (Patten et al. 2001). The restoration of the Florida Everglades, which involves reversing the undesired effects of low dams and water-control structures, is based on the acknowledged first step, which DOI agencies label "Get the water right." A similar overriding objective as a first step also fits well for the Platte River.

In addition to the different outcomes of applying the PHABSIM-IFIM and normative models, the two approaches are underlain by fundamentally different philosophies. IFIM deals with the issue of connecting river flows to physical characteristics of the river one species at a time. It defines the flows needed to create and maintain habitat for one species, then begins again to define the flows needed by a second species, and so on. The problem with that approach is that it obscures the complexity of the real world and instead treats the river as a machine that can be analyzed and "tooled" for the benefit of each species in turn. The normative-flow regime, in contrast, has as its philosophical basis the view of the river as an integrated system. The approach is to control water flows and pulses in a manner that is as close to the "natural" (in the case of the Platte River, the predam or pre-1909) discharges as possible. The normative approach recognizes that the existing river, its aquatic habitats, and its riparian systems are different partly as a result of river engineering (such as bridges) and partly as a result of upstream watershed changes. The idea underlying the normative approach is that if the more "normal" flows are restored, more "normative" habitats will logically follow, and these habitats will have a complexity sufficient to sustain a wide variety of species, including species that are endangered, threatened, and not listed.

The committee recognizes that the DOI agencies have used IFIM because at the time of their decisions it constituted the best available science—a circumstance that lends credibility to their management decisions. To maintain that credibility, however, the DOI agencies must shift their approach to one based on the normative flow regime because it now (2004) constitutes the best available science. In a recent, authoritative review of the relationships among stream flow, sediment, and habitat from the physical science perspective, Pitlick and Wilcock (2001) conclude that "restoration efforts that focus on site-specific issues or single-species enhancement are likely to fall short of their objectives." Similar sentiments come from the biological community (Poff et al. 1997). Poff et al. (2003) and Richter et al. (2003) provided the most recent statements on the normative-flow approach that can be employed on the Platte River. The normative-flow approach has also seen successful applications in river-restoration efforts in Australia and South Africa (Postel and Richter 2003). One important aspect of restoration of the Platte River concerns the woodland cover in and along the river. The Platte River presents a management conundrum. Riparian woodland established in the central Platte River between the 1930s and 1960s is now in its most productive and diverse stage and supports the majority of species on the floodplain. The clearing of large tracts of this woodland is recommended by USFWS to recover the target bird species. In contrast with other western U.S. rivers where river dewatering has stimulated the expansion of invasive trees of low wildlife value, the Platte's woodlands are dominated by native species, primarily cottonwood and willow, with high wildlife value. Because riparian areas are positioned at the convergence of terrestrial and aquatic ecosystems, they are hotspots of biodiversity and exhibit high rates of biological productivity in marked contrast with the larger landscape (NRC 2002a).

USFWS has chosen a conservation strategy that includes the removal of riparian woodland in the Platte River to produce more open-channel habitat for the three listed bird species (Figure 4-18). The clearing of wooded islands followed by periodic disking and mowing to keep vegetation short was begun in about 1980 by conservation organizations that used privately raised funds (Lillian Rowe Sanctuary–Audubon Society) and trust-fund earnings (Platte River Whooping Crane Trust). In addition, many wooded

FIGURE 4-18 Cleared area along central Platte River. Removal of woodland cover designed to improve long sight lines and open areas to benefit whooping cranes. Source: Photograph by W.L. Graf, May 2003.

islands were removed by being bulldozed to riverbed level. The practice has been effective in attracting larger flocks of roosting sandhill cranes (Faanes and LeValley 1993). Clearing was expanded more recently through expenditure of large amounts of public funds primarily from the Partners in Wildlife Program (a landowner-federal cost-sharing program) and the requirement that power companies clear woodlands to acquire new operating licenses. Exact measures are not available for the extent of woodland clearing in the restoration of the Platte River under present plans. As much as one-third of the 45-mile long intensive management segment may be cleared, with additional areas cleared by Partners in Wildlife. The context of this clearing is that it is focused in a particular segment of the 310-mile river. Clearing smaller parcels for Partners in Wildlife projects and private duck blinds may amount to 25% of the woodland in the less intensively managed river sections.

Clearing as a restoration strategy has both beneficial and adverse outcomes. The adverse consequences of clearing include loss of wooded nesting and migratory habitat for many species of songbirds; proliferation of invasive purple loosestrife due to soil disturbance and chopping of mature plants; possible oversupplying of sediment to downstream reaches, a cause of woodland expansion; loss of cottonwood and willow ingrowth needed to replace future senescent stands (cottonwood and willow are pioneer species that do not reproduce in established woodlands); and reduction in patch size of remaining woodlands. The expected beneficial consequences of clearing wooded areas include improved habitat for the three federally listed bird species of the central Platte as well as many other bird species that require more open habitat (Chapter 2, Appendix B). Other less defined benefits include re-establishing lost or reduced ecological processes that may be important to the proper function of a more natural river system.

Restoration sometimes focuses on benefits to the threatened and endangered species, but unintended detrimental consequences for other species ought to be minimized. In the case of the central Platte River, clearing of woodland to improve habitat for whooping cranes, for example, entails removal of forest environments for some songbirds. The "cost" to songbirds versus the "benefit" to cranes has not been carefully studied or determined. The effects of woodland clearing on other bird species are discussed in Chapter 2.

No quantitative assessments of the response of the listed bird species to clearing over the last 2 decades have been published, but some qualitative patterns are apparent. Sandhill cranes concentrate in wide, unobstructed channels, including areas that have been cleared. For example, in 1999, 66% of the night roosts for about 300,000 sandhill cranes were on reaches of river where forests had been removed (Platte River Whooping Crane Maintenance Trust 1999). The proportion of the whooping crane population that stops on the Platte River during its spring and fall migrations has increased since 1976 (Chapter 5); when cranes stop on the Platte, they often settle near or in cleared areas. In contrast, tern and plover populations have declined steadily in the central Platte since the late 1980s despite increasing cleared area. The reported downstream shifting of sandhill crane populations from the upper to the central Platte in the last 25 years (Faanes and LeValley 1993) has been attributed to upstream channel narrowing. Whooping cranes have also shifted their use patterns to the east (Stehn 2003). The channel area in the upper Platte has increased steadily since the 1950s (although not necessarily in the areas that the cranes have used or to the degree that roosting cranes will find useful), so there may be other factors in the downstream shifting. Other factors include the clearing projects that attract cranes, and the proximity to abundant supplies of waste corn located near suitable roost sites. The issue of shifting use patterns needs more study and analysis.

Few data have been collected in vegetation removal projects that illuminate the effectiveness of clearing. Thus, exactly what is lost and gained through woodland removal is often poorly known. Studies were initiated in the late 1990s at Cottonwood Ranch and Jeffrey Island to monitor the effects of clearing on vegetation, sediment supply, and channel structure. Several years of preclearing measurements, including an inventory of plants and birds, were made before the first clearing began in winter of 2003. It may be 5-10 years before results will be available from these studies to evaluate the practice of clearing. General studies are needed to develop broad standards to define the expected benefits of clearing to open-channel species; to assess the relative success of past clearing; to propose and test alternative woodland-removal methods; and to develop effective approaches to habitat management that best support ecological restoration to the extent possible. Historical data indicate a range of tree densities along the central and lower Platte River before settlement (see cover painting of this book). Therefore, when viewed as a whole, the river in its entirety should reflect a variety of tree densities.

Approaches to restoration in the central Platte have been under way for over 2 decades. Thus far, crane populations have increased during this period, and there is strong evidence that they followed cleared areas for roosting purposes (Faanes and LeValley 1993). Other factors, including availability of upland food sources, may also influence this geographic distribution of birds but to a lesser extent than does roost site availability (Su 2003; Iverson et al. 1987). Management has not stimulated increases in piping plover or interior least tern populations, and their numbers on the Platte have continued to decline. Restoration efforts have also benefited other waterbird species but are likely to have affected woodland species adversely. Restoration of the central and lower Platte River in the future can provide a context for a biologically diverse ecosystem that includes a variety of gradients from forest to open areas. In an adaptive management approach, the maintenance of some woodland along with the open areas will permit an assessment of the effects of restoration on all the species in the ecosystem. A monitoring system is essential to the success of such an ecosystem restoration.

The management of the central and lower Platte River through a partially restored flow regime (a normalized flow pattern) and reduced forest cover in some locations may cause a reduction in some native vegetation species (such as red cedar and hackberry) and some nonnative species (such as honeysuckle and buckthorn). Management of the ecosystem to benefit particular fauna inevitably will affect some species adversely, but the objectives of adaptive management, with its constant monitoring and redefinition of strategies, can minimize the unwanted effects. This approach will allow exploration of the responses of a variety of species to the managed changes and will provide an opportunity to learn more about the role of control factors other than the river, such as fire, predators, and human activities.

SUMMARY AND CONCLUSIONS

This section summarizes the committee's observations and conclusions about DOI's approach to understanding the physical processes and forms that underlie the habitats of threatened and endangered species in the Platte River Basin. It begins with a brief summary of the origins and major points in DOI's approach to river-flow management and habitat connections and then presents the committee's observations and suggestions. The section concludes with a list of specific recommendations.

USFWS has developed instream-flow recommendations through literature reviews, field observations, data collection and analysis, numerical modeling, workshops, and other approaches. Those processes and methods are scientifically valid, and the techniques applied in the Platte River continue to be used for many other rivers. DOI-recommended flow values appear reasonable, but their effects on this river system require further analysis based on empirical data collection and field observations. USFWS has already expended a great deal of effort to develop an effective flow-management plan, and more investigations are planned. According to USFWS (2002b), "these flow recommendations are intended to achieve the flow-dependent goal of rehabilitating and maintaining the structure and function, patterns and processes, and habitat of the central Platte River Valley ecosystem." Four types of flow recommendations were made: for species flows, annual pulse flows, peak flows, and program target flows. The values of the species flows for dry, normal, and wet years were based on a consultation process initiated in the 1980s and concluded with the discussions in the March 8-10, 1994, workshop, summarized by David Bowman (1994). The values of the annual pulse flows and peak flows for dry, normal, and wet years were presented by Bowman and Carlson (1994) and based on a workshop held on May 16-20, 1994. Flow values for the Platte River were based on expert opinions summarized in the two reports. According to USFWS (2002b), target flows are the discharges that the program activity seeks to establish through the water-control infrastructure to alter magnitude and timing of flows.

The central Platte River contains both meandering and braided reaches that represent complex hydrological and geomorphic conditions. The current state of knowledge is insufficient to predict its precise morphologic change due to flow volumes with confidence, and a great deal of field observation is needed to support the analysis. The analysis should experiment with series of flows designed to meet the variety of requirements related to vegetation growth, channel maintenance, sediment mobility, and ecosystem stability. It is essential that the field data be collected and analyzed to evaluate the actual effects of USFWS-recommended flows on the central Platte River, should they be implemented.

Monitoring river behavior requires careful design of field data collection. Specific data-collection characteristics—location, timing, goals, and level of detail—should be planned well before the occurrence of the targeted flow events. Field data can be expensive to collect, but timing is important because large flows are highly uncommon. If a research program does not collect data during and after a large flow, the next opportunity may not occur for many years. Perhaps USFWS could use its workshop technique to discuss various field needs of different disciplines. DOI flow recommendations would be most helpful if they were evaluated for suitability for listed and other regionally important species.

Regarding DOI interpretations of the interrelationships among flow, sediment, morphology, and vegetation, the committee has two sets of observations, one concerning the channel and history work by Murphy and Randle (2003) and the other concerning the sediment and vegetation work being undertaken by Murphy et al. (2001). First, undoubtedly, there are strong relationships among sediment, flow, vegetation, and channel morphology. Flow is also directly related to climate, water needs, reservoir storage, and diversion. Since the construction of the first large dam (Pathfinder in 1910), various water-resources developments have altered flow distribution and water consumption substantially, especially in the upper Platte River system. Active discussions between the Environmental Impact Statement team and the Parson team occurred in our August committee meeting in Grand Island, Nebraska, about the relative importance of climate and the water-resources developments for the change of river characteristics (for examples see Parsons 2003; Lewis 2003; Woodward 2003; Yang 2003; Murphy and Randle 2003).

Water-resources developments have diverted and returned flows into the North Platte River, South Platte River, and upper Platte River. Murphy and Randle (2003) have estimated consumptive uses of water, including sewage, and evaporation of reservoir water. Because the return flows from diversions (except Kearney Canal return) occur upstream near Overton, Nebraska, the relative flow effect of water-resources development is considerably greater on the upper Platte River than on the central Platte River. Regardless of climate change, water-resources development will continue to affect Platte River flows as long as there is a net irrigation water consumption and reservoir evaporation. The human controls on flows in the river are the most important controls on a daily, monthly, or annual basis, but the longer-term effects of climate change are a background control worthy of further investigation.

The mathematical modeling by Murphy et al. (2001) has some shortcomings that will challenge DOI investigators. Some of the required data are unavailable, and some of the modeling techniques are still in the development stage. The success of a numerical model depends on knowledge of flow roughness in the flow-momentum equation and of sediment transport rate in the sediment-continuity equation. Methods in field data collection to improve that knowledge have yet to be fully developed. The relationships of lateral variations of flow properties among the different subsections in a braided river are also difficult to determine.

There are four dimensions in any river system: the longitudinal direction, which is the main flow direction; the lateral direction, across the channel into the floodplain; the vertical direction, including surface-groundwater exchanges; and time. In a one-dimensional model, we investigate only the change in flow properties in the longitudinal direction with time and space by assuming no change in the other dimensions. Thus, in a strictly one-dimensional model, the lateral cross section is uniform because we assume no change of flow in the lateral direction. In a braided channel, a river cross section (perpendicular to the main flow direction) may include two flow channels with an island in between. Thus, Murphy et al. (2001) use a pseudo-one-dimensional model to include the possibility of having a channel cross section with nonuniform shapes. That approach is generally accepted, and many models use different schemes to represent sediment and lateral flow distribution among the various lateral cross sections. Several unproven assumptions have been used for the lateral distributions of flow and sediment in the current pseudo-one-dimensional model. The vegetation resistance should be determined from the field data instead of from other references. Some of the longitudinal intervals between two cross-sectional stations are too long to yield any reliable hydrogeomorphic relationships. If properly calibrated and validated, this model can give qualitative impressions of sediment and flow analyses, including the evaluation of the effect of vegetation removal and management.

Only a few two-dimensional mathematical models that can include sediment movements are under development. Two-dimensional models have limited application because two-dimensional flow data are often unavailable to calibrate them.

The committee recognizes six approaches for potential improvement of DOI investigations into ecosystem dynamics on the central and lower Platte River:

- Field data collection and methods for the monitoring of the effects of various flow recommendations and mechanical removal of vegetation must be carefully designed long before the occurrence of the targeted flow events or vegetation manipulations. Some kinds of data are also essential for calibrating and verifying the mathematical model.
- A risk-based hydrological model should be explored with various penalty functions (water not diverted to users as previously has been the case) at the water-demand points for optimization analysis of the flow-management plan of the river system. The effects of mechanical removal of vegetation should be included in the flow-management plan.

- For more detailed analysis, variations in flow velocity and flow depth are more important than flow discharge for evaluating ecological requirements because it is the depths and velocities that create habitats.
- Climate and water-resources developments can have strong influences on river flow distributions. Water-resources development affects river flows substantially in the upper Platte River, and its effects extend to the central Platte River. The relative importance of climate influences and water-resources development on channel characteristics should be analyzed and should encompass a record of several decades.
- Restoration of the central Platte River should include water processes and forms, control of invasive species, and some grazing and fire if research shows these phenomena to be important aspects of the pre-European river.
- More emphasis should be placed on the management of the Platte River as an ecosystem, rather than keeping the focus exclusively on listed species.

In summary, the committee's review of DOI's efforts to explain and model the connections among ecosystem components of water, sediment, morphology, and vegetation leads us to conclude that these efforts are underlain by valid science. Likewise, DOI's instream-flow requirements are grounded in scientific understanding of the system and in the technology of model construction that was state-of-the-art when the decisions and recommendations appeared. Science and engineering are making progress, however, and new technology is becoming available. New advances are needed because of the braided, complex nature of the Platte River, a configuration that is unlike that of the streams where others often apply the models. Current DOI model developments, including the emerging SEDVEG model, are likely to be helpful and useful in both understanding and managing the Platte River. DOI's determination of suitable habitat rests on the best available science. The committee also recognizes, however, that there has been no substantial testing of the predictions of DOI's modeling work,[1] and we urge that calibration of the models be improved and that monitoring of the effects of recommended flows and vegetation management be built into a continuing program of adaptive management. In such a system, monitoring can indicate whether recommendations and determinations are valid and can suggest further adjustments to the recommendations and determinations on the basis of observations.

[1]The committee did not consider USGS's in-progress evaluation of the models and data used by USFWS to set flow recommendations for whooping cranes.

5

Whooping Crane

The whooping crane is one of the world's most imperiled species. About 200 individuals exist in the single natural population, which migrates over a large portion of North America, from overwintering grounds on the gulf coast of Texas to breeding grounds in central Canada. During two annual migration passes through the American Midwest, individuals stop over along the central Platte River of Nebraska for periods of a day to several weeks. In support of the federal listing of the whooping crane as endangered, the U.S. Fish and Wildlife Service (USFWS) designated a portion of the central Platte River as critical habitat for the species.

This chapter is the response of the National Research Council Committee on Threatened and Endangered Species and the Platte River Basin to questions about habitat for the whooping crane: Do current central Platte habitat conditions affect the likelihood of survival of the whooping crane? Do they affect its recovery? Is the current designation of central Platte River habitat as critical habitat for the whooping crane supported by existing science? In addressing those questions, this chapter reviews relevant databases, examines the use of the central Platte habitat by whooping cranes, and evaluates the influence of the habitat on population parameters, including mortality, natality, and distribution.

HISTORICAL AND ECOLOGICAL CONTEXT

The whooping crane is native to a diverse array of ecosystems. Until its well-documented decline, it nested predominantly in the northern tall grass prairie of the upper Midwest and the eastern aspen parklands of Canada (Allen 1952). It also nested in such disparate regions as the taiga and sub-Arctic regions of the Northwest Territories, where the last remaining population has persisted to present times, and in the coastal marshes of Louisiana (Figure 5-1), where a nonmigratory population bred until 1939 and was extirpated from the wild in 1950 (Gomez 1992; Lewis 1995). Wetland ecosystems historically used by wintering whooping cranes were diverse as

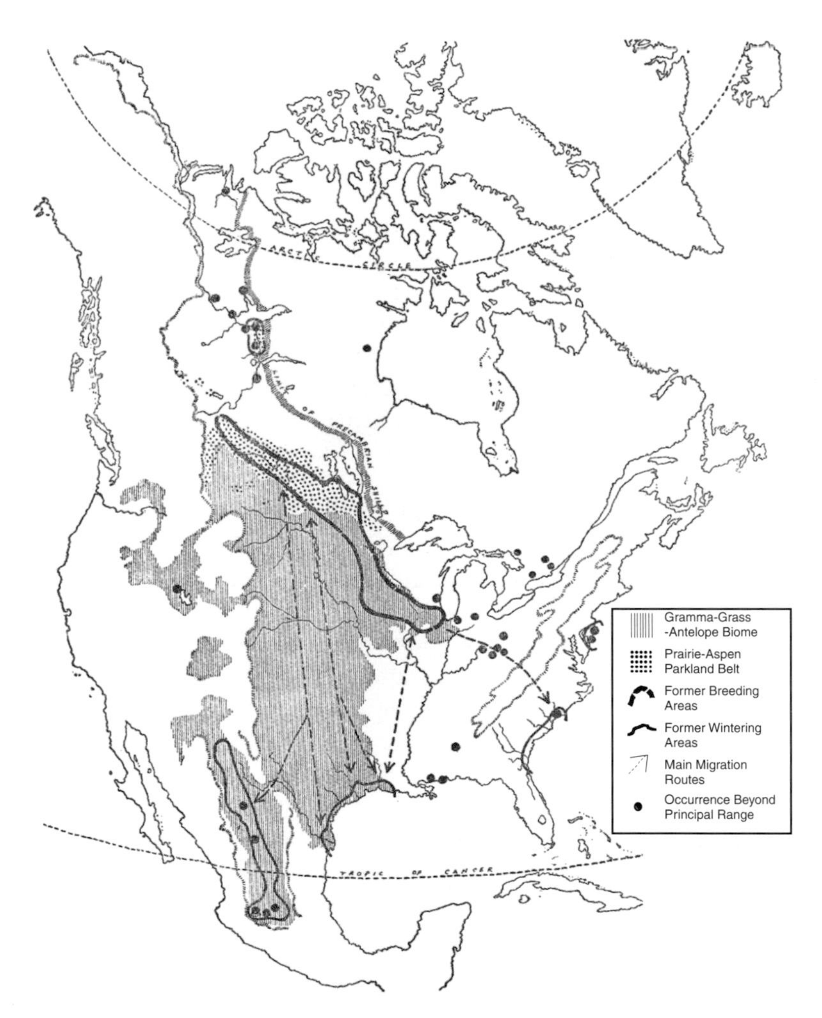

FIGURE 5-1 Historical distribution of whooping cranes in North America. Source: Allen 1952. Reprinted with permission; copyright 1952, National Audubon Society.

well, ranging from the high plateau wetlands of central Mexico to the varied coastal wetlands that extended from Texas to South Carolina (Nesbitt 1982; Allen 1952). Moreover, migratory populations of whooping cranes followed a variety of flyways between breeding and wintering areas (Figure 5-1). Although whooping cranes used and continue to use a wide range of environs, they primarily have depended on highly productive wetland ecosystems for nesting, overwintering, and migratory stopover (Hayes and Barzen 2003).

Whooping cranes from two breeding-wintering populations are thought to have migrated through Nebraska (Allen 1952). Few historical records of whooping cranes summering in Nebraska exist and no records of breeding are available, so it is assumed that the state has been used predominantly by migrants. A whooping crane population that bred in the upper Midwest (and perhaps Manitoba) was thought to have wintered in Louisiana and to have migrated through eastern Nebraska (Figure 5-2a).

The disappearance of cranes from eastern Nebraska coincided with the extirpation of the upper Midwest-Louisiana migratory population. No historical whooping crane records (either observations or specimens) are reported for the Platte River downstream of Chapman. A population that bred further west of Manitoba and wintered in Texas (whose remnant is the extant, migrating population) is thought to have passed through the region of the central Platte River and other portions of central Nebraska (Allen 1952). The last remaining wild population of whooping cranes now breeds in and near Wood Buffalo National Park (WBNP, Canada), migrates through the Great Plains (including Nebraska), and winters along the Texas coast (Figure 5-3). In 1860 or 1870, the population may have been as large as 1,300 or 1,400 or as small as 500 or 700 individuals (Allen 1952; Lewis 1995). By 1912, the Aransas-Wood Buffalo migratory population (AWP), wintering in what is now Aransas National Wildlife Refuge (ANWR), was only 36 birds; and by 1941, the number had declined to 15 (Allen 1952).

Causes of the decline in the population were primarily overhunting and loss of natural habitat to agriculture (Allen 1952). Subsequent population recovery has been attributed to enforcement of the Migratory Bird Act of 1916 (which made it illegal to shoot whooping cranes) and the creation of WBNP and ANWR. WBNP was established in 1922 as the primary summer area, and ANWR in 1937 as the primary wintering area (Lewis 1995). Federal protection for the whooping crane was conferred in 1967, and critical habitat was designated in 1978. USFWS established a species recovery team in 1975 for the United States, and Canada followed suit in 1985. A recovery plan for the whooping crane was published in 1980 (Olsen et al. 1980; Smith 1986; Lewis et al. 1994) and was last revised in 2003 (CWS and USFWS, unpublished material, May 2, 2003). A memorandum of understanding signed between the Canadian Wildlife Service and USFWS in

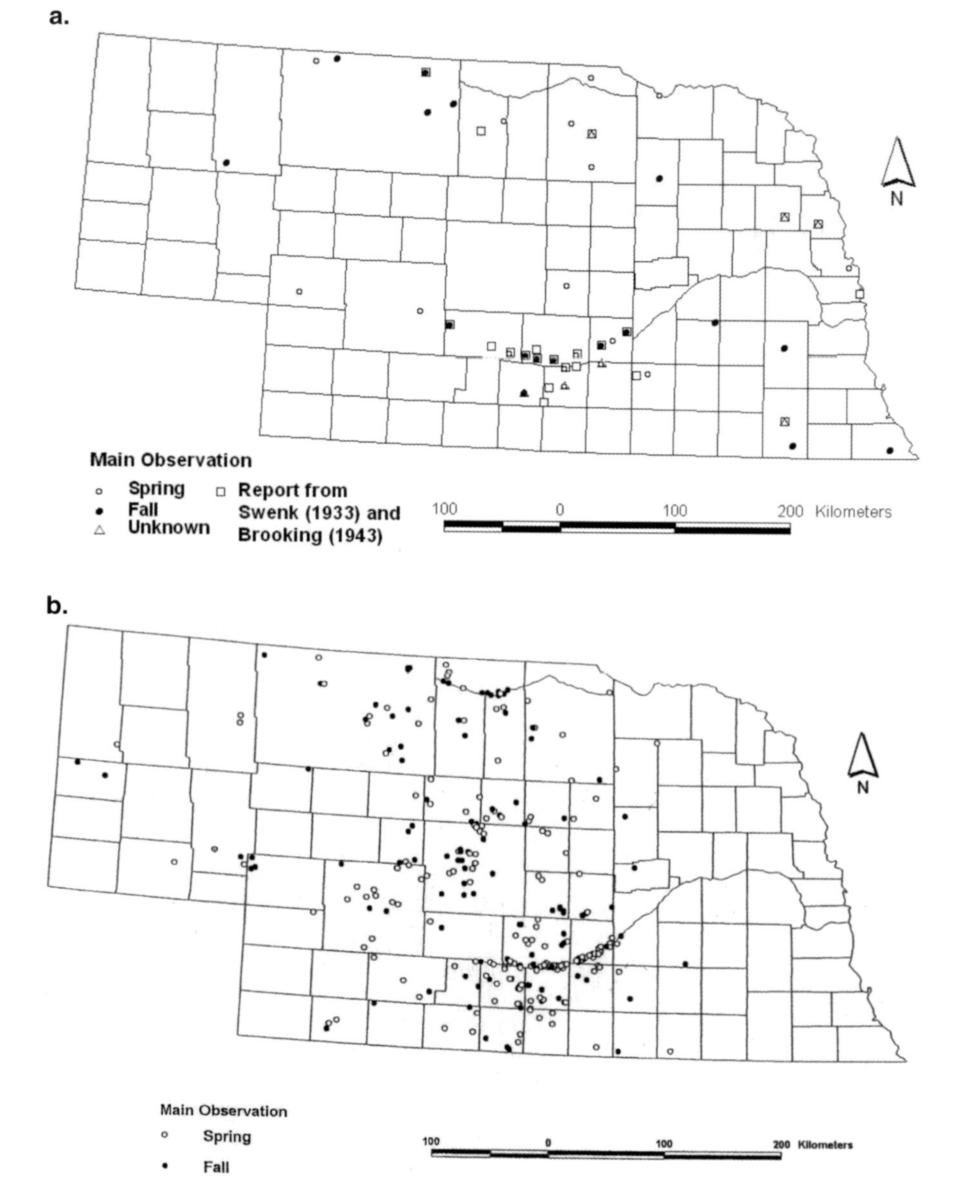

FIGURE 5-2 Distribution of whooping crane sightings in Nebraska. (a) 1820-1941, (b) 1943-1999. Sources: (a) Modified from Allen 1952. Reprinted with permission; copyright 1952, National Audubon Society. (b) Austin and Richert 2001.

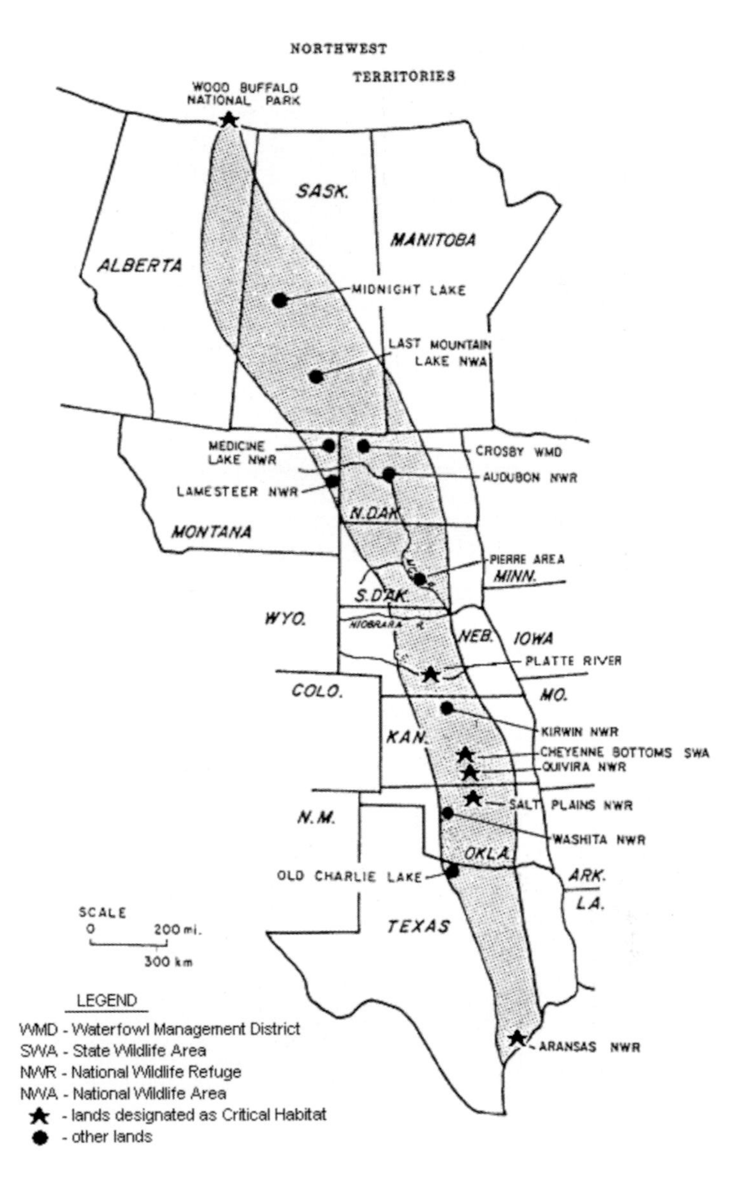

FIGURE 5-3 Map showing present (2003) home range of migrating whooping cranes and central position of Platte River, Nebraska. Source: CWS and USFWS, unpublished material, 2003.

1985 further assisted coordination of monitoring and conservation efforts (Smith 1986).

Apart from the work of Allen, little was known about whooping cranes at the time of their listing in 1967. The AWP consisted of 43 birds; by the time critical habitat was designated in 1978, it had increased to 75 (CWS and USFWS, unpublished material, May 2, 2003). Since 1938, investigators at ANWR have counted birds in the AWP each winter (Lewis 1995). Surveys of breeding birds on the nesting grounds in WBNP began soon after the breeding site was discovered in 1954. In 1977, researchers at WBNP began a banding program to capture and tag flightless young with colored leg bands (Kuyt 1979). In migration areas, a coordinated effort to collect, verify, and maintain records of migrating whooping cranes began with the fall migration of 1975 (interview with USFWS personnel, Aug. 13, 2003) and was expanded in 1977 (Olsen et al. 1980). Using newly acquired data and the few papers published since Allen's work, Johnson and Temple (1980) completed the first comprehensive analysis of migration habitat for the whooping crane.

When USFWS designated critical habitat for the species, surprisingly few data existed beyond the historical data (reviewed below) and winter estimates of the population size (Johnson and Temple 1980). USFWS identified the central Platte River as critical to protect "so that it would not be lost to development before its overall significance to whooping cranes could be determined" (Johnson 1982, p. 34). The goal of the recovery plan was to ensure protection of key stopover points along the flyway that had the highest historical and current use by cranes. That string of protected areas along the north-south migration route would assist the species by decreasing the distance between stopover locations. The central Platte River became part of a series of stopovers protected as critical habitat that included Quivira National Wildlife Refuge, Salt Plains National Wildlife Refuge, and Cheyenne Bottoms State Wildlife Area (Figure 5-3).

DIFFICULTIES IN DEFINING CRITICAL HABITAT FOR THE WHOOPING CRANE

Endangered species with low population numbers are especially problematic for researchers. First, researchers cannot determine the importance of various habitat types for the species by simply investigating present use patterns. Inferences about the importance of various habitat types depend on assessments of what habitats are available (Manly et al. 2002), random selections of habitat for evaluation (Pereira and Itami 1991), and evaluations of nonused habitats (Faanes et al. 1992). The species is defined as showing preference or avoidance when it uses a particular habitat type disproportionately or when habitat use differs substantially from random selection. When an otherwise preferred habitat is not fully occupied, absence of animals from it does not necessarily mean that the habitat is of poor quality or not desired. Habitats of high quality may be nonused simply because there are not enough animals in the population to occupy all available habitats.

Traditional habitat-use studies are constrained when used for low population numbers. For example, the AWP presently (2002) numbers only 185 (CWS and USFWS, unpublished material, May 2, 2003), and this small population cannot be assumed to be at carrying capacity or to have saturated any habitat that it uses during its annual cycle (Lewis 1995). That caution is relevant to both the historical analysis provided by Allen (1952) and more current analyses (Austin and Richert 2001).

Habitat use by whooping cranes varies during different parts of their annual cycle of activities: summer breeding period, nonbreeding period in summer, fall staging, fall stopover, winter, and spring stopover (Lewis 1995; Kuyt 1979, 1992; Howe 1989). Migrating birds use staging habitats for several days or weeks to acquire and store energy or nutrients that are required for future stages in the annual cycle (Melvin and Temple 1982). In contrast, cranes use stopover habitats to meet immediate needs for energy and nutrient provisioning (such as water and protein) for up to several days while they wait for appropriate weather conditions to continue migration (Melvin and Temple 1982).

Habitat limitations during any part of the annual cycle can limit a population independently (although, given present data, this does not appear to be the case for whooping cranes). For example, if breeding habitats are limiting, populations can grow more slowly or decline even if all other habitats used during the annual cycle are abundant. Allen (1952) and Lewis (1995) argued that protecting breeding habitats (WBNP) and winter habitats (ANWR) has been critical to survival and recovery of the whooping crane. Others (EA Engineering, Science and Technology, Inc. 1985; Lutey 2002; R. Brown, Colorado Water Conservation Board, unpublished material, August 11, 2003) have also suggested that inasmuch as the AWP continues to increase, stopover habitats used by whooping cranes during migration cannot be limiting the species. No data indicate that any specific habitat related to a particular stage of the annual cycle is generally limiting the AWP (CWS and USFWS, unpublished material, May 2, 2003). Some data suggest that changing winter habitat conditions may become limiting in the future or are limiting now in some years (Chavez-Ramirez 1996; Stehn 2001). Improved understanding of the roles of the various habitats used by the crane population requires long-term, multiyear assessments of the whooping crane population that compare historical and modern data.

Long-term studies would also illuminate time lags that affect the AWP. Natality, for example, occurs only during spring, although factors that influence natality may occur on habitats outside the breeding area (Chavez-Ramirez 1996). Adult mortality occurs throughout the year but primarily during migration (Lewis et al. 1992). Changes in population numbers thus take a full year to become measurable. The lifespan of whooping cranes is 30-50 years (Mirande et al. 1991, 1997; Brook et al. 1999), so population dynamics can take many years to become apparent. Given current population growth rates and assuming no substantial change in population measures, the AWP will take 20 years (Reed 2003) to reach the minimal AWP goal of 1,000 birds and 175 pairs—numbers that would have to be maintained for an additional 10 years to achieve downlisting.

Because of those long response times for the crane population, conservation strategies must take into account planning horizons of 30 years or more. Critical habitat for the species must not only serve the needs of the present small population but be adequate to accommodate the needs of the larger population anticipated for meeting the delisting requirements. Extrapolation of future needs will have to take into account the possible influence of a variety of factors, including adjustments in agricultural policies and land-use patterns. Changes in agricultural land use, for example, have already affected sandhill cranes (Krapu 2003), and such changes may become more relevant for whooping cranes as the AWP changes.

CURRENT DATABASES

Data on whooping cranes made available to the committee varied in years covered, variables included, and geographic region covered. The committee carefully reviewed them for errors and requested updates from those who created or maintained the databases to provide the most accurate, complete, and current information for this report.

Historical Sightings by Allen (1952)

In the 1940s, when it was clear that whooping cranes were close to extinction, Robert Allen documented the distribution, abundance, and ecology of the species. His work, published by the Audubon Society in 1952, provided the first thorough investigation of historical records of whooping cranes. Allen interviewed many people who had resided in the whooping crane home range during the late 1800s, and he captured valuable information that would otherwise have been lost. It is not possible to derive quantifiable conclusions from those interviews, but they allow some informed conclusions about the ecological relationships between the species and its habitat.

Allen concluded that Nebraska was important for migrating whooping cranes and that the central Platte River (which he referred to as the Big

Bend region) was an important habitat for migrating whooping cranes. He based his conclusions on three lines of evidence: hunting data, observational data, and qualitative information from interviews. Nebraska had more reported whooping crane hunting kills than any other region in the United States and Canada (45 of 333, Allen 1952, p. 76, Table J); within the state, more kills were in the central Platte than anywhere else (Figure 5-4). Cranes vulnerable to hunting are those not flying at altitudes typical of migrating birds; migrating birds would have been too high to be killed with a shotgun (Kuyt 1992). The distribution of cranes shot on the central Platte River reflects historical use of local habitats, much of it on the river itself (Allen 1952, p. 102). Hunting records are particularly reliable because mistakes in species identification are infrequent in contrast with sightings.

Allen also reported historical observations from the central Platte River (Figure 5-2a). Some of the historical observations, however, may have been erroneous. Inaccurate observations from Swenk (1933) and Brooking (1943) led some scientists to conclude that Allen's compilation of observations was

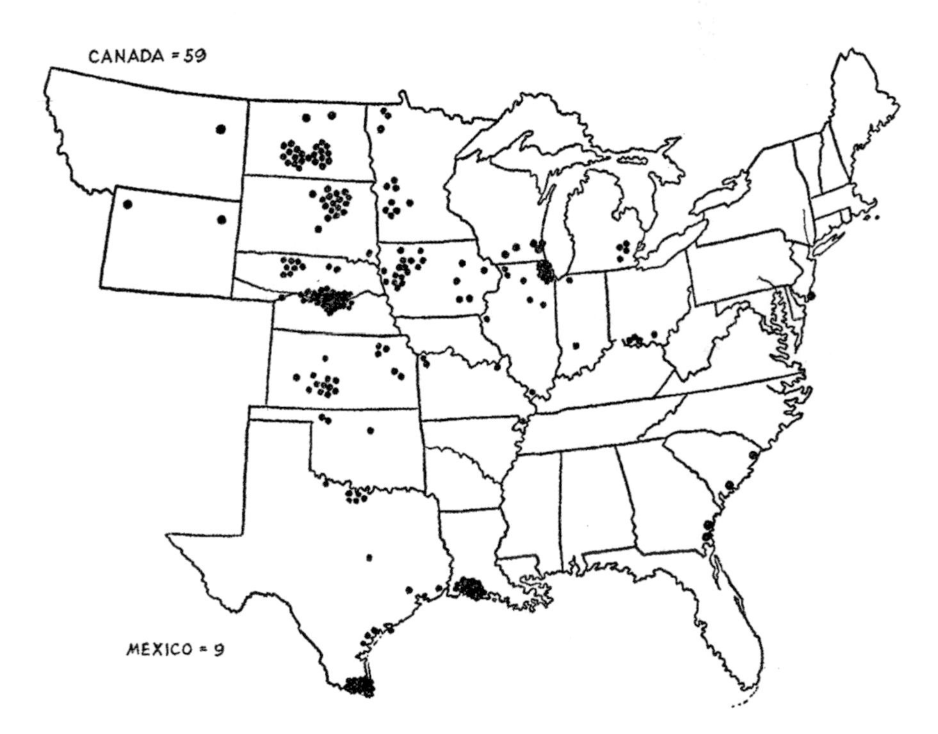

FIGURE 5-4 Available kill record locations in United States. Each black dot equals one hunter-killed whooping crane. Whooping cranes killed in Canada are not mapped. Source: Allen 1952. Reprinted with permission; copyright 1952, National Audubon Society.

"burdened with remaining uncertainty" (EA Engineering, Science and Technology, Inc. 1985). To evaluate the historical record, the committee remapped Allen's original records and distinguished the observations of Swenk (1933) and Brooking (1943) from other records (Figure 5-2a). Whooping crane locations reported from other observers or from museum and hunter-shot specimens (where identity is not in question) did not vary greatly from the records in Swenk (1933) or Brooking (1943). There is little evidence to suggest, therefore, that whooping crane use in Nebraska as summarized by Allen (1952) was inaccurate.

Allen interviewed residents who lived along the Platte River during the late 1800s and early 1900s. The veracity of interviews and their qualitative data can be determined only through recommendations of the interviewer or through assessment of the completeness (and believability) of the descriptions. The following interview is representative (Allen 1952, p. 79):

> While in Nebraska a few years ago I talked with Loren Bunney, State Conservation Agent at Ogallala. He told me that he had lived as a boy in Harlan County, south of Holdredge, between the Republican and Platte Rivers. He started hunting as a boy of ten, 'some fifty years ago,' and, even at that time (about 1897), whooping cranes were a rare sight, although the area in which he lived was in the main pathway of their migration. When flocks of sandhill cranes went through he always 'looked them over for the big white ones.' Now and then a few were seen.

That interview, coupled with other similar reports, suggests that whooping cranes did not frequently stop south of the Platte and crossed the region during migration. Residents' observations corroborate the lack of hunter-killed whooping cranes from the area (Figure 5-4) and suggest that cranes may not actually have stopped as frequently south of the Platte River on migration.

Confirmed Sightings in 1976-2002

A coordinated effort to collect, verify, and maintain records of migrating whooping cranes on a consistent basis was initiated in fall 1975 (Wally Jobman, USFWS, pers. comm., August 13, 2003) and expanded in 1977 (Olsen et al. 1980). Before then, records of whooping crane sightings were collected haphazardly. USFWS has closely monitored migration information since 1975 to identify appropriate survey periods for whooping cranes on the Platte. In spring, for example, Grand Island USFWS staff coordinate their survey efforts with ANWR staff that monitor when whooping cranes initiate migration. Observers at each stopover site also communicate as the birds continue moving north. Providing further continuity, Wally Jobman has coordinated whooping crane sightings for the Platte River region and elsewhere since 1978. On the central Platte River, ground surveys rely on reports from the public and from government staff and coordinated surveys. About two-thirds of all whooping crane sightings originate with the general public, and agency or other qualified biologists attempt to verify all sightings (Wally Jobman, USFWS, pers. comm., August 13, 2003). Large numbers of tourists visit the central Platte River each spring to observe migration, and many hunters visit the region each fall. In both seasons, potential observers are numerous and well distributed along the central Platte River. Nongovernment observers have contributed observations at the Rowe Sanctuary since 1975, as have the Whooping Crane Trust staff since the early 1980s. Aerial surveys in the 1980s found only one group of three whooping cranes that ground surveyors had not already located. Aerial surveys were re-established in 2001 and have been helpful in confirming sightings but not in discovering new birds. The Nebraska Wildlife Federation initiated its Whooper Watch Program in 2002. Most birds (besides those seen at the Rowe Sanctuary) are first reported in upland areas and then watched to determine where they roost at night and monitored for as long as they stay in the area of the central Platte River. Overall, because most sightings of whooping cranes in the central Platte originate in observations of the general public (ubiquitously spread along the central Platte in spring and fall), the search effort for whooping cranes has probably varied little among years since the 1980s.

Whooping crane sightings that originate with the public undoubtedly underestimate stopovers on the Platte River because many stopovers are of short duration and because birds are not always detected when they first arrive (Kuyt 1992). Fall detection of cranes is also unreliable because hunters can be reluctant to report whooping crane sighting for fear that their hunting areas will be closed until the birds depart. A federal waterfowl production area (the Funk Basin, a wetland complex in the nearby Rainwater Basin) was temporarily closed in 1993 while whooping cranes stayed there (Wally Jobman, USFWS, pers. comm., August 13, 2003).

In a larger-scale effort, Austin and Richert (2001) summarized whooping crane sighting data collected throughout the United States from 1943 to 1999, analyzing their contribution to whooping crane migration ecology, and evaluated their usefulness. They identified data-comparability concerns due to unequal effort at locations throughout the Northern Plains. For example, many observers watch for whooping cranes in the Platte River Basin, especially during spring, when thousands of tourists come to the Platte River to see migrating sandhill cranes, but comparatively few people watch for them in the Sandhills to the north. Because of such variability in search effort among regions, it is not possible to compare frequencies of sightings among regions. Howe (1989) demonstrated that effect with cranes that were followed with radiotelemetry. Austin (pers. comm., USGS, August 29, 2003), however, did not feel that those biases were important when comparing the frequency of sightings among years for any one site.

The committee used confirmed sightings in 1976-2002 in all central Platte analyses (Appendix C) but followed Austin and Richert (2001), who used data collected in 1943-1999, when referring to nationwide sightings. Before analysis of crane records for the Platte River, the committee examined carefully the data and omitted records where double-counting was suspected. Data in Appendix C, therefore, differ slightly from input data used in analyses (Appendix D). Data from 1976-2002 were used because consistent survey efforts were unavailable before then.

Crane Sightings on the Platte River

The pattern of present-day whooping crane use in Nebraska (Figure 5-2b) is similar to historical whooping crane use (Figure 5-2a). In addition to a growing whooping crane population, the frequency of Platte River sightings and the number of whooping crane use-days spent on the Platte have increased over the last 50 years (Figure 5-5). Douglas H. Johnson (USGS, Northern Prairie Wildlife Research Center, Jamestown, ND) analyzed these data to define the portions of the increase related to overall population increase as opposed to other causes (Lutey 2002). Johnson's analysis divided the number of cranes observed by the total known population and

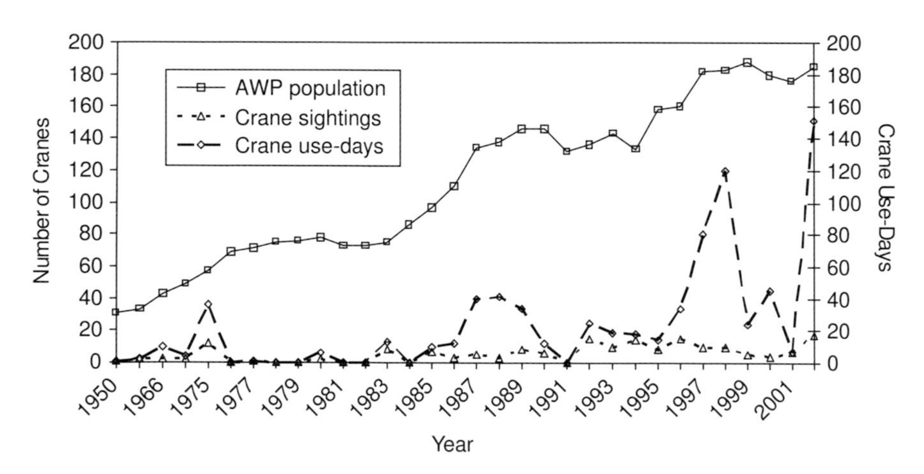

FIGURE 5-5 Population numbers of the Aransas-Wood Buffalo population (AWP) of whooping cranes, number of whooping crane sightings on the Platte River, Nebraska, and the number of sighting days (number of birds times number of days seen) on the Platte River, 1950 to 2002. The population has slowly increased during the past half century. Source: CWS and USFWS, unpublished material, 2003.

compared the result, called the sightings ratio, with calendar dates (Lutey 2002). If increases in whooping crane sightings at the Platte were due solely to increases in the whole AWP, the ratio should not change appreciably, because the numerator (number of central Platte sightings) and the denominator (number of birds in the AWP) should change in parallel. Sightings for the spring and fall migrations were combined to estimate an annual rate of use for the central Platte River. Few color-banded or radio-tagged birds have been resighted on the Platte River across seasons or years (USFWS, unpublished material, 2003; Howe 1989), so there is little chance of double-counting by combining migration seasons.

For the purposes of the committee's report, Johnson redid the analysis to reflect the changes made to the central Platte whooping crane sightings file described above (Appendix D) and to update the dataset with the four most recent years of observations (D. Johnson, USGS, pers. comm., August 29, 2003) because the analysis in Lutey (2002) could be updated with four additional years of data. Estimates of whooping crane numbers at ANWR (the denominator) are within one or two birds (less than 1% error; Tom Stehn, USFWS, pers. comm., February 12, 2004). The same is true for the ratio involving crane use-days.

From 1976 to 2002, the sightings ratio increased ($R^2 = 0.245$, $p = 0.009$; Figure 5-6a). The sightings ratio represents the proportion of the AWP seen at the Platte River in any one year. If the regression model is used to describe the relationship, it indicates that the proportion of the AWP observed annually (fall and spring combined) on the central Platte River increased from 1.2% in 1976 to 4.1% in 1989 (the median date) and to 7.0% in 2002.

There are few data to explain the increase in sightings on the central Platte River, which is greater than the increase in the AWP. It may be that as the AWP continues to increase, visitation rates of the entire population to the central Platte River change. Subadults, birds that have adult plumage but are incapable of breeding (Lewis 1995), and breeding birds may use stopover sites differently. Austin and Richert (2001) noted a trend toward a nonrandom distribution of different social groups (families vs subadults, or nonbreeders) on riverine roosts, most of which were in Nebraska.

Another possible explanation of the differential increases in sightings on the central Platte River, as opposed to the AWP, is that crane-management activities on the river now concentrate the movement of cranes through the management areas and effectively narrow the migration corridor (Austin and Richert 2001; Kuyt 1992). Over the last 20 years, whooping cranes' use of night roosts has shifted to the east toward Chapman (and east of the critical habitat area) at the same time that clearing activities have occurred in that same stretch of river (Stehn 2003). For habitats outside the central Platte River, declines in migration-habitat quantity or quality could also force more birds to use the river. Few color-marked whooping cranes have used the Platte River repeatedly, and few radio-tracked birds have used stopover sites repeatedly (Howe 1989), so there are no data to identify the proportion of marked birds that change their distribution as they move through Nebraska. The eastward shift by whooping cranes has paralleled a shift in habitat use from west to east along the central Platte by sandhill cranes that has occurred over the last 40 years (Faanes and LeValley 1993).

Crane Use-Days on the Platte River

Examination of the confirmed-sightings database suggests that cranes have increased the duration of stopover times on the central Platte River in recent years. Repeating the analysis with updated information to compare use-days with whooping crane populations over many years, as discussed in Lutey (2002), resulted in an increase in the use-days ratio over time (R^2 = 0.3594, p = 0.001; D. Johnson, USGS, pers. comm., August 29, 2003). The pattern of increase in the use-days ratio was similar to, but somewhat stronger than, that in the sightings ratio (Figure 5-6a,b). Whooping cranes are spending more time on the central Platte River than can be explained by increases in the size of the AWP alone.

These data do not reveal why whooping cranes may be using the Platte River for longer periods. If subadults, as opposed to adults, are not using the Platte River for greater lengths of time because of their different ecological needs, other hypotheses can be raised. As the AWP grows, there will be more birds in the population that have different resource or habitat needs. Allen (1952, p. 94) related an interview that described whooping cranes as using the central Platte River for longer periods when overall population numbers were higher. Behaviors related to lifestage may change bird use of stopover areas as the population grows.

Individually Marked Birds

The information gained by studying individually marked birds differs from data acquired by studying groups of unmarked birds. For example, Austin and Richert (2001) use sightings of unmarked whooping cranes to measure uses of migration habitats and conclude that high proportions of whooping crane sightings come from wetlands and agricultural fields. Whether the same birds use both habitats or some birds use only wetlands and some only agricultural fields is unknown. Only through the study of individually marked birds can habitats used by birds be linked. For this report, the committee examined two primary databases that contain information on individually marked birds: color-banded birds and radio-tracked birds.

a.

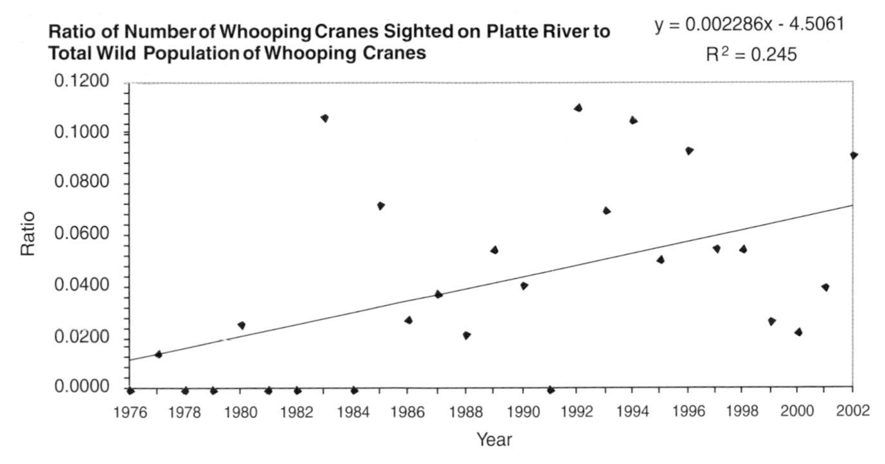

b.

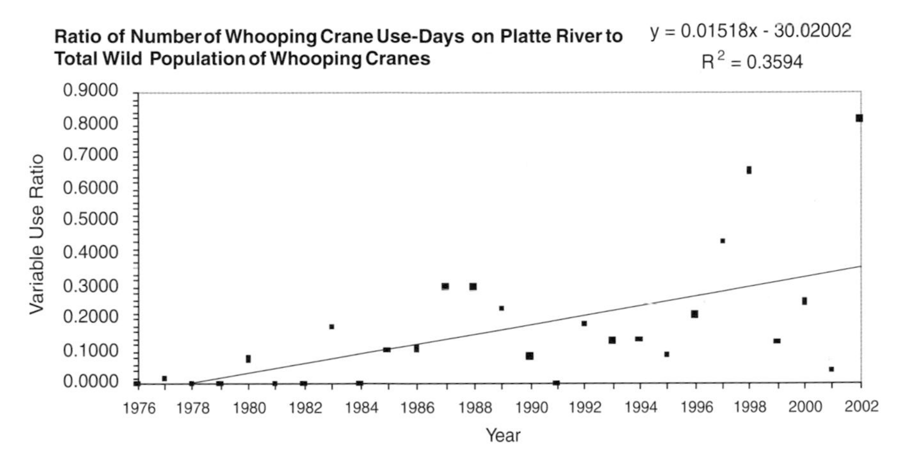

FIGURE 5-6 (a) Number of whooping cranes sighted at Platte River divided by total number of birds in Aransas-Wood Buffalo population (AWP) of whooping cranes in same year (sightings ratio). (b) Number of whooping crane use-days (number of birds times number of days birds spent on Platte River) divided by AWP in same year (use-days ratio). Data collected prior to 1975 were not utilized in these analyses because data collection effort was extremely variable among years before this period. Data for 2003 were not yet complete during committee's deliberations.

Color-banded Birds

Beginning in 1977, researchers captured whooping cranes on breeding grounds and attached color bands to their legs (Kuyt 1979). That program ended in 1988 when 133 birds had been banded (Kuyt 1992) because initial research questions had been answered and mortality associated with capture, although low, was of concern (Tom Stehn, USFWS, pers. comm., August 15, 2003). Seventeen color-banded birds (12.8% of the total number of marked birds) have been seen at least once on the central Platte since 1977 (USFWS, unpublished material, 2003). Only two of the 17 have been seen on the central Platte in more than one year. That low proportion of repeat sightings of the same birds on the central Platte River suggests that little fidelity to this stopover site occurs.

Radio-tracked Birds

In the early 1980s, after extensive experimentation on sandhill cranes, scientists fitted whooping cranes with radio-transmitters and tracked them during fall and spring migration (Kuyt 1992; Howe 1989) for the first time. The key nature of this study made these migration data a basis of many reports and papers on whooping cranes and migration (including this report).

Using radio-tracking data, Pitts stated that of "27 migration passes by individual birds monitored across Nebraska, no stopovers occurred on the Platte" (EA Engineering, Science and Technology, Inc. 1985, pp. 3-19). Referring to these same data, Lingle stated "18 whoopers were tracked on 3 southbound and 2 northbound migrations" (G. Lingle, University of Nebraska, unpublished material, March 22, 2000). Although Pitts did not have full access to radio-tracking data at the time of his report, Lingle did. Twenty-seven migration passes come directly from Howe (1989, p. 5), who referred to the number of marked and unmarked whooping cranes observed during the migration studies. They do not, however, refer to the number of birds tracked on migration, because some marked birds traveled with other birds and behaved largely as other members of the flock.

A whooping crane family of two adults and one chick, for example, travel as a group until the chick is permanently forced from the family group (Lewis 1995). For whooping cranes, eviction of the chick occurs during late spring migration or after arrival on breeding grounds (Kuyt 1992) and therefore entails most of two migration periods. Because they travel together, habitat choice by family groups represents a sample of one and not three. Individuals of one family group violate the assumption of spatial independence (Millspaugh et al. 1998; Alldredge and Ratti 1992) because their movements and habitat choice are biologically dependent on each other.

Five individual whooping cranes with radiotransmitters were tracked across Nebraska over 4 years (1981-1984, Kuyt 1992). These five birds were tracked over Nebraska in seven fall and five spring migration passes. Because there is little evidence of fidelity for stopover sites among years or between seasons (color-banded data above and Howe 1989), multiple migration passes by the same bird are probably best considered as independent. The effective sample size of whooping cranes studied, for habitat selection purposes, is 12 (not 18 or 27) or the number of migration passes made by radio-tracked birds over Nebraska. Specifically, one migration pass was tracked in 1981, three in 1982, five in 1983, and 3 migration passes in 1984 (Kuyt 1992). From 1981 to 1984, no tracked whooping cranes stopped on the central Platte.

In comparison, no whooping cranes (tracked or otherwise) were seen on the Platte River during 1981, 1982, and 1984, whereas eight sightings occurred in 1983 (Appendix C). The predicted portion of whooping cranes expected to stop at the Platte River in 1981-1984 (Figure 5-6a) would be 2.2-2.9% of the AWP, so the chance that 1, 3, 5, or 3 (the number of migration passes in 1981, 1982, 1983, or 1984) radio-tracked migrating birds would stop at the Platte would be low. Failure of tracked whooping cranes to stop at the central Platte during the early 1980s, therefore, does not necessarily suggest that the central Platte is unimportant to whooping cranes.

Two other factors might reduce the likelihood that whooping cranes with radiotransmitters will stop on the central Platte. First, radio-tracked bird use of Nebraska was more frequent in spring than in fall (Table 2 in Howe 1989, p. 6) but only five of 12 passes occurred in spring (Kuyt 1992). Disproportionate sampling, coupled with a small sample size, would increase bias against observing stopovers on the central Platte. Second, Austin and Richert (2001) describe a weak tendency for family groups stopping over in Nebraska to be found more frequently at locations away from the Platte River. Of the 12 migratory passes tracked, eight passes were made by family groups (Kuyt 1992), so additional bias against observed stopovers on the central Platte was possible.

Radio-tracking data have provided much needed insight into habitat use and behavior during migration. These data suggest that cranes move rapidly through stopover areas (Kuyt 1992) and use a variety of habitats for foraging and resting during stopovers (Austin and Richert 2001; Howe 1989). It is apparent that weather events and migration behavior related to weather and climate may determine where a crane might stop during migration as much as the quality of a potential stopover habitat (Kuyt 1992). Accordingly, it is difficult to assess how important individual stopover areas may be to individual cranes or to the AWP as a whole. Pitts (EA Engineering, Science and Technology, Inc. 1985) and Lingle (G. Lingle, University of Nebraska, unpublished material, March 22, 2000) used radiotelemetry data from migrating cranes to argue that habitats on the central Platte River are not critical to whooping cranes, inasmuch as birds that stop at the Platte could readily use other habitats if the Platte River were not available and most whooping cranes appeared not to use the central

Platte River at all. Conclusions of Pitts and Lingle should be taken with caution, however, because they are derived from data on just five whooping cranes that made a total of 12 flights over Nebraska at a time when the size of the AWP was less than half what it is now (Figure 5-5). Furthermore, the probability of any tracked crane stopping at the Platte was biased by the nonrandom grouping of experimental birds by social status (a chick with parents vs a nonbreeding bird) and by season (fall vs spring). More data on radio-tracked (or satellite-tracked) birds are clearly needed.

Howe (1989, Table 2), in contrast, describes Nebraska as an important area for whooping cranes in spring. Richert's (1999) analysis of whooping crane habitat on multiple geographic scales suggests that use of individual regions (such as the Platte River and the Rainwater Basin) is interdependent and depends on regional weather cycles. If that is true, assessment of the importance of individual habitat areas must be carried out over sufficient periods to incorporate potential variations in conditions. The radio-tracking study examined whooping crane habitat use over a relatively short period (4 years) and should be expanded temporally to include wider ranges of environmental conditions. In general, however, Howe's depiction of Nebraska as an important geographic region for crane stopovers, especially in spring, is important in understanding the overall migration ecology of whooping cranes. Assessment of central Platte River habitats should not be made independently of events in other regions.

CHARACTERISTICS OF CRITICAL HABITAT

Relative importance of a given habitat to the population of a species of concern is typically measured by the degree to which that habitat is preferred by individuals of the population. Habitat suitability indexes therefore depend on habitat use in relation to habitat availability.

Assessments of the suitability of specific habitat characteristics for whooping cranes or the degree to which alternative habitats may be available to cranes now using areas of diminished quality begin by addressing issues of scale: within local habitats, within the home range of individuals, within the home range of the population, and within the entire geographic range of the species (Johnson 1980). Researchers who fail to identify their scale of analysis may miss important relationships. For whooping cranes, pertinent questions about habitat use in the central Platte River are connected to roosting (Johnson 1982; Austin and Richert 2001), although foraging in nearby wetlands is also important (Lingle 1987). A number of reports have examined habitat selection for whooping cranes on all four scales. Within habitats, night roosts have a number of physical characteristics believed to be important, including water depth and substrate profile (Johnson 1982; Faanes et al. 1992; Faanes 1992; Ziewitz 1992). Within a home range, whooping cranes can select for riverine vs palustrine wetland roost sites (Lingle et al. 1991). Within the home range of a population, they can select for stopover areas between larger regions, such as the central Table Playas as opposed to the central Platte (Richert 1999). Finally, within the geographic range of the species, they can select the state of Nebraska as a major location for stopovers in spring but may prefer Kansas and Oklahoma in fall (Howe 1989).

Austin and Richert (2001) summarized habitat evaluations at all known migration stopovers in the United States for whooping cranes from 1943 to 1999. On migration stopovers in Nebraska, whooping cranes typically use riverine habitats to roost at night and feed in upland (usually agricultural) or wetland habitats during the day. Those findings represent descriptions of habitat that whooping cranes have actually used but do not represent use of habitats in relation to their availability or to randomly chosen habitats, so interpretation of the data on which they are based is limited.

Foraging sites in wetlands tended to have characteristics similar to those of roost sites described below. Water where whooping cranes foraged was shallower, on the average, than in the wetland as a whole. Foraging areas also tended to be largely devoid of tall (>1 m) vegetation and had open horizons (Austin and Richert 2001). Physical characteristics within agricultural habitats used by foraging cranes were not evaluated, but these habitats also tended to be open and have short (or no) vegetation.

Several habitat assessments have focused on roost habitat characteristics. Johnson (1982, p. 40) listed 10 important physical characteristics for whooping crane roost sites in riverine areas; his measurements included water depth, flow velocities, channel width, vegetation, and other variables. Since Johnson's study, upriver and downriver visibility (Faanes 1992) and protective deep channels (Faanes et al. 1992) have been added to the list of important physical characteristics and have yielded new habitat suitability models (Ziewitz 1992; USFWS, unpublished material, June 16, 2000).

Roost characteristics for used habitats are different from those for nonused habitats. Faanes et al. (1992) examined river cross sections and compared characteristics of sites that were used by whooping cranes with those of nonused sites along the same transects. Johnson's (1982) roost characteristics were found important, as were additional characteristics of channel profile. Roost locations for whooping cranes were more likely to be surrounded by deeper water than would be expected by chance alone. Few other studies have examined habitat selection on this geographic scale of selection.

Even though substantial data (in both quantity and quality) are available and reasonable habitat suitability models have been developed, the size of the AWP remains the chief limitation on analysis and interpretation. Such a small number of birds ensures that some high-quality sites inevitably are nonused because there are not enough birds to fill them. For example, Faanes et al. (1992) described selected river profiles in habitats preferred by roosting whooping cranes. However, the pattern of preference observed might not be the same if the number of whooping cranes using the central Platte were substantially larger. It is difficult to assess how much habitat in the central Platte will be adequate to support a viable delisted population of whooping cranes. Many reports indicate that the presence of forests is inversely related to the desirability of roosts. Whooping cranes use roosts where the width of unvegetated channel and the nonobstructed view of areas upstream and downstream of the roost are extensive (Faanes 1992; Faanes et al. 1992). Researchers, however, have yet to define specifically what constitutes a nonobstructed view for a whooping crane. Is the presence of a few trees enough to significantly alter use of an otherwise desirable (open) roost site? More experimentation could be done with habitat modification of forests in the river valley (see Chapter 4). Debates about the appropriateness of forests in the Platte River Valley are mired, in part, because only two extremes (forest vs open) are presented where once an entire gradient of vegetation communities existed.

On a scale expanded to evaluate habitats within a home range, potential additional variables that might distinguish used from nonused habitat for whooping cranes include distance to nearest human development, distance from roost to feeding areas, and variety of habitats used (Johnson 1982; USFWS, unpublished material, June 16, 2000; Austin and Richert 2001). No studies of habitat use by whooping cranes according to availability of those variables have been made; thus, their relative importance has not been quantified. Foraging whooping cranes use both agricultural and wetland habitats throughout their migration (Howe 1989; Austin and Richert 2001), and this behavior differs from that of sandhill cranes, which strongly prefer agricultural habitats in Nebraska (Iverson et al. 1987). Lingle et al. (1991) found that whooping cranes spent roughly equal foraging time in agricultural and wetland habitats in Nebraska. In the specific area of the central Platte, no extensive habitat-use studies have been done for foraging whooping cranes except that several authors characterize whooping crane foraging activities as parallel to sandhill crane preference for foraging in agricultural areas (e.g. Lingle 1987; USFWS, unpublished material, June 16, 2000).

From analysis on the scale of the Platte River Valley, more is known about use of roost habitats by whooping cranes. The river channels (including their bars and islands) are the most important habitats: 77% of all whooping cranes confirmed in the valley in 1975-2003 used the channel areas (USFWS, unpublished material, June 11, 2003; Appendix C). Of whooping crane groups that stayed in the Platte River Valley for more than 1 day, 91% used channel areas as roost habitat (USFWS, unpublished material, June 11, 2003; Appendix C). Within the valley, few whooping cranes have roosted off-channel; when doing so, most have used wetlands associated with the Platte River (Howe 1989).

Riverine roost sites are important whooping crane habitats throughout Nebraska (Austin and Richert 2001), not just on the central Platte. In Nebraska, and elsewhere in the flyway, wetlands not associated with the river also serve as important roosting habitats for whooping cranes in the AWP (Austin and Richert 2001). Night roosts used by whooping cranes tend to be isolated from human disturbance and are within 1 mi of foraging sites. The relative availability of nonriparian habitats is not known. In general, wetlands are important to migrating whooping cranes for both foraging and roosting. In the central Platte, use of riverine habitats by whooping cranes is well documented and appears important to their survival. Preference studies have not been done on this scale and, although useful, will be limited by the same problem of small population size as noted above.

When the scale of analysis was expanded to include the entire flyway for the AWP, one-third of the habitats used by roosting whooping cranes were riverine, a different distribution of use from Nebraska, where 59% of the roosting sites are riverine (Austin and Richert 2001). Telemetry studies did not find riverine habitats important to roosting whooping cranes, however, because the few tracked birds primarily used palustrine wetlands (Howe 1989). Bias occurred in both studies. Austin and Richert (2001) identified the limitations to analysis of habitat-use evaluations for migratory whooping cranes. Comparisons among sites are difficult because efforts to observe whooping cranes along the entire 4,000 km migration route are uneven. Some regions are subject to underreporting (Howe 1989). Still, the high incidence of use of riverine habitat by whooping cranes in Nebraska, compared with other states, is striking. Migration stopover sites in Nebraska include the central Platte River, Rainwater Basin, central Table Playas, and northern Sandhills (Richert 1999; USFWS, unpublished material, June 16, 2000; Austin and Richert 2001). The quality of both palustrine and riverine habitats in Nebraska is high (Stahlecker 1997) and, accordingly, these wetlands are used extensively by whooping cranes (Austin and Richert 2001).

With the opportunistic and nontraditional patterns of habitat selection exhibited by whooping cranes, it is difficult to predict habitat preferences on the scale of home ranges. Although nonhabitat variables (such as weather events) can influence habitat selection greatly (Kuyt 1992), the overall pattern of habitat use in Nebraska suggests that the general region of the central Platte (including the Rainwater Basin) is important to migrating whooping cranes. On large time scales, as wet and dry periods cycle through the region, the interaction among the sites will probably be important (Richert 1999). Whooping cranes will need stopover habitat during droughts when wetlands in the Rainwater Basin are dry and during wet years when the Platte River is above flood stage. Maintaining a complex of wetlands over a large geographic area will help to mitigate the effects of climatic variations.

Few data relevant to use of habitats by migrating whooping cranes on the largest geographic scale exist. However, telemetry data reported by Howe (1989) suggest differences in use of stopover areas between fall and spring migrations. Whooping cranes appear to spend more time in Nebraska in spring than they do in fall (Howe 1989, Table 2, p. 6). In fall, many whooping cranes tend to stop in a Saskatchewan staging area for several weeks before continuing their migration, presumably to acquire fat reserves for migration. It takes cranes about two daily flights to reach that region in Saskatchewan after leaving their breeding grounds in WBNP (Kuyt 1992). After leaving Saskatchewan, the birds migrate rapidly and reach Nebraska in 2-3 days and Texas in 5-7 days more (Kuyt 1992). During fall migration from 1981 to 1984, several radio-tracked whooping cranes flew over Nebraska without stopping in the state at all (Kuyt 1992). In spring, however, cranes do not use a staging area (Kuyt 1992). For spring migration, Nebraska is about the same distance from ANWR as Saskatchewan is from WBNP. Longer periods spent in Nebraska in spring may therefore be a function of proximity to ANWR as a starting point. Collectively examining habitat selection for whooping cranes at all four geographic scales suggests that Nebraska, and the central Platte in particular, provide critical habitat for this species. Climatic variations and development pressures likely over the next 30 years in which whooping cranes will need to recover (assuming that current population trends continue) will further emphasize the importance of providing stopover habitats for whooping cranes on the central Platte.

WHAT DO WHOOPING CRANES GAIN FROM CENTRAL PLATTE HABITATS?

Two basic ecological needs of migratory bird species are met by migration (stopover or staging) habitats: food and a safe resting place. Energy and nutrients are acquired and stored for use during future phases of the annual cycle (see Alisauskas and Ankney 1992). The reserves can be used to provision future flight, allow birds to persist through periods of food shortage, or help to meet reproductive needs when exogenous food resources are insufficient. Because energy and nutrient reserves are acquired in one stage of the annual cycle and used in another, mortality and natality can be influenced by events that occur in other stages of the annual cycle, producing a cross-seasonal effect (Weller and Batt 1988; Fretwell 1972). Birds require safe environs at each habitat they visit. Protection from natural predators and relative isolation from human activities contribute to the utility of stopover sites. The central Platte River has historically filled both needs for whooping cranes during their migrations.

Energy and Nutrient Storage

When whooping cranes leave breeding areas at WBNP, they often fly for 2 or more days to reach a staging area in Saskatchewan, where they remain for a few days to a month (Kuyt 1992); 2-3 weeks is the average period of occupancy in this fall staging area. Once whooping cranes leave the Saskatchewan staging area, they typically move rapidly to ANWR, stopping for just short periods at individual stopover sites (Kuyt 1992). Although no direct data exist, the pattern of migration suggests that birds leave their breeding area with varied body conditions (such as amount of stored fat), stop in Saskatchewan and feed intensively on grains to build fat reserves, and then continue with migration once sufficient fat is stored. A similar pattern of fat storage (more directly measured) occurs in migrating sandhill cranes in fall (Krapu and Johnson 1990).

Presumably, whooping cranes then arrive at ANWR lean, having used substantial amounts of fat reserves during fall migration. These measures are also indirect and are inferred from similar patterns of migration physiology of sandhill cranes. Sandhill cranes complete long migrations with substantially reduced fat reserves (Krapu et al. 1985).

Chavez-Ramirez (1996) used behavioral descriptions and energetic models to conclude that whooping cranes regain fat reserves in winter areas before beginning their spring migration. They migrate back to WBNP relatively quickly, using stopover habitats en route but no staging areas. According to data from radio-tracked whooping cranes, not all delays at stopover sites were related to weather delays (Kuyt 1992, pp. 33-35). Presumably, energy and nutrient reserves acquired in winter are used to migrate quickly from Texas to Canada. Once in breeding areas, whooping cranes initiate nesting soon after arrival. Little food is available when whooping cranes arrive at WBNP, because ice and snow cover typically does not disappear until after nests have been initiated. Stored fat reserves, in addition to their importance for migration, are probably used for egg formation and incubation. Similar patterns occur in waterfowl, in which links between stored reserves and reproduction have been studied better (Alisauskas and Ankney 1992).

As with sandhill cranes, cross-season effects probably are important for whooping cranes. Chavez-Ramirez (1996) argued that variations in winter habitat conditions are related to reproductive effort in the following spring and that winter body condition could affect winter survivorship. If winter habitat conditions influence mortality or reproduction, it may be that migration-habitat conditions affect whooping cranes in a similar way.

Stehn (2001) has suggested that habitat quality at ANWR may be declining. The amount of fresh water entering ANWR is important ecologically because it influences overall productivity in the estuary, especially for such species as blue crabs, which form the primary food of whooping cranes during winter. Declining fresh-water flows have been linked to the declining abundance of blue crabs (Stehn 2001). Fresh water for the estuarine refuge comes from the Guadalupe River. Today, 29% of the river's flow is appropriated, with additional diversions likely to be necessary to support growing urbanization in southeast Texas (Stehn 2001).

Whooping cranes that leave ANWR with insufficient stored fat will either migrate to WBNP, facing the potential for added mortality and reduced productivity (Chavez-Ramirez 1996), or increase stopover times on migration to acquire fat reserves. Responses by individual birds are likely to vary. For example, nonbreeding subadults may be able to stop longer on migration than can breeding adults because they do not need to initiate nests. The potential for cross-seasonal effects means that if deteriorating habitat conditions in ANWR are severe and long-lasting, whooping cranes may not be able to acquire sufficient fat reserves and will have to acquire them in other habitats, such as the central Platte River.

Few data are available for testing the hypothesis that spring migration habitats influence energy reserves, reproduction, or mortality in whooping cranes. Some birds stay longer on spring stopover areas than do others (Kuyt 1992; Howe 1989), and more whooping cranes stop at the Platte River for long periods than did 25 years ago (Figure 5-6b). If fall staging areas are important for whooping cranes, as argued by Pitts (EA Engineering, Science and Technology, Inc. 1985; Howe 1989), the situation may be reversed in spring as birds move from winter to spring migration habitats. Other species of waterfowl acquire energy and nutrient reserves on spring staging areas before arrival on breeding grounds in the manner suggested for whooping cranes (Barzen and Serie 1990). Although relevant data are sparse, the hypothesis warrants further investigation. Documented migration patterns may not predict future patterns of movement in that habitat conditions change over time at the numerous stopover and staging areas. The changes might be related to the crane population, to climatic adjustments, or to changes in altered agricultural practices. It is possible that spring migration areas could be used in new ways by whooping cranes. This possibility cautions against underestimating the ecological value of stopover habitats to whooping cranes, especially stopover habitats on the central Platte River. Satellite telemetry studies over multiple years would help greatly in testing these ideas as they have done with sandhill cranes (Krapu 2003).

Safe Stopping Areas

Population monitoring at ANWR since winter 1938-1939 is useful for assessing the dynamics of the AWP (CWS and USFWS, unpublished material, May 2, 2003). Accurate yearly counts of the birds arriving on the winter grounds coupled with low summer adult mortality and counts of birds at the end of winter provide estimates of mortality at several stages of the annual cycle. As much as 80% of whooping crane deaths appear to occur during migration, even though this period constitutes less than 20% of the birds' annual cycle (Lewis et al. 1992). Although the total mortality is reasonably well known, causes of death are more problematic. Of 133 adults and subadults that disappeared away from winter areas between April and November (mortality on breeding areas is assumed to be low) from 1950-1987, the cause of death is known in 13 cases, five of which were due to collisions with power lines (Lewis et al. 1992).

Because most deaths of adult whooping cranes occur during migration, mortality may be linked to the quality or quantity of stopover habitats. A significant portion of the whooping crane population stops at the Platte River during migration (Figure 5-6a,b). If the Platte River were no longer available, whooping cranes would probably shift their use to other habitats in Nebraska. Would these shifts alter mortality? Migrating whooping cranes use wetlands near the central Platte River in the Rainwater Basin south of the river or in the central Table Playa to the north (Richert 1999). Those off-stream wetlands apparently serve as good migration habitats for whooping cranes, except in droughts or during waterfowl disease outbreaks to which these wetlands are prone (Friend 1981). The Platte River ecosystem is not prone to disease outbreaks, because, in contrast with off-stream wetlands, water is flowing rather than stagnant. The Platte River also does not dry out as frequently as do surrounding wetlands; therefore, the Platte River provides safer conditions than do surrounding wetlands at least in some years (Johnson 1982).

Crowding of waterfowl and cranes and short-term use of unusual habitats may also affect crane mortality. An experimentally released pair of whooping cranes unexpectedly migrated from their release area in Florida (where they were part of a nonmigratory flock) and spent the summer in Michigan (M.A. Hayes, International Crane Foundation, Baraboo, WI, unpublished material, 2002). The pair began fall migration normally, but the male disappeared (and presumably died) during a snowstorm on the first day of migration. The female migrated successfully to Florida in 11 days, during which she used many stopover habitats similar to those described by Howe (1989). Some stopover habitats this whooping crane used, however, were unusual. For example, she used an abandoned coal mine pond in Ohio and a forested area (probably including a stream) in a mountainous area of Virginia.

Even though this bird survived the migration, did the probability of mortality increase when poor quality habitats were used? An incremental increase in mortality due to poor conditions on stopover habitats would be difficult to document with a small population. Higher mortality due to nocturnal predators has resulted from use of inappropriate habitat with an experimentally reintroduced flock of whooping cranes in Florida. Once crane behavior was modified, and better roost habitat was used, mortality decreased (Nesbitt et al. 1997).

Mortality and Population Viability Analysis

The impact that variations in mortality have on the AWP can be examined through population viability analysis (PVA). The sensitivity of a target population to changes in mortality can be assessed with simulations that use measured population characteristics. The biology of whooping cranes, however, makes the species a challenging subject for PVA. In the wild, only one functional population of whooping cranes exists, and its members use the central Platte River only to stop during migration (Lewis 1995). Individual birds typically stay over for just a few days, but on occasion a bird stays for as long as several weeks. About 7% of the migratory population now uses the Platte River each year, but many more, if not all members of the species, use the Platte River at some time during their lives (Lutey 2002). Furthermore, specific habitat use by birds during migration and selective use of habitats by particular age groups (subadult or adult), reproductive status (breeding or nonbreeding), and sex are poorly known. In this context, relating the fitness of individual birds or the likely persistence of the population to central Platte River habitat conditions is tenuous.

Nonetheless, three PVAs for whooping cranes are available and warrant consideration in conservation planning for the species. All use the same database for the Platte River population but use different years of data. Mirande et al. (1991, 1997) published PVAs that were done with VORTEX, and Brook et al. (1999) published PVAs that compared the output of different modeling packages, including VORTEX, RAMAS (Age, Stage, Metapop), INMAT, and GAPPS. All are age-based stochastic simulations. According to Brook et al. (1999), analysis of census data yields no evidence of density dependence; even if it did, current populations are probably well below carrying capacity, so density dependence would be unimportant. The Mirande et al. (1991) PVA concluded that the population is likely to grow at a low, steady rate ($r = 0.046$) and has less than a 1% chance of going extinct in the next 100 years. The model included density-dependent effects. Some simulations included catastrophes that affected reproduction and survival. In sensitivity analysis, increasing recent mortality by 50% (an increase in the annual mortality rate from 9.4% to 14.1%) would cause the population to go extinct, but an increase of 25% in mortality would not. The Brook et al. (1999) analysis provided similar results: all model applications predicted population growth. However, model predictions differed in details, and differences were caused by a variety of features, such as inclusion of an inbreeding depression, stochastic variation in monogamy, and different projections for the same feature (Brook et al. 1999).

Because there is no full-time resident or breeding population of whooping cranes in the Platte River Basin, conclusions about the role of the central Platte River in the persistence of the population must be based on some stringent assumptions, including the assumption that the loss of Platte River habitat will result in some increase in mortality or other contribution to decreased fitness of the population. Accepting that assumption, the PVA of Brook et al. (1999) with VORTEX is a reasonable model to address several issues pertinent to the committee's charge, including lengthening the time frame of the analysis, mortality vs persistence, lifespan reduction, and nutrition.

First, does lengthening the goal for population persistence from 50 years (used by Brook et al. 1999), which is too short for conservation planning, to 200 years substantially alter optimistic predictions of the model? The modified Brook et al. model was run with an initial population size of 185. No other modifications were made except lengthening the time frame for viability to 200 years. The population parameter values used in the model were gathered over decades from a population that was growing. If the same parameters are used, starting the model with a population size that is closer to the current size, the prediction is for population viability under current circumstances—a 99% probability that the population will persist for 200 years.

Second, how much would adult mortality have to increase for the likelihood of population persistence to decrease significantly? Model inputs were adjusted to increase mortality by 1% increments up to 5%, reflecting loss of or further decline in quality of Platte River habitats. The increments were added to the existing 9.4% mortality for all whooping crane age classes. At its current population size the whooping crane is fairly well buffered against environmental and demographic variation that could cause it to go extinct. If, however, there is a 3% increase in mortality, the model predicts that the population would become nonviable, and the probability of its persistence for 200 years would decline to 86% (Figure 5-7).

Third, given that the whooping crane lifespan is not known, what is the effect on population persistence of shortened lifespans that could result from losses of key habitats? Loss of habitat or decreases in its availability on the central Platte River could result in a decrease in the vigor of individual birds, shortening lifespans. However, model runs with lifespans reduced from 50 years to 40 and to 30 years showed that reduction in lifespan had no statistically significant effect on population persistence (Table 5-1).

The PVAs suggest relatively stable populations well into the future under current conditions, but the high sensitivity of population persistence

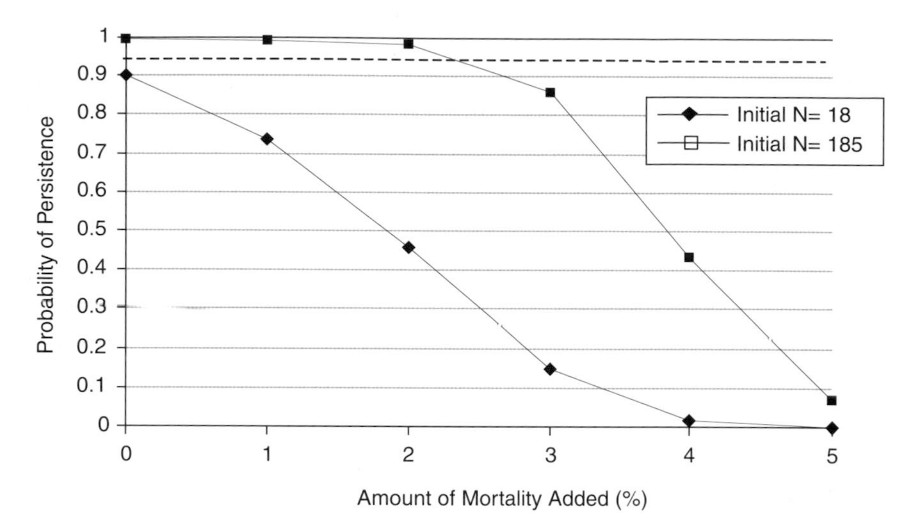

FIGURE 5-7 Results of population viability analysis scenario 2. Viable population numbers are those above 0.95 probability of persistence for 200 years. Source: Reed 2003.

to rather minor changes in mortality suggests a note of caution. Reductions in foraging or resting habitat that manifest in decreased vigor of individual cranes could have a substantial effect on the fate of the entire population. Without central Platte River stopover habitats, alternative habitats elsewhere in the migratory pathway will be necessary to provide resources for dispersing whooping cranes.

TABLE 5-1 Results of Scenario 3 Simulations

Initial Size of Population	Extinction Probability	Of Populations that Do Not Go Extinct, Mean Population Size at Target Time	Of Populations that Go Extinct, Mean Time to Extinction, years
Baseline: maximal age, 50 years			
18	0.098	1,305	51
185	0.005	1,360	133
Maximal age, 40 years			
18	0.113	1,267	55
185	0.008	1,345	117
Maximal age, 30 years			
18	0.126	1,253	60
185	0.006	1,344	80

Source: Reed 2003.

If mortality increased only enough to slow growth, rather than to cause a population decline, adverse effects on the AWP could still occur. A net loss of alleles will continue to occur in the AWP through genetic drift (a loss of genetic material from a population due to small population size) until the population becomes large enough for the rate of genetic mutations (which produce new alleles) to roughly equal the rate of drift (which loses alleles). The longer it takes for the AWP to reach a size at which drift and mutation rates are balanced, the more genetic material will be lost from the population. When the AWP went through a bottleneck in 1941, reaching a low of 15 birds, an estimated 66% of the mitochondrial DNA diversity in the population was lost (Glenn et al. 1999). Further loss of genetic diversity because of a slowed recovery could be problematic.

SUMMARY AND CONCLUSIONS

This chapter began with three questions related to whooping cranes: Do current Platte River habitat conditions affect the likelihood of survival of the species? Do they affect recovery of the species? Is the current designation of central Platte River habitat as critical habitat for whooping cranes supported by science? The chapter has reviewed information collected by the committee from the literature and from unpublished DOI documents concerning the species, from testimony given by agency and other specialists, from a visit to DOI facilities in Denver and Grand Island, and from field examinations.

The committee concluded that current habitat conditions along the central Platte River adversely affect the likelihood of survival and recovery of the whooping crane population. Geographically, the central Platte River occupies a critical position along the migration route of the species, between the wintering grounds in coastal Texas and the summer breeding grounds in north central Canada. The river and its closely associated lands provide useful roosting areas for birds in the midst of migrations of thousands of miles. The portions of the river that are not heavily wooded provide open areas separated from the banks by channels—an arrangement that provides security for the roosting birds. Nearby wetlands and agricultural areas provide important forage for the birds and allow them to obtain needed nutrition to support their continued migration.

The committee concluded that there are no apparently suitable alternatives to replace the central Platte River in its function as habitat for migrating whooping cranes. The Rainwater Basin, south of the river, includes numerous small wetland basins and patches, but these areas periodically dry completely, whereas the Platte River flows relatively continuously. Because the Platte draws its water from distant mountain watersheds, its flows are less susceptible to periodic drought on the plains than are the ephemeral wetlands of the Rainwater Basin. Nearby rivers also lack the consistency of the Platte, and the Niobrara and Loup Rivers and smaller streams do not offer alternative habitat with the same qualities as the Platte.

On the average, about 7% of the whooping cranes currently use the central Platte River as a stopover during migration, but there is substantial fluctuation in this fraction from year to year. The loss of the Platte River habitat would have potentially serious consequences for the species. If mortality increases by only 3%—a likely scenario if the Platte River habitats become unavailable, because most crane deaths occur during migration—the entire migrating population of less than 200 is likely to become unstable. The general total population of migrating whooping cranes is slowly increasing from its low of only 15 in 1941, and the proportion of the population that uses the central Platte River as a stopover each year is also gradually increasing. The river therefore directly affects species recovery.

The committee also concluded that the current designation of central Platte River habitat as critical habitat for whooping cranes was supported by the science of the time of the designation and that it is supported by present scientific understanding. When DOI agencies designated the critical habitat in 1978, they relied on information that was available then. Agency personnel used information that was available in the refereed scientific literature, agency reports, and additional observations made by agency personnel. Agency personnel made the designation by using procedures that the biological community commonly recognized, and internal peer review strengthened confidence in the designation. There have been no developments in scientific knowledge that invalidate the 1978 decisions.

Given the small numbers of migrating whooping cranes, some parts of the designated habitat inevitably will not be used in some years. Because both the population and the percentage that stops on the Platte River are slowly increasing, it is likely that the occasionally nonused portions of the critical habitat will be increasingly important. The general utility of the central Platte River for roosting whooping cranes also changes from one year to the next, depending on the hydrological conditions of the river, which are subject to some change. The committee also acknowledges that conservation actions at locations outside the central Platte River, particularly in overwintering and breeding habitats, will have important effects on continued use of Platte River habitats, and that a wide variety of management responses—including captive rearing, reintroductions, and translocations—will continue to play critical roles in the conservation of the species as long as populations continue to persist at low levels.

Finally, in addition to addressing the three questions related to whooping cranes, the committee identified gaps in knowledge and data that should be addressed. The committee strongly urges managers of the central Platte River to adopt a multispecies approach to their research and decisions.

Solving problems related to whooping cranes without reference to other species will probably lead to additional problems.

Knowledge and data on how the various species may interact with each other through habitat manipulation are especially important. When managers clear woodland for whooping crane habitat, for example, there appear to be favorable outcomes for the crane population, but there is little knowledge about the effects of such actions on other species, so continued monitoring and measurement efforts are required. The committee also recognizes the importance of long-term records and data in reaching conclusions about whooping crane use of the central Platte River. Because of the annual fluctuations in the river processes and in crane use, trends of only a few years are not likely to be informative regarding longer-term, decade-long trends, so analysis and prediction require datasets on birds and environmental conditions that exceed a few years.

6

Piping Plover and Interior Least Tern

Antithetical though it may seem from their vernacular name, shorebirds are common in North America's vast interior. Sandpipers, willets, curlews, and their many relatives are found across the Great Plains and central Canada north to the high Arctic. Why piping plovers and interior least terns, now recognized as endangered and threatened species, reside in Nebraska, so far from ocean and sea, is immediately apparent to the astute observer. Great Plains rivers—with their braided channels, gravel bars, and sandy shores—offer conditions for rearing young that are remarkably similar to the conditions that support plovers and terns on distant coastal shores. But those conditions, which have been sustained for millennia by dynamic fluvial geomorphic processes, are changing rapidly. Impoundment and exports of Platte River Basin waters have altered the river's hydrograph and caused the disappearance of landscape features that sustain the two species. As their habitat has been lost, populations of piping plovers and interior least terns have declined dramatically along the central Platte River.

The following chapter explores three specific questions posed to the committee: Do the present habitat conditions on the Platte River affect the likelihood of survival and recovery of the piping plover? Do they affect the likelihood of survival and recovery of the interior least tern? Is the currently designated critical habitat on the central Platte River for the piping plover supported by existing science? Observations in this chapter also support conclusions related to a more general question concerning gaps in knowledge regarding the two species.

PIPING PLOVER

The piping plover (*Charadrius melodus*) is a small shorebird that has threatened or endangered status throughout its range (USFWS 1988; Haig 1992; Thompson et al. 1997). The species is distinguished from other, smaller plovers by a single black neck band that is present during the breeding season and a short, stout bill (orange during breeding), pale gray back and wings, white belly, and orange legs (Figure 1-4). The name "piping" refers to the bird's distinctive flute-like vocalizations.

Distribution

The U.S. Fish and Wildlife Service (USFWS) and Nebraska Game and Parks Commission (NGPC) collect and maintain data on plover nesting. Methods used to monitor plovers in the Platte River are the same as or similar to those used by biologists elsewhere in the range of this species in North America. Protocols for monitoring breeding pairs, estimating productivity, and reporting results have been formalized.[1] Since the Northern Great Plains (NGP) population was listed, Platte River nesting records have been consistently collected and maintained.

The general method used to obtain estimates of breeding pairs throughout the piping plover range is to walk toward potential nest habitat or approach it by boat (Plissner and Haig 2000). Potential habitat is identified on the basis of historical records (plovers are very site-faithful if habitat is suitable) and knowledge of habitat characteristics. In the central Platte, USFWS and NGPC biologists obtain estimates by using an airboat (Erika Wilson, USFWS, pers. comm., July-September 2003; John Dinan, NGPC, pers. comm., May-September 2003); they do not go on the land, because some of it is private property. At sandpits and Lake McConaughy, monitors approach plover habitat on foot (Mark Peyton, CNPPIR, pers. comm., September 2003; Jim Jenniges, Nebraska Public Power District, pers. comm., May-August 2003). USFWS's surveys by river use two observers and are conducted in May and early June, depending on river conditions. At potential habitat sites, the monitors stop and check the landscape with a telescope and record the presence or absence of plovers. Monitors also record the Global Positioning System (GPS) location of the site. If plovers are present, the number of individual birds is counted, and their behavior in riverine habitat is recorded. Plovers typically vocalize and are usually visible to monitors because of their flight or rapid movements on the ground. If monitors are close to a nest, plovers may perform "broken-wing

[1]Governance Committee, Executive Director's Office, Tern and Plover Monitoring Protocol Implementation Report, unpublished material, Feb.12, 2001.

behavior" to distract the observer. Monitors record the presence or absence of birds at a particular site and note behavior that indicates nesting. If there is evidence of nesting or if the team believes that the birds will nest, they return daily or several times per week to determine whether chicks have hatched. Visits are made until adults leave or it is decided that no chicks are present, typically June-July. At the end of each season, data on pair estimates and nesting success are sent to NGPC, where the Nebraska piping plover database is maintained. Since records were first recorded for piping plovers, the methods have become more formalized, particularly after the methods section of the Cooperative Agreement Tern and Plover Monitoring Protocol was prepared.

The piping plover breeds only in North America. Its total population was estimated at nearly 6,000 adults in 2001 (Susan Haig, USGS, pers. comm., July-August 2003). The Great Lakes population is recognized as endangered under the Endangered Species Act (ESA), and the NGP and Atlantic Coast populations are recognized as threatened. Birds nesting on the Platte River are part of the NGP population. Piping plovers breed on open beaches along the Atlantic Ocean and Great Lakes, on alkali flats, on islands in broad prairie rivers, and along reservoir shorelines in the NGP. They winter along the Atlantic Coast from Virginia to southern Florida, in the Caribbean, and along the Gulf Coast from Texas and Mexico.

Historical distribution and nesting records exist for piping plovers in Nebraska back at least to the 1800s (EA Engineering, Science and Technology, Inc. 1988; Lutey 2002). Before listing, the species was reported from 32 of 97 counties in the state. Information on population status is more recent: statewide estimates were not conducted until after the "population" was listed in 1986. The presettlement plover breeding-population size in Nebraska is unknown.

The breeding range of the NGP population includes southern Alberta, southern Saskatchewan, and southern Manitoba; extends south to eastern Montana, North Dakota, South Dakota, southeastern Colorado, Iowa, and Nebraska; and extends east to Lake of the Woods in north central Minnesota. Most of the breeding pairs in the U.S. portion of the population's range are in North Dakota, South Dakota, Montana, and Nebraska. USFWS conducted international winter and breeding censuses for piping plovers in 1991, 1996, and 2001. Trend data indicate that the NGP piping plover population declined by 15% from 1991 to 2001.

Plovers breed in Nebraska on sandbars, along reservoir shorelines, in commercial sand mines, and at other artificially created sites along three major rivers (Figure 6-1, 6-2, and 6-3). In the northeastern corner of the state, along the border with South Dakota, nesting occurs along about 64 km of the Upper Missouri River and 153 km of the lower Niobrara River

FIGURE 6-1 Sandy low bars along central Platte River serve as nesting areas for piping plovers and interior least terns. Source: Photograph by W.L. Graf, May 2003.

(Figure 6-4). Farther south, plovers are found along about 390 km of the central and lower Platte River, from the Missouri River west to Lexington. Breeding also occurs at Lake McConaughy in western Nebraska and on the Middle Loup and Loup rivers in central and eastern Nebraska (L.C. Wemmer, USFWS, unpublished material, February 11, 2001). Census efforts during the breeding season estimated 398, 366, and 300 pairs of piping plovers in Nebraska in 1991, 1996, and 2001, respectively (Figure 6-5) (Susan Haig, USGS, pers. comm., July-August, 2003). Those estimates do not include plovers on the portion of the Missouri River that is shared with South Dakota. Current estimates indicate that nesting pairs of piping plovers in Nebraska make up 10-12% of the NGP population, and about 9% of Nebraska's breeding piping plovers nest on the central Platte River (Dinan, NGPC, pers. comm., May-September 2003); about 1% of the NGP population of piping plovers nest on the central Platte River.

Lutey (2002) reported that the numbers of piping plovers observed on the Platte River declined from 1987 to 1998, a period during which the number of breeding pairs averaged just one for the South Platte, 35 for the North Platte (at Lake McConaughy), four on the upper Platte, 29 on the

FIGURE 6-2 Sand mines along margin of central Platte River serve as nesting areas for piping plovers and interior least terns, but are not as suitable as sand masses in river. Source: Photograph by W.L. Graf, May 2003.

central Platte, and 50 on the lower Platte. John Dinan (NGPC, pers. comm., May-September 2003) reported a 61% decrease in piping plovers nesting on the central Platte River from 1991 to 2001, so piping plovers appear to have declined over the last decade at regional, state, and local levels. Declines at all levels are attributed primarily to human activities, including direct and inadvertent harassment of birds and nests by people, domesticated animals, and vehicles; destruction of shoreline habitat as a consequence of development projects; increased predation due to human presence in less-visited beach areas; and water-level regulation policies that result in changes in nesting habitat (Haig and Elliott-Smith, in press; Haig 1992). Causes of the declines recorded in the NGP population include predation of eggs and chicks, habitat destruction and degradation that result from channelization of rivers and modification of river flows, disturbance by humans and pets, contaminants, and inadequate regulatory mechanisms (L.C. Wemmer, USFWS, unpublished material, February 11, 2001).

To understand habitat use by plovers in the Platte River, it is important to establish whether birds are sedentary—that is, use one or a few sites during their lifetime—or disperse extensively. That is important because it

FIGURE 6-3 Sandy shore of Lake McConaughy provides nesting areas for piping plovers and interior least terns when the reservoir is low enough to expose beaches. Source: Photograph by W.L. Graf, May 2003.

is a key to understanding local population dynamics, population persistence, and the spatial scale on which management of the Platte River will affect regional populations. Movements of plovers that hatched along the Platte River or were first banded as adults nesting along the Platte have been studied by Lingle (1993c). He reports resightings of 329 plovers banded from 1985 to 1989 between Lexington and Grand Island and observes that in the year after banding, 43% of the plovers banded as adults returned to the same site, but only 18% of the birds banded as chicks returned to the site where they hatched. Those observations suggest that missing birds either died or moved to other locations. An unknown number of birds were probably present in the study area but not detected. Without human observers present to look for banded birds, it is not possible to infer where the surviving banded birds eventually nested. Lingle did, however, obtain several dispersal records worth noting: seven birds were observed at other nest locations in the Platte River Valley, including an adult banded near Kearney that nested 155 mi downstream 2 years after banding and a banded bird captured at Lake McConaughy, 300 mi from its banding site on the lower Platte River. Lingle's observations indicate that plovers from the Platte River are capable of moving far between breeding seasons and show the difficulty in making inferences

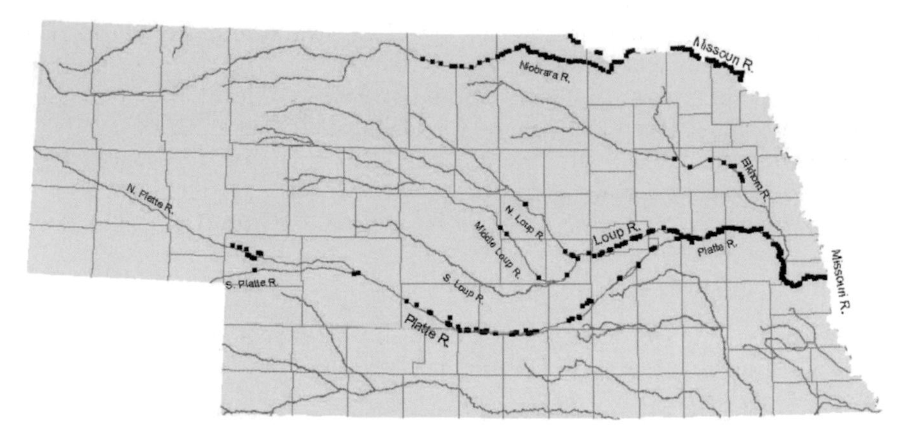

FIGURE 6-4 Piping plovers and interior least terns distribution in Niobrara, Loup, and Platte Rivers, Nebraska. Source: Dinan 2003.

about local populations, movements of individuals, and patterns of habitat use without information on the behavior of marked birds. They also suggest strongly that piping plovers in Nebraska are part of a single demographic unit of interacting individuals.

Records on timing of breeding-season activities are maintained by the Wildlife Division of the NGPC (Dinan, NGPC, pers. comm., May-September 2003). Piping plovers typically arrive at the Platte River in middle to late April and depart by late August. During that interval, they locate a mate,

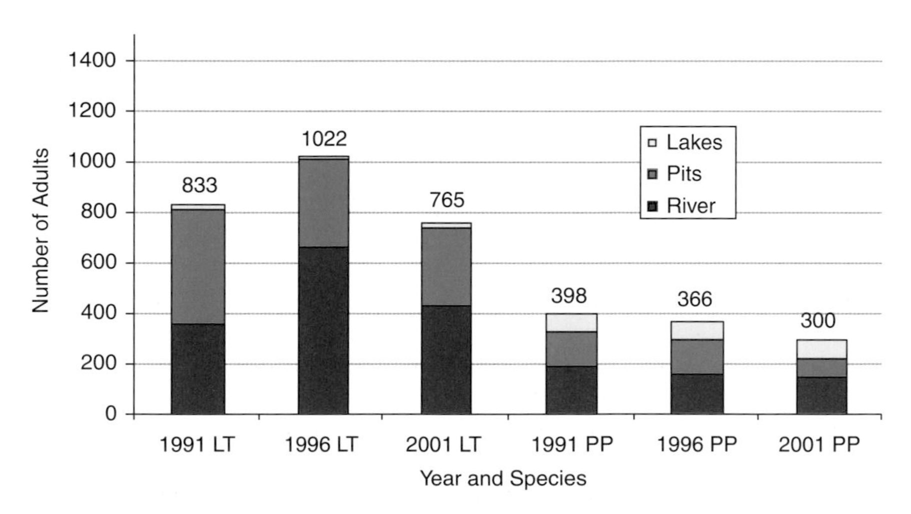

FIGURE 6-5 Estimated piping plover (PP) and interior least tern (LT) population numbers represented in pairs during 1991, 1996, and 2001 breeding season for Nebraska, excluding Missouri River. Source: Dinan 2003.

construct a shallow nest on sand or gravel substrate, and, between early May and early July, lay eggs and incubate them for about 28 days. First hatching typically occurs in late May or early June but, depending on spring water flow, may extend into late June and July. Renesting can occur if the entire clutch is lost. Chicks require more than 21 days of growth and development after hatching before they are able to fly. After young birds fledge, they continue to feed and mature along the Platte River until they begin autumn migration in August.

Influence of Current Central Platte Conditions on Survival of Piping Plovers on Platte River

Plovers, like all species, have specific requirements for survival. USFWS identifies among the requirements space for population growth and normal behavior; nutritional or physiological requirements (such as food, water, air, light, and minerals); cover or shelter; sites for breeding, reproduction, and rearing of offspring which are protected from disturbance; and habitats representative of the historical distribution of the species (Fed. Regist. 67 (176): 57638 [2002]). Those survival requirements have been used by USFWS to identify the primary constituent elements (PCEs) of critical habitat in accordance with Section 3(5)(A)(i) of the ESA. PCEs for the NGP population of piping plovers are habitat components (physical and biological) essential for the biological needs of courtship, nesting, sheltering, brood-rearing, foraging, roosting, intraspecific communication, and migration (Fed. Regist. 67 (176): 57638 [2002]).

According to the NGP population recovery plan (USFWS 1988) and critical habitat final rule (Fed. Regist. 67 (176): 57638 [2002]), PCEs on the Platte River include sparsely vegetated channel sandbars, sand and gravel beaches on islands, temporary pools on sandbars and islands, and interface zones with the Platte River (Figure 6-6). Those habitat features are directly influenced by dynamic precipitation cycles and longer-term climate patterns. Habitat area, abundance and availability of insect foods, brood and nesting cover, and lack of vegetation are all linked to weather and climate. Variability in flow can cause high rates of turnover of naturally variable sandbars. Flowing water creates diverse habitats for feeding, nesting, and brooding. Habitat variables were quantified by Ziewitz et al. (1992) during the 1988 breeding season and used to prepare USFWS's target species suitable habitat document (USFWS, unpublished material, June 2000). Table 6-1 summarizes habitat characteristics important to the species.

Studies on the habitat requirements of the piping plover across its breeding range have been few. However, the study by Ziewitz et al. (1992) quantified a number of variables (such as channel width, sandbar area, mean nest elevation, and maximal nest elevation) present at the piping

FIGURE 6-6 Area of central Platte River channel near Shelton with many primary constituent elements for the piping plover. Channel sandbars lack woodland cover, temporary pools on bare sand are abundant, and interface zones are plentiful. Source: Photograph by W.L. Graf, May 2003.

plover nesting sites along the Platte River. The effort was an important contribution to understanding of the piping plover habitat requirements along the central Platte River. Using an airboat, Ziewitz's research team traveled the reach of the Platte River from Lexington to its confluence with the Missouri River in the middle of June. They located as many nest sites as possible and (to minimize disturbance to breeding birds) marked the nests for future study. Dates of hatching and egg laying were estimated by floating eggs in a container of water. Age is determined by the position of the egg as it floats in the water; older eggs float higher in the water column (Westerskov 1950). Ziewitz et al. (1992) used aerial videography to record the Platte River over the summer. They analyzed scenes at the 26 sites occupied by plovers and univariately compared physical characteristics of occupied and unoccupied sites. The study showed that more plovers nested on the lower Platte River than the central Platte, birds nested on river segments that were wider and had greater areas of sparsely vegetated sandbars, channels were wider and sandbars higher along the lower Platte River than on the central Platte, nests on the lower Platte had greater clearance above the river's water line when initiated than did nests on the central Platte, and the average elevation of nests was greater on the lower Platte, which suggested that nest habitat was limited on the central Platte. Several other studies (e.g., Ducey 1988; Faanes 1983) attempted to quantify vegetation at piping plover nest sites, documenting that occupied sites tended to be sparsely vegetated or to lack vegetation early in the nesting season. No studies are available that quantify plover habitat characteristics during periods of especially low water flow.

Current information suggests that no suitable piping plover habitat now exists along the central Platte River (Erika Wilson, USFWS, pers. comm., July-September 2003; John Dinan, NGPC, pers. comm., May-September 2003). Flow has been so low that no new sandbars are being created. Sandbars that once were suitable are now unsuitable because no recent scouring has occurred to prevent establishment of vegetation, and many river reaches have been nearly dry in recent years. In 1988, about 25 pairs nested on the central Platte River and in sandpits; Ziewitz et al. (1992) concluded that reproduction by piping plovers was being compromised by low flows that were punctuated by sudden water peaks in July, which led to nest flooding. Lingle (1993a) attributed the July peaks to a combination of heavy local precipitation and releases of water to the river via the Johnson-2 return during local thundershowers. He reported that nest success was higher on the lower Platte River, where sandbar heights are greater than on the central Platte. Plover nesting in 1988-1996 varied from 20 to 42 pairs on the central Platte (1-15 pairs on the river, 13-30 pairs in sandpits). However, since 1996, fewer than five pairs per year have nested on the central Platte. Chick productivity at river sites along the central Platte declined to about 1.3 chicks per pair in 1997 and 1998; since 1999, there has been no successful reproduction.

In addition to riverine habitat, piping plovers nest in commercial sandpits on the central Platte River. Another important adjacent source of alternative habitat is Lake McConaughy to the west. Large numbers of plovers nest along the shoreline of Lake McConaughy when water levels are low. For example, high water elevation, 3,262 ft above mean sea level (msl), along the lake in 1986 resulted in narrow beaches (<25 m) and restricted habitat availability. In May 1991, in contrast, lake elevation was extremely low, 3,245 ft msl, and resulted in beach widths of 400-800 m and abundant nesting habitat (Wingfield 1993). It is important to consider potential habitat at Lake McConaughy because plovers produced at this reservoir may nest along the central Platte River, central Platte plovers may nest at Lake McConaughy (Wingfield 1993), and relationships between nesting along the river and the lake are probably related to water levels and habitat conditions at both locations. Finally, it is important to emphasize that lack of knowledge about population fidelity (tendency to return to same place to nest) and movement on all spatial scales (including both the

TABLE 6-1 Habitat Characteristics Important to the Piping Plover

Piping Plover Habitat	Observed Measurements of Habitat Parameters (OMHPs)	Preliminary Goals for Habitat Management	References
Riverine habitat			
Channel or sandbar characteristics			
Channel width	975-1,554 ft	≥900 ft, initially	Ziewitz et al. 1992 Kirsch 1996
Sandbar area (early June, at nest initiation)	0.03-3.58 acres	Variable	Ziewitz et al. 1992 Kirsch 1996
Mean elevation above 400-cfs stage (early June, at nest initiation)[a]	0.2-2.0 ft (mean, 0.4 ft)	Low, ephemeral sandbars; high enough to provide dry, bare sand during nesting season	Ziewitz et al. 1992
Maximal Elevation above 400-cfs stage (early June, at nest initiation)[a]	0.4-4.4 ft (mean, 2.7 ft)	Low, ephemeral sandbars; high enough to provide dry, bare sand during nesting season	Ziewitz et al. 1992
Vegetation			
Cover: at nest site	<5-20% (early in nesting season)	Sparsely vegetated or unvegetated	Schwalbach 1988 Ducey 1983
On sandbar	<10-25%		Faanes 1983
Height (at nest site)	<2 ft (early in nesting season)	Same as OMHP	Schwalbach 1988 Prindiville-Gaines and Ryan 1988

Forage			
Expanse of wet or moist substrate		—	Corn and Armbruster 1993
Sandpit habitat			
Sand and gravel operation characteristics			
Availability of forage	Within 1 mi	—	Corn and Armbruster 1993 Nordstrom and Ryan 1996
Total surface area	1.5-197 acres (mean, 18.3 acres)	—	Sidle and Kirsch 1993
Water surface area	1-150 acres (mean, 27.4 acres)	—	Sidle and Kirsch 1993

[a]To compare water-surface-elevation datum from different nest sites to a common elevation datum, original data were adjusted to yield elevation measurements as if all central Platte River data were collected at a discharge of 11.3 cms (400 cfs).
Source: USFWS, unpublished material, June 16, 2000.

interior NGP and the Platte River) compromises the ability to determine habitat use by central Platte River plovers.

Several studies have concluded that artificial habitats cannot provide the full complement of essential habitat requirements for piping plovers over the long term and therefore cannot substitute for riverine habitat (Corn and Armbruster 1993; Lingle 1993b; Sidle 1993). For example, prey-capture rates for plovers and densities of invertebrates were higher at river-channel sites than at sandpit sites (Corn and Armbruster 1993). (A similar conclusion was drawn for interior least terns [Sidle 1993; Wilson et al. 1993].) USFWS believes that sandpits do not meet the requirements of critical habitat (Fed. Regist. 67 (176): 57638 [2002]). Although plovers will nest in active sandpits, use of this habitat is temporary and terminates when vegetation becomes too dense. Moreover, water is often distant from sandpits, and adults must travel farther than a mile to forage. Finally, because sandpit sites are not isolated on islands, nests there are more vulnerable to predation. A number of investigators (including Erika Wilson, USFWS, pers. comm., July-September 2003; John Dinan, NGPC, pers. comm., May-September 2003; Lackey 1997 and Plettner 1997 in abstracts from 1997 Platte River Basin Ecosystem Symposium, Kearney, NE, February 18-19, 1997) have reported that despite some limitations, productivity in the sandpits can be high if the pits are intensively managed to exclude predators and human disturbances. No studies have examined whether survival from fledging to first breeding is higher in natural or in artificial habitats.

The contribution of alternative habitat to the survival and recovery of piping plovers can be summarized as follows: sandpits provide refuge and nesting substrate when water is high on the river, but they do not appear to provide the complete array of essential habitat elements required by piping plovers.

The presence of the habitat components required for survival influences whether piping plovers are able to nest, hatch eggs, and raise their offspring to fledging. According to USFWS and NGPC biologists (Erika Wilson and John Dinan), only a few plovers have attempted to nest on the river (as opposed to sandpits and other artificially created habitat) in the central Platte over the last 5 years. All nests apparently have failed to produce young; the failure is attributed to low water flow and lack of dynamic ecosystem processes appropriately timed to create new islands or sandbar habitat. It is believed that if water is too high or sandbars are too low during the breeding season, nest habitat (potentially including eggs and chicks) will be inundated and plovers will be unable to find appropriate locations to breed. If water is too low, because of water management or climatic conditions, scouring does not occur and vegetation develops on sandbars and small islands in the Platte River; in addition, reduced water flow fails to create the new sandbars that are necessary for nesting (Figure 6-7).

FIGURE 6-7 Portion of central Platte River channel near Shelton. Typical low water condition fails to create new sandbars, and existing sandbars are becoming stabilized with vegetation. Source: Photograph by W.L. Graf, August 2003.

The primary causes of breeding failure in the central Platte River for the last decade appear to be nest flooding and reductions in habitat after periods of very low stream flow. Lingle (1993a,b) studied causes of nest failure and mortality of plovers along the central Platte from 1985 to 1990 and reported that the greatest causes of plover nest failure on the river were flooding (61%) and predation (19%). In contrast, predation and abandonment accounted for most (42% and 21%, respectively) failures at sandpits.

In summary, current conditions in the central Platte River appear to be compromising the continued existence—that is, the survival—of the NGP population of the piping plover. Supporting that conclusion are observations that few plovers have attempted to nest on the central Platte River in the last 5 years, no eggs have hatched or chicks fledged in the river during the last 5 years, hydrological conditions necessary for development and maintenance of nesting habitat have not occurred in the central Platte in the last 5 years, and loss of habitat along the river appears to be forcing birds to use alternative sites that are less secure from predators and other sources of disturbance and death and do not provide suitable habitat for reproduction. The loss of central Platte River habitat has reduced breeding options for the already-vulnerable regional population.

Recovery Goals for the Northern Great Plains Piping Plover Population

According to USFWS's (1988) recovery plan for the Great Lakes and NGP piping plover populations, the NGP population can be considered for delisting when four recovery criteria are met: the number of birds in the population reaches 1,300 pairs, essential breeding and winter habitats are protected, the Canadian recovery objective of 2,500 birds for its prairie region is reached, and the 1,300 pairs in the United States are maintained in a specific distribution for 15 years (assuming at least three major censuses during this period). For Nebraska, the recovery goal is 465 pairs; the goal for the Platte River is 140 of those pairs. The NGP recovery plan has been revised twice since 1986, but neither revision has been formally approved by USFWS. The most recent draft revision proposes increasing the recovery goal for NGP from 1,300 to 2,400 breeding pairs (L.C. Wemmer, USFWS, unpublished material, February 11, 2001).

Lutey (2002) reported a 1987-1998 average of 84 pairs of piping plovers breeding on the Platte River including birds nesting in sandpits along the river. If Lake McConaughy birds are included, the average is 119 pairs. Current numbers therefore are about half those needed to meet the recovery goal for the Platte. It is important to note that the Lake McConaughy plovers usually are not included in the recovery goal for the Platte River, because the lake is not considered part of the Platte River (Erika Wilson, USFWS, pers. comm., July-September 2003; John Dinan, NGPC, pers. comm., May-September 2003). The Lake McConaughy birds, however, are included in Nebraska's total number of breeding pairs and can contribute to Nebraska's goal of 465 pairs as part of NGP population recovery. It should be noted here that re-establishment of a more natural Platte River hydrograph alone will not ensure recovery of the piping plover. The continued presence of reservoir and sand pit breeding sites may provide sink habitat situations that draw plover pairs into circumstances where survival of offspring is low, and this in turn could lead to reduced fitness and regional declines in population numbers. Ameliorating direct losses and harassment of birds and nests caused by recreation and other human activities on the river will be a necessary component of recovery. Losses of adults, juveniles, and eggs to predation are likely to increase and lead inevitably to increases in introduced species associated with human activities on the river. And, although the contribution of contaminants to current mortality is unknown, population losses from that source will probably need to be better understood before downlisting or delisting of the species.

Population Viability Analysis for Piping Plovers

Assertions regarding population sizes necessary for recovery can be tested with viability modeling. Population viability analysis (PVA) uses demographic models to predict the probability that a population will go extinct within a specific period under specific circumstances. PVAs are often used to derive estimates of the minimal number of individuals needed to ensure long-term population persistence. They also are used to simulate effects of various environmental phenomena and population characteristics on trends in population size. Three PVAs have been published for the NGP population of piping plovers (Ryan et al. 1993; Plissner and Haig 2000; Larson et al. 2000, 2002). The PVA by Plissner and Haig (2000) found that a 36% increase in productivity (offspring fledged per pair per year) or an 8-10% increase in adult and immature survival rates, was needed to ensure a 95% probability that the population will persist for 100 years. That level of reproduction would require a mean reproductive success of 1.7 chicks/pair per year, a scenario in which the population would decline but would not go extinct. To ensure population stability, Plissner and Haig reported that an annual mean productivity of 2.0 fledglings/pair was required. That assessment has been considered conservative by other biologists because of the low reproduction rate used in the model; see Larson et al. (2000, 2002). However, Plissner and Haig and Larson et al. agree that the NPG metapopulation (a set of local populations connected by migrating individuals) will not persist unless reproductive success is increased substantially.

Conclusions about effects on recovery of piping plovers are essentially the same as reported above for current conditions on survival. Plissner and Haig (2000) suggested that at least 1.7 chicks (preferably 2.0 to provide a buffer against stochastic events) must fledge per pair per year to meet recovery goals for the NGP population; the estimate may be reduced somewhat if mortality decreases. Ryan et al. (1993) suggested 1.13 chicks as sufficient for recovery, but they assumed an unreasonably high rate of juvenile survival; Larson et al. (2000, 2002) report a populationwide fledging rate of 1.1 fledglings/pair per year to stabilize simulated populations. Reproductive success for the central Platte River has been zero for the last 5 years, and it is unlikely that this is being offset by groups of birds elsewhere in the population's range. On a more local scale, relatively high productivity was reported at Lake McConaughy, especially in 2002 and 2003 (1.7 chicks/pair) (Mark Peyton, CNPPID, pers. comm., August 2003). Although Lake McConaughy is not considered part of the Platte River—and, just as important, does not provide secure habitat, because it depends on Kingsley Dam operations—Lake McConaughy plovers currently contribute substantially to the Platte River population. However, without the ability to differentiate individual birds from different source populations, the relationship between birds that nest at Lake McConaughy and birds that use the Platte remains unknown.

The committee performed an elementary PVA on the NGP and Platte River piping plover populations and found, not surprisingly, that losing Platte and Loup Rivers habitat has little effect on the persistence of the

NGP metapopulation. This finding is, however, a result of current low productivity by plovers on the Platte River. That is, when a declining population is removed from a group of stable or declining populations, it has little effect on overall persistence of the group. Indeed if the areas of the Platte and Loup Rivers, in their current conditions, are acting as sink habitats, removing them completely as potential breeding sites could actually increase the persistence probability of the NGP population. However, if habitat management succeeds in increasing habitat value to a point where it allows the proportion of nests that produce offspring to double, the Platte and Loup River population would have a moderate probability of persisting, and in turn again contribute to the regional metapopulation persistence. Current conditions can therefore be viewed as adversely affecting recovery. The analysis was based on a number of assumptions about the spatial structure of piping plovers in the NGP population; before the contribution of birds from the central Platte or the Platte River system can be accurately evaluated, movements of birds among habitat areas and their uses of those areas must be better understood.

Although insufficient data are available to construct a PVA for piping plovers on the Platte River that has a high level of resolution and predictive power, the committee used the VORTEX model 9.3 (Lacy et al. 2003) created by Plissner and Haig (2000) for the NGP to address several circumstances. First, to assess the effect of removing the Platte and Loup River birds from the model, resident birds were removed from the larger NGP population, and then the apparent carrying capacity (number of individuals that can be supported permanently in a given area) of the Platte River was decreased. Second, Platte and Loup River birds were separated from the Nebraska population to examine the effect of reducing plover productivity in the Platte River to zero. Finally, the percentage of nests producing fledglings in the Platte and Loup system was doubled to assess the effect of reducing nest failures.

No information exists regarding the population structure of piping plovers in Nebraska. Consequently, birds were modeled as two populations: one on the Missouri and Niobrara Rivers, and one on the Platte and Loup Rivers. Population recovery goals (USFWS 1988) were used to estimate carrying capacity. Carrying capacity for the Niobrara portion of the Missouri and Niobrara population is 100 birds, and carrying capacity for the Platte and Loup is 610 (Platte, 280; Loup, 330). Counts from 2001 were used to estimate current population sizes as 84 birds on the Niobrara and 216 on the Platte and Loup. Carrying capacity of those rivers is substantially higher than current populations; recovery goals are based on many sources of information (e.g., knowledge of distribution and abundance, survey data, historical population data, and loss of viable habitat) and include assessment of the potential to increase the number of breeding pairs at occupied sites and to establish pairs at unoccupied sites (L.C. Wemmer, USFWS, unpublished material, February 11, 2001).

Although breeding sites are dynamic—they are lost and created—it was assumed that on the average the number of sites and dispersal among sites was consistent across years. As data become available to evaluate that assumption, a more sophisticated PVA can be developed.

Tables 6-2 and 6-3 summarize the input values used for the initial model. Movement among other populations and the Platte and Loup

TABLE 6-2 Summary Input for the Piping Plover PVA Initial Model

	Great Lakes	Manitoba and Lake of the Woods	Northern Prairie	Platte and Loup	Niobrara	Colorado
Carrying capacity	300	250	4,700	610	100	40
Initial population size	48	73	2,786	216	84	13
1st year mortality	56.8	56.8	56.8	56.8	56.8	56.8
Mortality (SD)	11.36	11.36	11.36	11.36	11.36	11.36
Adult mortality	34	34	34	34	34	34
Mortality (SD)	6.8	6.8	6.8	6.8	6.8	6.8
% nests failing	49.68	79.9	51.75	58	58	72.4

Source: Reed 2003.

TABLE 6-3 Migration Rates Among Piping Plover Populations

	Great Lakes	Manitoba and Lake of the Woods	Northern Prairie	Platte and Loup	Niobrara	Colorado
Great Lakes	—	0.001	0	0	0	0
Manitoba and Lake of the Woods	0.001	—	0.01	0	0	0
Northern Prairie	0	0.01	—	0.01	0.01	0
Platte and Loup	0	0	0.01	—	0.02	0.005
Niobrara	0	0	0.01	0.02	—	0.005
Colorado	0	0	0	0.005	0.005	—

Source: Reed 2003.

population and Niobrara population is the same as Plissner and Haig used for the Nebraska populations (half going to each population); migrations between the Platte and Loup population and the Niobrara population used the same rule as Plissner and Haig. Also assumed were no inbreeding depression, no correlation in environmental variance, no catastrophes, no additional mortality associated with dispersal, no density dependence, monogamous mating, males and females breeding at the age of 1 year with a maximum of 10 years, an equal sex ratio, maximal clutch of four eggs, and a stable age distribution. Other model assumptions followed those of Plissner and Haig.

The first model run reduced carrying capacity for the Platte and Loup population (by 280 birds) without altering initial population size. Birds that did not emigrate died. In the second model, carrying capacity was restored, but reproductive success was reduced to zero. The last model also restored carrying capacity, and nest failures were halved (the percentage of nests failing was reduced by half), with proportions of nests producing one, two, three, and four eggs kept the same.

Three key results should be noted: removing the plover carrying capacity of the Platte and Loup Rivers slightly increased the rate at which the NPG metapopulation went extinct (Table 6-4), reducing Platte and Loup reproductive success to zero increased the time to metapopulation extinction and reduced population persistence time in the Platte and Loup Rivers system by two-thirds (from 28 years to 10) (Table 6-4), and reducing nest failure rate by half (increasing the 42% of nests producing at least one fledgling to 84%) had a large effect on population persistence. For the NGP metapopulation, the probability of persisting 200 years increased from 0% to 16.3%. For the Platte and Loup population, persistence probability went from 0% to 93.7%, and the mean time to extinction (when populations went extinct) increased from 28 years to 107 years. When populations persisted, their mean population size at 200 years was 456 (Table 6-4).

The analyses indicate that losing Platte and Loup River habitats has a measurable effect on regional metapopulation persistence. However, because all populations are declining, the effect is slight. If management actions could produce as much as a doubling of the proportion of nests that produce offspring (and the proportions of nests producing one, two, three, and four fledglings remain the same), the Platte and Loup population of plovers has a high probability of persisting. When the proportion of nests producing chicks, for example, is low, it is possible to double the proportion of nests that produce chicks by intensive nest-site management (e.g., nest exclosures, beach closure, and volunteer plover patrols) (Wemmer 2000), as recommended (Larson et al. 2002) for the Great Lakes and NGP populations of piping plovers.

TABLE 6-4 Summary Statistics from PVA Similar to that of Plissner and Haig (2000), with Nebraska Birds Separated into Platte and Loup and Niobrara Populations

	Default	Scenario		
		Platte-Loup Habitat Removed	Zero Reproductive Success in Platte-Loup	Doubled Nests Producing Fledglings in Platte-Loup
Northern Great Plains metapopulation				
Probability of surviving 200 years	0.00	0.00	0.00	0.163
Mean final population size if persisting	—	—	—	116
Mean years to extinction if it goes extinct	27	22	24	40
Population growth rate (r)	–0.167	–0.168	–0.214	–0.123
Niobara population				
Probability of surviving 200 years	0.00	0.00	0.00	0.025
Mean final population size if persisting	—	—	—	3
Mean years to extinction if it goes extinct	22	22	22	23
Population growth rate (r)	–0.144	–0.145	–0.148	–0.114
Platte and Loup Rivers population				
Probability of surviving 200 years	0.00	—	0.00	0.937
Mean final population size if persisting	—	—	—	456
Mean years to extinction if it goes extinct	28	—	10	107
Population growth rate (r)	–0.145	—	–0.232	0.078

Source: Reed 2003.

Evaluation of Science Supporting Piping Plover Critical Habitat Designation

Critical habitat for the NGP breeding population of the piping plover was designated in 2002, including 19 critical habitat units totaling 183,422 acres (74,228 ha) and portions of four rivers totaling approximately 1,208 river miles (1,944 km) (Fed. Regist. 67 (176): 57638 [2002]). One unit (Unit NE-1) was designated that encompasses about 440 mi (707.9 km) of river habitat on the Platte, Loup, and Niobrara Rivers.

The critical habitat designation for the NGP breeding population of piping plovers identified 405.5 km (252 mi) of the Platte River from the Lexington bridge east to the Platte's confluence with the Missouri River. In designating critical habitat, USFWS is required to use the best scientific and commercial data available.

The critical habitat final rule for the NGP piping plover population (Fed. Regist. 67 (176): 57638 [2002]) reports that:

- The best scientific and commercial data available were used.
- Conservation of the NGP breeding population of piping plovers undertaken by local, state, tribal, and federal agencies operating within the range of the species since listing in 1986 was reviewed.
- Steps necessary for recovery of the NGP piping plover population were identified in the NGP plover recovery plan (USFWS 1988).
- Available information pertaining to the habitat requirements of the species (including new material since preparation of the recovery plan in 1988) was reviewed. Specific sources of information included data in reports submitted during Section 7 consultations and by biologists holding Section 10(a)(1)(A) recovery permits, the 1994 technical agency review draft revised recovery plan for piping plovers breeding on the Great Lakes and NGP (USFWS, unpublished material, Dec. 23, 1994), research published in peer-reviewed articles and presented in academic theses and agency reports, annual survey reports, regional geographic information system coverages, and personal communications from knowledgeable biologists.

The committee found that scientific knowledge supported designation of critical habitat for the piping plover on the Platte River by DOI agencies in 2002. USFWS used methods that are comparable with those successfully used for designation of critical breeding habitat for the Great Lakes population of piping plovers. The only major difference between the two efforts was that substantial data on habitat use and breeding biology of nesting piping plovers were not available in the Great Lakes case, but were available in the case of the central and lower Platte River. USFWS's use of the data in designating the Platte River critical habitat resulted in greater confidence in the designation than would be possible without such data (Figure 6-8). Use of the central and lower Platte River by plovers is better known than use of most of the designated critical habitat elsewhere. For the NGP population of piping plovers, the PCEs of critical habitat are the essential habitat characteristics for the biological needs of courtship, nesting, sheltering, brood-rearing, foraging, roosting, intraspecific communication, and migration (Fed. Regist. 67 (176): 57638 [2002]). In the case of the central and lower Platte River, these characteristics are sparsely vegetated channel sandbars, sand and gravel beaches on islands, temporary pools on sandbars and islands, and interface zones with the river (USFWS 1988; Fed. Regist. 67 (176): 57638 [2002]).

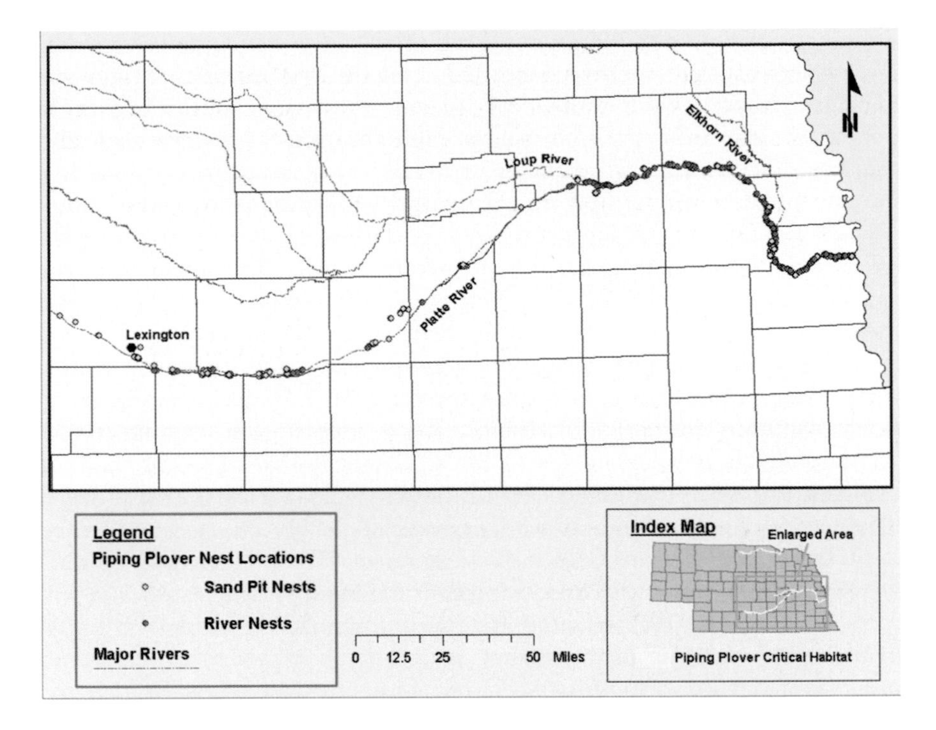

FIGURE 6-8 Piping plover nesting locations along critical habitat section of central Platte, 1987-2003. Source: J. Runge, USFWS, unpublished material, 2004.

The hydrology and geomorphology of the Platte River, driven by drought and flood cycles, create and maintain these habitat components. The water flow, sediment, vegetation, and channel morphology influence the abundance and availability of insect foods and brood and nesting cover. The variability of flow in the channel causes recycling of sediment in ever-changing forms, including sandbars and beaches that form, are destroyed, and reform in new configurations. The dynamic, sandy surfaces of these forms remain little vegetated and provide diverse habitats for feeding, nesting, and brooding. USFWS specified the "presence of dynamic ecological processes" (produced by the river processes) as the overarching PCE—an accurate, reasonable decision supported by published studies and agency reports (e.g., Faanes 1983; Haig 1992; Ziewitz et al. 1992; Lingle 1993a; Fed. Regist. 67 (176): 57638 [2002]).

In addition to the general systematic properties of the central and lower Platte River, USFWS investigated four more-specific characteristics required for designation of critical habitat. The designated habitat must be currently or recently used for breeding, or have a documented history of occupation by breeders, be deemed potential breeding habitat and fall within geographic boundaries of distribution, and be included in habitat complexes essential to the conservation of the species. The committee concluded that USFWS followed a specific process in addressing those points, as reviewed in the following paragraphs, and the committee found that the procedure resulted in scientifically valid designation of critical habitat for the piping plover on the central and lower Platte River.

USFWS began its designation process by using recovery plans (USFWS 1988; USFWS unpublished material, December 23, 1994) to identify the specific recovery needs of NGP piping plovers (including birds nesting on the central Platte River). The plans provided records of geographic nesting locations at the time the documents were prepared and included descriptions with quantitative data on habitat variables at some breeding locations. USFWS obtained and used more recent breeding-survey data to create a complete and current (as of 2001) database that reflected the best available knowledge of historical and current breeding by piping plovers. The database included inputs from census and survey efforts coordinated with USFWS, plover nest records maintained by the Nebraska Wildlife Division every year since 1984, and prelisting records. The result is a picture of plover distributions along the Platte River that constitutes the best available information on the subject. Recent methods for assessing plover use have increased in sophistication, particularly by including GPS, so the later records may be more accurate than the earlier ones. An important decision by USFWS was to designate the entire reach between Lexington and the Missouri and Platte confluence as critical habitat. The alternative was to attempt to identify individual parcels along the reach that had been used in the past by nesting plovers. For some plover nest habitat in North America, that parcel-by-parcel approach is possible, and it has been done to some extent in the Great Lakes region. However, given the ephemeral nature of plover habitat along the major prairie rivers including the Platte, sites are not predictable or stable. Extensive literature on plover habitat justifies the decision to designate the entire reach, but there will be some parcels in the designated area of critical habitat that do not meet all the requirements every year. USFWS (Fed. Regist. 67 (176): 57638 [2002]) correctly pointed out that because of the nature of the northern Great Plains, some of the designated habitats will not have the PCE every year but must have them over time to be considered critical habitat.

The decision to designate the entire reach was a practical solution. USFWS initially declined to designate critical habitat on the central and lower Platte River because it deemed the task too difficult and essentially infeasible. The agency did not designate critical habitat for the interior least tern for the same reasons. In response to legal action, the agency undertook the task at a relatively coarse scale for the Lexington to Missouri confluence segment of the river, but could not resolve a finer resolution down to parcel scale. The committee recognizes that the extensive investigation required to designate critical habitat at the parcel scale was not possible within a reasonable period because recent land-use changes—including the conversion of abandoned sand mines to urban housing, shoreline development around Lake McConaughy, and expansion of the use of the river bed by off-road vehicles—are proceeding too quickly.

INTERIOR LEAST TERN

The least tern (*Sterna antillarum*) is the smallest of North American terns. Adults weigh about 40-45 g and are distinguished from other small terns during the breeding season by a black cap and eye stripe, prominent white forehead, gray back and wings, and white underside. The least tern's bill is orange or yellow and has a dark tip (Figure 1-5). The rarity of the least tern and loss of its nesting habitat have led the USFWS to designate as endangered, threatened, or a species of concern across most of its range.

Distribution

Least terns breed along coastal beaches and major interior rivers of North America. They winter on marine coastlines of Central America and South America (Thompson et al. 1997). The size of least tern populations are difficult to estimate accurately because of the wide geographic distribution of the species and a weak understanding of population limits. Thompson et al. (1997) estimated that at least 55,000 individuals could be found across all breeding sites in the United States during the 1980s and 1990s. Birds nesting on the Platte River are part of an "interior population" that was listed as endangered nearly 2 decades ago (Fed. Regist. 50: 21784 [1985]).

Historical distribution and nesting records exist for interior least terns in Nebraska since the 1800s (EA Engineering, Science and Technology, Inc. 1988; Lutey 2002). Before listing, nesting and migration records were reported for 56% of the counties in Nebraska, but estimates for the size of the statewide population are not available until USFWS listed the population as endangered in 1985.

Historically, the interior least tern bred along the Colorado, Red, Rio Grande, Arkansas, Missouri, Ohio, and Mississippi river systems in a vast region covering much of the mid-continent region of the United States (Thompson et al. 1997). In this region, interior least terns often nest near piping plovers and for this reason are thought to have similar habitat requirements. Although the boundaries of the breeding range of least terns in the 1990s were similar to its boundaries a century earlier, the distribution of least terns in that area is now much more fragmented than previously, especially distribution of the interior population (Thompson et al. 1997). Kirsch and Sidle (1999) examined the status of the interior population after listing and reported that in 1995 the rangewide population exceeded the recovery goal of 7,000 pairs, primarily because of a tripling of tern numbers along the lower Mississippi River. They concluded that the increased population size was not a result of increased reproduction but probably reflected both more-thorough surveys and possible immigration of terns from the Gulf Coast. Kirsch and Sidle (1999) noted that the number of breeding pairs in most interior locations, including the Platte River, had not reached recovery goals.

Breeding terns in Nebraska typically co-occur with piping plovers on sandbars, on reservoir shorelines, in commercial sand mines, and at other artificially created sites along three major rivers (Figures 6-1, 6-2, and 6-3). In the northeastern portion of the state, nesting occurs along about 64 km (40 mi) of the upper Missouri River shared with South Dakota and along 153 km (95 mi) of the lower Niobrara River. Farther south, terns are found along 386 km (239 mi) of the central and lower Platte River from Lexington downstream to the confluence with the Missouri River. Breeding also occurs at Lake McConaughy in western Nebraska and on the Middle Loup and Loup Rivers in central and eastern Nebraska (Figure 6-4). Missouri River habitat is shared with South Dakota and population estimates for Nebraska typically do not include the Missouri River (John Dinan, NGPC, pers. comm., May-September 2003).

The structure of the tern population using the Platte River habitat is partly determined by the movement and distribution of birds during their lifetimes. Lingle (Lingle 1993c) examined the movements of interior least terns that had hatched along the Platte or that were first encountered as adults nesting along the river (1993c). From 1984 to 1989, Lingle banded 704 birds at locations between Lexington and Grand Island. He found that 29% of the terns banded as adults returned to their banding site, 26% returned to the site where they had hatched, and the rest died or moved to other locations. Emigrating Platte River terns were observed as far away as Stafford, Kansas: a tern chick banded in 1987 was found nesting at Quivira National Wildlife Refuge in 1990. An adult banded at Quivara was recorded on a nest near Kearney (dispersal distance, 170 mi or 274 km). Birds banded on the central Platte were found on the lower Platte and vice versa. The study strongly indicates that the interior population is open on spatial scales beyond that of the central Platte River, and that terns from the Platte River are capable of moving substantial distances between breeding seasons. Tern use of the Platte River is seasonal. Interior least terns arrive on the Platte River in early to middle May. After arrival, they locate mates, construct shallow nests on sand or gravel substrate, and lay eggs that are incubated for about 20 days. Egg-laying occurs from late May to middle

July, depending on river levels and the ability of birds to locate suitable habitat. Birds renest if an entire clutch is destroyed. First hatching typically occurs in June but can extend to late July. Chicks are able to fly after about 20 days. After fledging, young birds feed and mature in the Platte River ecosystem until autumn migration in late July to early September. A mated pair of interior least terns, along with their offspring, remain on the Platte for about 4 months.

Influence of Current Central Platte Conditions on Survival of Interior Least Terns on Platte River

All species, including interior least terns, have specific requirements for survival, including space for population growth and normal behavior; nutritional or physiological requirements (such as food, water, air, light, and minerals); cover or shelter; sites for breeding, reproduction, and rearing of offspring; and habitats that are protected from disturbance or representative of the historical geographic and ecological distributions of the species (Fed. Regist. 67 (176): 57638 [2002]). Ensuring that those requirements are met is a key focus of conservation efforts. In addition to habitat for nesting, interior least terns require adjacent habitat for foraging. Unlike plovers, interior least terns do not obtain food in their nesting territories.

To fledge young along the Platte River, interior least terns require bare or nearly bare alluvial islands or sandbars, favorable water levels during the nesting season, and food. Island habitat is critical because nests that are on islands have lower rates of predation than nests on the mainland. Interior least terns forage in open river channels with pooled or slow-flowing water less than 15 cm deep. In general, terns forage within 0.8 km (0.5 mi) of riverside nest sites, but they may forage up to 2.4 km (1.5 mi) from sandpit colonies. Habitat variables were quantified by Ziewitz et al. (1992) and used to prepare a description of suitable habitat (USFWS, unpublished material, June 16, 2000) (Table 6-5).

A number of independent investigators have studied breeding habitat of least terns across their broad geographic range in North America (Thompson et al. 1997). Several studies have focused on interior least terns in Nebraska (e.g., Sidle and Kirsch 1993; Ziewitz et al. 1992; Kirsch 1996; Ducey 1988; Faanes 1983; Corn and Armbruster 1993). One study quantified environmental variables at interior least tern nest sites along the Platte River (Ziewitz et al. 1992). That effort focused on both terns and plovers and is widely cited in documents from which water-management recommendations were established for the central Platte River. Study methods for interior least terns were similar to those described above for piping plovers.

Using aerial videography, Ziewitz et al. (1992) obtained records of the river during the summer of 1988. They analyzed scenes at 26 sites occupied

TABLE 6-5 Habitat Characteristics Important to the Interior Least Tern

Interior Least Tern Habitat	Observed Measurements of Habitat Parameters (OMHPs)	Preliminary Goals for Habitat Management	References
Riverine habitat			
Channel and sandbar characteristics			
Channel width	975-1,554 ft	≥900 ft, initially	Ziewitz et al. 1992 Kirsch 1996
Sandbar area (early June, at nest initiation)	0.3-3.58 acres	Variable	Ziewitz et al. 1992 Kirsch 1996
Mean elevation above 400-cfs stage (early June, at nest initiation)[a]	0.2-2.0 ft (mean, 0.4 ft) above stage at 400 cfs	Low, ephemeral sandbars; high enough to provide dry, bare sand during nesting season	Ziewitz et al. 1992
Maximal Elevation above 400-cfs stage (early June, at nest initiation)[a]	0.4-4.4 ft (mean 2.7 ft) above stage at 400 cfs	Low, ephemeral sandbars; high enough to provide dry, bare sand during nesting season	Ziewitz et al. 1992
Vegetation			
Cover: at nest site On sandbar	<5-20% (June and July) 28% (May and June)	Sparsely vegetated or unvegetated	Schwalbach 1988 Ducey 1988 Faanes 1983
Height (at nest site)	<2 ft (May and June)	Same as OMHP	Schwalbach 1988 Ducey 1988
Forage			
Proximity to forage fish	0-2 mi Typically within 1 mi	Within 1 mi	Smith and Renken 1990 Faanes 1983
Size of forage fish	2-10 cm	Same as OMHP	Atwood and Kelly 1984 Dugger 1997

Sandpit habitat			
Sand and gravel operation characteristics			
Proximity to river and forage	0-2 mi	—	Smith and Renken 1990 Sidle and Kirsch 1993 Wilson 1991
Total surface area	1.5-197 acres (mean, 18.3 acres)	—	Sidle and Kirsch 1993
Water surface area	1-150 acres (mean, 27.4 acres)	—	Sidle and Kirsch 1993

[a]To compare water-surface-elevation datum from different nest sites to a common elevation datum, original data were adjusted to yield elevation measurements as if all central Platte River data were collected at a discharge of 11.3 cms (400 cfs).

Source: USFWS, unpublished material, June 16, 2000.

by plovers and terns and compared physical characteristics of occupied and vacant sites (every 3 mi along a river transect). The study showed that more terns nested on the lower Platte River than on the central Platte and that birds nested on river segments that were wider and had larger areas of unvegetated or sparsely vegetated sandbars than on vacant sites. Channels were wider and larger and sandbars higher on the lower Platte River than on the central Platte, and nests on the lower Platte had greater clearance above the river water line when initiated than did those on the central Platte. The average elevation of nests was higher than that of vacant sites available on the lower Platte, but lower than that of vacant sites available on the central Platte River. On the basis of those results, Ziewitz and colleagues concluded that habitat availability may be limiting populations on the central Platte River. Current reports describe riverine habitat as nonexistent along the central Platte (Erika Wilson, USFWS, pers. comm., July-September 2003; John Dinan, NGPC, pers. comm., May-September 2003). Recent flows are so low that new sandbars are not being created and old sandbars have become unsuitable because no scouring has occurred to prevent vegetation from colonizing. In 1988, about 60-65 pairs of interior least terns nested on the central Platte River; Ziewitz et al. (1992) contended that later reproductive success and nesting activity were compromised by low flows, which were punctuated by sudden water peaks in July that flooded nests. Tern nesting efforts between 1988 and 1996 varied between about 60 and 90 pairs per year on the central Platte River. From 1997 to 2001, the number of pairs attempting to nest ranged downward from 45 to 15. From 2001 to 2003, the number of pairs on the river was below 12. Since 1991, interior least tern productivity on the river has been no greater than two chicks fledged per pair; no chicks fledged in six of those years, and none during the last five.

Reproductive success of interior least terns appears to be influenced primarily by water-flow regimes. The birds require sandbars or small islands that are relatively free of vegetation and are at sufficient elevation for eggs and chicks not to be flooded. Sandbars and islands also provide isolation from mammalian predators. The relationship between water levels and site occupancy is well documented by numerous authors throughout the breeding range of least terns in North America. In a study by Lingle (1993b), predation and flooding accounted for 74% of nest failures in Platte River habitat. Predation was the greatest source of nest failures in sandpits, followed by human disturbance and weather.

USFWS and NGPC collect and maintain data on interior least tern nesting. Methods used to monitor terns in the river are the same as or similar to those used by biologists elsewhere in the range of the species in North America. Protocols for monitoring breeding pairs, estimating productivity, and reporting results have been formalized. Since the interior population of the least tern was listed, nesting records for the Platte River have been among the most consistently collected and maintained. Kirsch and Sidle (1999) reviewed the status of the interior population and stated that "methods, level of effort, and degree of coordination that biologists use on the upper Missouri, the Platte, and lower Mississippi rivers are needed throughout the interior breeding range"; that supports the idea that data used to recommend conservation and management strategies for interior least terns on the Platte River have been drawn from appropriate scientific approaches. Because piping plovers and interior least terns typically co-occur on the Platte River, the species are surveyed at the same time, and similar methods are used, as described earlier in this chapter.

Our findings are essentially the same as those reached for the piping plover. According to discussions with USFWS and NGPC biologists (Erika Wilson and John Dinan, respectively), who manage the interior least tern database and survey for nesting birds, only small numbers of terns have attempted to nest along the central Platte River over the last 5 years. All have failed to produce young. Failure is attributed to low water flow and lack of dynamic ecosystem processes appropriately timed to create new island and sandbar habitat without flooding nesting pairs. Therefore, current conditions on the central Platte River appear to be compromising the continued existence (survival) of the interior population of least terns. That conclusion was based on the following observations: few terns have attempted to nest on the central Platte River in the last 5 years, no eggs have hatched or chicks fledged along the river during the last 5 years, water levels deemed necessary to produce nesting habitat have not been reached on the central Platte in the last 5 years, and loss of habitat along the river is forcing birds to use alternative sites that tend not to offer adequate protection from predation and flooding and do not provide habitat conditions needed for survival of adults and chicks. In addition, loss of central Platte River habitats from the geographic range of the interior population of least terns reduces breeding options for an already vulnerable regional population.

Influence of Current Central Platte Conditions on Recovery of Interior Population of Least Terns

For many threatened and endangered species for which a recovery plan has been prepared, a population goal is set that must be reached before the species or population can be removed from the federal list. Additional criteria are usually stipulated for recovery. For the interior population of the least tern, the recovery plan (USFWS 1990) states that "in order to be considered for removal from the endangered species list, 1) interior least tern essential habitat will be properly protected and managed and

2) populations will have increased to 7,000 birds." The population goal is broken down into five river systems—Missouri River, Mississippi and Ohio Rivers, Arkansas River, Red River, and Rio Grande. The Platte River is included in the Missouri River system, for which specific recovery guidelines have been promulgated: the number of birds in the Missouri River system must increase to 2,100 adults; essential breeding habitat will be protected, enhanced, or restored; and breeding pairs will be maintained in a specific distribution (listed by state and river) for 10 years (assuming that a minimum of four censuses have been conducted). The recovery goal is 1,520 adults in Nebraska, including 750 adults along the Platte River. Lutey (2002) reported the average numbers of interior least tern pairs at sites along the Plate River from 1987 to 1998—South Platte River (one), North Platte (Lake McConaughy) (six), upper Platte (13), central Platte (74), and lower Platte (187)—for a total of 550 individuals, not counting six pairs at Lake McConaughy. From 1991 to 2001, interior least tern nesting on the central Platte River decreased by 47% (John Dinan, NGPC, pers. comm., May-September 2003).

Like recovery of the piping plover, least tern recovery will require more than the return of a more favorable Platte River hydrograph. Human activities not directly associated with habitat loss, including harassment of adults and young, and destruction of nests also contributed to reducing local populations. Losses of least terns, their offspring, and nests to predation from natural and exotic sources will need to be evaluated and most likely mitigated in some form. Also, as in the case of plovers on the Platte River, the role of contaminants in compromising breeding success needs study to assess its effects on least tern populations.

Population Viability Analysis for Least Terns

Two PVAs have been completed for least terns: one for birds in the Platte River (Boyce et al. 2002), and one for the California least tern (Akçakaya et al. 2003). The models are completely different in structure and sensitivity to altered vital rates. The model by Boyce et al. (2002) determines extinction probability from time-series data on population sizes, so the only parameters for the model are variance in population growth rate (as determined by changes in population size across years), initial population size, and carrying capacity. In contrast, the model by Akçakaya et al. (2003) is a matrix-based, stochastic simulation model of a metapopulation that uses RAMAS and includes age-specific vital rates.

The Boyce et al. model predicted a high probability of persistence of least terns on the central and lower Platte River—an almost 100% probability of persisting in at least one location for 100 years. In addition, it found that increased connectivity between the populations increased persistence probability. At least two problems, however, plague this type of PVA approach. First, sensitivity to altered vital rates cannot be evaluated, because data entering the model come solely from population size. How much immigration contributes to persistence, for example, cannot be assessed, even though it could be essential to population persistence. Second, there is some concern that this type of model is not able to predict persistence very far into the future. Work by Fieberg and Ellner (2000) and Ludwig (1999) suggests that even with perfect knowledge of population size, estimates of extinction probabilities based on this approach are accurate only to 20% of the length of the time series. That means that the 11 years of least tern data used by Boyce et al. might validly predict only 2 years into the future. The authors' recognition that the central Platte River birds contribute substantially to the joint persistence of the populations is correct in the context of the model, but the model makes implicit assumptions about each of the local populations, for example, that Platte River birds are self-replacing (the population is not maintained in part by immigration), which probably is not valid. Consequently, it is difficult to determine what this model offers for predicting the contribution of Platte River birds to the persistence of their own population, or that of the regional population.

The model by Akçakaya et al. (2003) is designed for terns in California but is more useful in a Platte River application. The model treats clusters of colonies or individuals within 5 km (3.1 mi) of one another as elements of the same population. It includes age structure, year-to-year changes in survival and fecundity, regional catastrophes (such as strong El Niño-Southern Oscillation events), and local catastrophes (reproductive failure due to predation). The period used to assess risk of extinction and substantial population declines is 50 years. Lacking local data, they used data on dispersal from a Massachusetts population (Atwood 1999) and included spatial and temporal autocorrelation. Although Akçakaya et al. had no evidence of density dependence in reproductive success, they assumed that reproductive success declined when populations were at high and low population densities. The high-density effect is programmed to occur near carrying capacity, and an Allee effect occurs when a colony has five or fewer females. Sensitivity analysis was carried out by rerunning the model using low, medium, and high values for parameters and quantifying the effects on population viability. The model predicted that during normal (noncatastrophe) years, the finite rate of increase (λ) is ~0.994, which converts to a slow annual decline (-0.6%). In a revised model, $\lambda = 1.07$, so slow average population growth was predicted. Using medium values of all parameters, the model predicts zero risk of extinction or substantial decline in the next 50 years. Using lower vital rates, the model predicts a higher risk of decline but a low risk of least tern extinction in California over the next 50 years. The model was sensitive to survival and fecundity rates and moderately sensitive to carrying capacity. It appears that reproductive success, philopatry (tendency to return to hatching site to nest), and site fidelity are all greater for California birds than for Platte River birds.

A PVA commissioned by the committee sheds some light on several points that are important for the recovery of Platte River interior least terns. Analyses suggest that the least tern interior population is likely to persist for 200 years and that the Platte and Loup River birds contribute minimally to its persistence. Reproductive success is zero or very low on the Platte, so birds breeding on the Platte are probably coming from outside Nebraska. Therefore, Platte River habitat is serving as a "sink"; the population there is not replacing itself with local reproduction and could be drawing breeding birds into habitat not suitable for reproduction. Those conclusions, however, are based on a hypothetical spatial structure of least tern populations; before population persistence and the importance of the Platte River birds to the larger interior population can be accurately evaluated, the movements and habitat use of terns in the interior population must be understood.

The committee's conclusions about effects on survival of interior least terns and piping plovers were similar. As in the case of the PVA for the piping plover, the committee used available data to ask several basic questions about the persistence of the Platte River interior least tern population. Again, it is important to note that data are not sufficient to describe in terms of parameters for a PVA with robust predictive capacity. Nonetheless, the committee built a simple three-population model for the interior United States: Niobrara River (eventually adding the Missouri River birds to this population would be reasonable), Platte and Loup Rivers, and all birds outside Nebraska. The model allowed the committee to consider the fate of the interior population after excluding Platte River birds, assuming a completely isolated population of 920 adult birds. The committee also considered the fate of Platte River terns with and without immigration and assuming increased reproductive success.

For the least tern PVA, the committee again used VORTEX (9.3). There is no information about the population structure of interior least terns in Nebraska (or, for that matter, the interior United States). Although breeding sites are dynamic—they are lost and created—it had to be assumed that on the average the number of sites and dispersal among sites were consistent. As data become available to evaluate that assumption, a more sophisticated PVA can be developed. Consequently, the PVA is exploratory, so the model and its results should be viewed as hypotheses to be tested.

The model used initial population sizes (N) and carrying capacity (K) from the recovery plan (Table 6-6). Survival data came from Thompson et al. (1997) or from California populations (Table 6-7). All populations are

TABLE 6-6 Summary Input for Interior Least Tern PVA Initial Model

	Interior United States Excluding Platte and Loup Rivers and Niobrara River Populations	Platte and Loup Rivers	Niobrara River
Initial population size (K)	7,000	920	200
Initial carrying capacity (N)	5,900	700	200

Source: Reed 2003.

assumed to have the same life-history parameters: age at first breeding, 2 years; longevity, 20 years; maximal clutch size, three eggs; adult survival, 85%; survival from fledging to age 1 year, and from age 1 year to age 2 years, 56% (Akçakaya et al. 2003); environmental variance for reproduction and survival, 20% of the parameter value; no inbreeding depression; no catastrophic environmental events; no change in carrying capacity over time; and no cost to dispersal. It appears that the percentage of nests that fail each year on the Platte River is increasing, so the base model was run with all populations experiencing a 20% failure rate, then with the Platte

TABLE 6-7 Some Parameter Values for California Least Tern PVA (Akçakaya et al. 2003) and Parameter Values from Platte River Data

Model Parameter	California PVA	Platte River Data
Reproduction (fledglings/female per year)	0.6964	0.47 (Kirsch 1996 as cited in Thompson et al. 1997)
Survival, hatching to fledging (normal years)	0.6237	0.504 (Smith and Renken 1993)
Survival, hatching to first breeding (normal years)	0.16	Unknown
Survival, from 1 to 2 years	0.5627	Unknown
Survival, from 2 to 3 years, and 3 to 4 years	0.81	Unknown
Survival after age 4 years	0.92	Unknown
Philopatry	>50%	26% (Lingle 1993c, in Thompson et al. 1997)
Site fidelity	70%	28% (Lingle 1993c, in Thompson et al. 1997)

Source: Reed 2003

and Loup River population experiencing increasing failure rates. Fledglings per female surviving per year is lower than in California (Table 6-7) at 0.47 (presumably, this includes nest failures) (Kirsch 1996 as cited in Thompson et al. 1997). The fledgling survival per female as reported by Kirsch is very low; to duplicate this low estimate in the model output required use of high nest failure rates in the model. It was assumed that of the birds whose reproductive efforts did not fail, 70% had one offspring, 28% had two, and 2% had three. Those assumptions generated predicted rates of reproductive success higher than observed (just over one fledgling per nest), so the model was run with increased failure rates. In addition, the base model was run with the Platte and Loup population and carrying capacity removed from the model.

Dispersal rates are unknown, but philopatry and site fidelity appear to be lower in Nebraska than in Massachusetts or California (Akçakaya et al. 2003), so it was assumed that dispersal in Nebraska is higher than in Massachusetts or California. Lower site fidelity is expected in Nebraska because riverine sandbars are less stable than coastal beaches. High immigration from the greater interior metapopulation would inundate model results, so it was held to 1%. Migration between Nebraska populations was set at 30%; it was set at 10% between Nebraska and other interior sites. The model was run with a variety of dispersal values and with the nest failure rate set at 20%.

The current recovery plan has a goal of 920 adults in the Platte River and Loup River combined. A PVA assuming a completely isolated population of this size was run with the base-model parameters, first in a version with immigration and then in a version of the model that assumed increased reproductive success. To model immigration, the population was supplemented with 10 birds (five adult males and five adult females) each year. In a nonimmigration model, nest failure rate was decreased to 10% (from 20%); in another scenario, failure rate was sustained at 20%, but it was assumed that of the birds that did not fail, 50% had one offspring, 42% had two, and 8% had three. Finally, a last scenario used variances in reproductive success and survival that decreased to 10% (from 20%).

Most scenarios run with the model showed poor least tern population persistence probability (Table 6-8); all were characterized by multiple extinction and recolonization events among smaller populations.

The Platte and Loup River population contributed substantially to the persistence of the greater interior metapopulation as it was modeled. That was demonstrated by the increased extinction probability associated with increasing nest failure in the Platte and Loup population and after removing the population and its carrying capacity entirely (Table 6-8). Model results indicate that the interior metapopulation is more adversely affected by

TABLE 6-8 Results of the Platte and Loup Habitat Removed Scenario PVAs, Varying Nest Failure Rate on Platte and Loup Rivers, and Removing Birds of Platte and Loup Rivers and Carrying Capacity (Fixed Dispersal Rate of Interior United States into Nebraska of 0.0005 and Between Nebraska Populations of 0.01)

Scenario	% Nest Failure in Platte and Loup Rivers	Metapopulation			Platte and Loup Rivers			Niobrara River		
		P(ext)[a]	Size[b]	Mean Time[c]	P(ext)[a]	Size[b]	Mean Time[c]	P(ext)[a]	Size[b]	Mean Time[c]
(1) Base	20	0.729	216	114	0.757	57	108	0.745	51	107
	40	0.788	232	110	0.808	56	103	0.811	52	103
	60	0.840	184	104	0.861	45	95	0.855	41	95
(2) Platte and Loup Rivers gone	20 (overall failure)	0.761	280	112	—	—	—	0.803	52	100
(3) Base, 20% failure	Dispersal[d]									
	1	0.729	216	114	0.757	57	108	0.745	51	107
	3	0.658	111	127	0.665	59	123	0.669	55	123
	5	0.342	96	125	0.287	68	125	0.414	64	128

[a]Probability of extinction.
[b]Mean N at 200 years for those which lasted.
[c]Mean extinction time (years) for those which went extinct.
[d]Immigration rate from interior United States into Nebraska; fixed rates out of Nebraska (10%), and between Nebraska populations (30%).
Source: Reed 2003.

Platte and Loup habitat serving as a sink than by an outright loss of Platte and Loup River habitat. That is reinforced when immigration from outside Nebraska is increased, causing a drain from the larger regional population to the smaller, sink population. The same pattern was found by Plissner and Haig (2000) for piping plovers.

Under current conditions, an isolated population of 920 adults in the Platte River and Loup River combined would have a poor probability of persisting (Table 6-9). Increasing reproductive success substantially did not have much effect on persistence, but decreasing variance in reproductive success and survival improved persistence (Table 6-9). Under the last scenario, populations almost always persisted, and population sizes tended toward 80% of carrying capacity, on the average, after 200 years. Adding immigrants equivalent to an influx of 10 adults, or 1% of the local carrying capacity, improved persistence probabilities almost to zero probability of extinction (Table 6-9).

Under the simulated model structure and conditions, the least tern metapopulation is unlikely to persist for 200 years. The Platte and Loup River population contributes to the persistence of the metapopulation but is a negative factor if reproductive success is very low and immigration into Nebraska is high. Clearly the model is sensitive to variance in reproductive success and survival, so it is important to determine the best estimates for those parameters.

Immigration into the population also had a considerable effect in increasing population persistence. That supports the need to understand the spatial structure of the tern populations in this area.

TABLE 6-9 PVA Results for Single Platte and Loup River Population with Different Amounts of Immigration

Scenario	P(ext)[a]	Size[b]	Mean time[c]
Isolated, base model	0.973	107	78
10 adult immigrants/year	0.000	781	—
Decrease nest failure to 10%	0.921	193	84
Reduced partial brood loss[d]	0.854	235	88
Last two combined	0.795	363	91
Same, but variance in reproduction and survival decreased to 10%	0.003	786	123

[a]Probability of extinction.
[b]Mean N at 200 years for those which lasted.
[c]Mean extinction time (years) for those which went extinct.
[d]50% had one offspring, 42% had two, and 8% had three.
Source: Reed 2003.

SUMMARY AND CONCLUSIONS

This chapter has reviewed the necessary background to address three questions regarding the piping plover and interior least tern populations of the central and lower Platte River. First, the committee found that habitat conditions on the central and lower Platte River affect the likelihood of survival and recovery of the NGP population of piping plovers. The decline in the river's plover population has been coincidental with the loss of its preferred habitat, especially in the central Platte River, where suppressed variability in flow has led to reductions in sandbars and beaches and indirectly to increased woodland and reduced open sandy areas. The piping plover population along the Platte River has consistently declined since 1996, and improvement in its numbers is likely to be closely tied to habitat conditions on the river that respond to hydrological adjustments. Breeding along the central Platte River is mostly in artificially created habitats that are not sustainable on a multidecade basis. Recovery requires a reversal of present trends by rejuvenation of a more natural regime of river flows, sediment processes, vegetation, and channel morphololgy.

Second, the committee found that, for the same reasons, the current habitat conditions on the Platte River are likely to affect the likelihood of survival and recovery of the interior least tern. The interior least tern has habitat preferences that are highly similar to those of the piping plover, so what affects one bird population affects the other. Because terns seek sandy, beach-like habitats and because these habitats are becoming less common along the Platte River as a result of flow, sediment, vegetation, and morphology changes, the tern population associated with the river is far below recovery goals. That sufficient suitable habitat stimulates a healthy tern population is demonstrated by components of the least tern population in other subunits of its range.

Third, the committee found that the USFWS designation of critical habitat for the piping plover is based on the best available scientific knowledge at the time of designation (2002). The agency followed reasonable and established procedures in the designation process, drawing on information from peer-reviewed journals and unpublished information, including agency reports and surveys. It consulted with practicing biologists who had specific useful information. USFWS adopted a prudent strategy in deciding on a general designation from Lexington to the confluence with the Missouri River in the absence of substantial time and support to refine the designation to a finer, parcel-by-parcel scale.

As in investigations related to the whooping crane, research into issues related to piping plovers and interior least terns revealed some gaps in data and knowledge. The integrated habitat use by plovers and terns shows the need for knowledge about how multiple species use the same habitat space and how they interact with other listed and nonlisted species. There is also a need to integrate the Platte River populations of plovers and particularly of terns with their larger populations. Platte River terns, for example, appear to range widely (up to at least 170 mi away from the river), so understanding of the Platte population cannot focus solely on the river. Finally—as is the case with conclusions related to time trends in climatic data, hydrological data, and crane population numbers—conclusions related to trends in the populations of plovers and terns must take into account the longest periods possible rather than focusing on a few years of record.

7

PALLID STURGEON

This chapter addresses the question, Do current habitat conditions in the lower Platte River affect the likelihood of survival and limit (adversely affect) the recovery of the pallid sturgeon? The segment of the lower Platte River of interest in this question extends downstream from the confluence of the Elkhorn River with the lower Platte to the mouth of the lower Platte at its confluence with the Missouri River.

Sturgeons (family Acipenseridae) comprise an ancient group of fishes that are found in the northern hemisphere. Some species are anadromous, and others are freshwater; all spawn in freshwater rivers and streams. The tail fin is heterocercal (having unequal lobes), and the vertebrae are cartilaginous. All species have five prominent rows of bony scutes on the body that run from behind the head to the tail. Four barbels are on the ventral surface of the snout in front of the protrusible mouth.

Three species of sturgeon have been collected from the Platte River. The most common and widespread is the shovelnose sturgeon (*Scaphirhynchus platorynchus*). It is found in large rivers throughout the Mississippi River and Missouri River drainages. It is now confined in most years to the lower Platte and its tributaries upstream to the downstream-most dam or obstruction. However, before construction of mainstream dams, it was collected as far west as Casper, Wyoming, in the North Platte River (Baxter and Stone 1995). The lake sturgeon (*Acipenser fulvescens*) has a wide distribution from the Mississippi River and St. Lawrence River drainages north into Canada but has rarely been collected from the lower

Platte River and is uncommon in the Missouri River bordering Nebraska. In July 2000, a specimen was captured near Schuyler, Nebraska; that constitutes the furthest upstream documented locality in the Platte River. The pallid sturgeon was described from specimens captured in the Mississippi River near Grafton, Illinois, by Forbes and Richardson (1905). Specimens lack the spiracle found in the genus *Acipenser*, are light in color, and have a caudal peduncle that is fully armored and laterally compressed (Figure 1-5). The breast or belly of the pallid sturgeon is naked; this distinguishes it from the shovelnose sturgeon, whose belly is covered with scale-like plates. Keenlyne et al. (1994) measured shovelnose and pallid sturgeon from the upper Missouri River drainage and found overlap in morphometric characters between the two species. Sheehan et al. (1999) developed an index with meristic and morphometric characters to distinguish pallid and shovelnose sturgeon. Differences between species of the genus *Scaphirhynchus* have been debated over the last several years (Phelps and Allendorf 1983), but recent genetic studies are providing valuable insight into the evolutionary relationships of these species and have confirmed the genetic distinction between pallid and shovelnose sturgeon (Campton et al. 2000; McQuown et al. 2000; Simons et al. 2001; Birstein et al. 2002). There is no way to distinguish male and female pallid sturgeon visually.

A common denominator in the distribution of sturgeon species appears to be access to flowing freshwater, at least for spawning and initial development. That is especially true for members of the genus *Scaphirhynchus* (also known as river sturgeons), which spend virtually their whole lives in riverine environments. Sturgeons are long-distance travelers. Tagging studies have documented movements of well over 100 km. Sturgeon eggs are demersal and adhesive. Shortly after hatching, the larvae become buoyant and drift with the current.

SPECIES DISTRIBUTION

Pallid sturgeon were found in the Mississippi River from near Keokuk, Iowa, downstream to the Gulf of Mexico, and in the Achafalaya River in Louisiana. The main part of their range is the Missouri River, from its confluence with the Mississippi River upstream to Fort Benton, Montana (Figures 7-1 and 7-2). In addition, pallid sturgeon have been documented from downstream reaches of several major tributaries of the Missouri River. In Kansas, they are known only from the lower 65 km of the Kansas River, mostly during the time of floods. In Nebraska, they have been captured up to 46 km upstream of the mouth of the Platte River. In Montana, their distribution extends 113 km up the Yellowstone River. Damming, water diversions, flood control, and channelization have modified rivers throughout much of its range (Keenlyne 1989). Today, the distribution of pallid

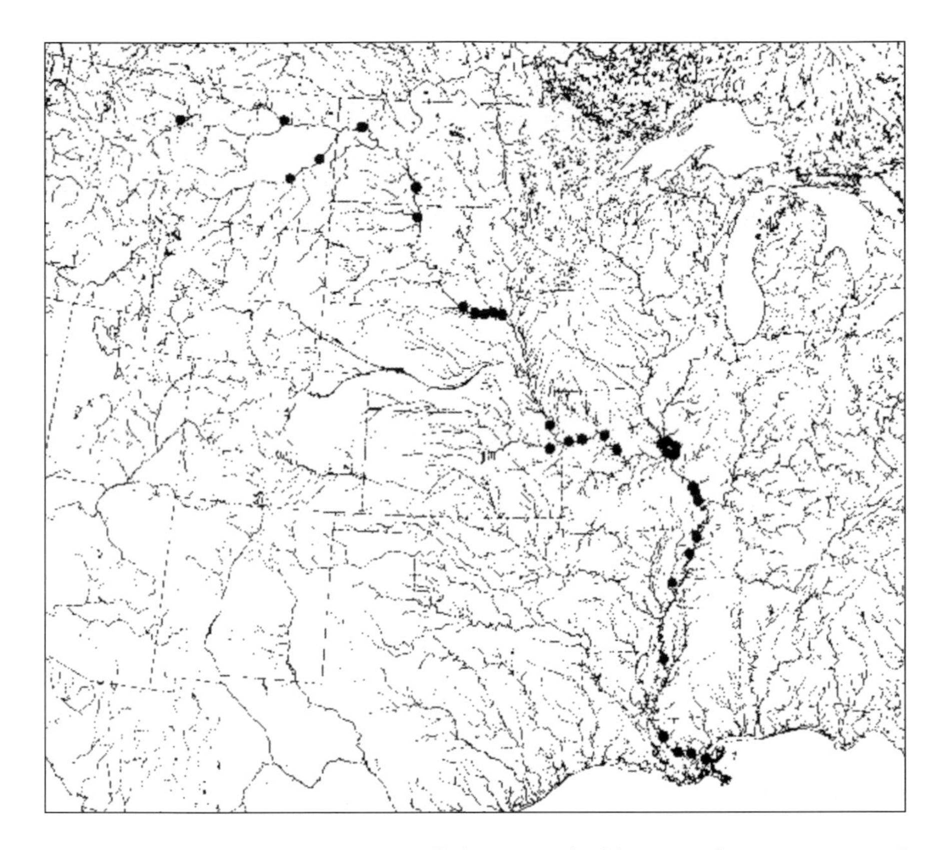

FIGURE 7-1 Map showing where pallid sturgeon had been caught in Missouri and Mississippi Rivers up to 1980. Each dot represents one location. Source: Lee 1980. Reprinted with permission; copyright 1980, North Carolina State Museum of Natural Sciences.

sturgeon in the Missouri River drainage is disjunct, with a northern subpopulation in Montana that is associated with the Yellowstone River confluence and another subpopulation downstream of the Gavins Point Dam near Yankton, South Dakota. In this lower section of the Missouri River drainage, the Platte River is the only tributary from which pallid sturgeon have been captured regularly in the last 20 years.

Pallid sturgeon were probably more abundant in the past, but no accurate records have been tabulated. At the time when pallid sturgeon were described, Forbes and Richardson (1905) stated that the ratio of pallid to shovelnose sturgeon in the catch at Grafton, Illinois, was 1:500, but at the mouth of the Missouri River the ratio was 1:5. Carlson et al. (1985) collected 4,355 sturgeon, of which 11 were identified as pallid sturgeon, for a ratio of 1:396. Watson and Stewart (1991) found one pallid sturgeon among

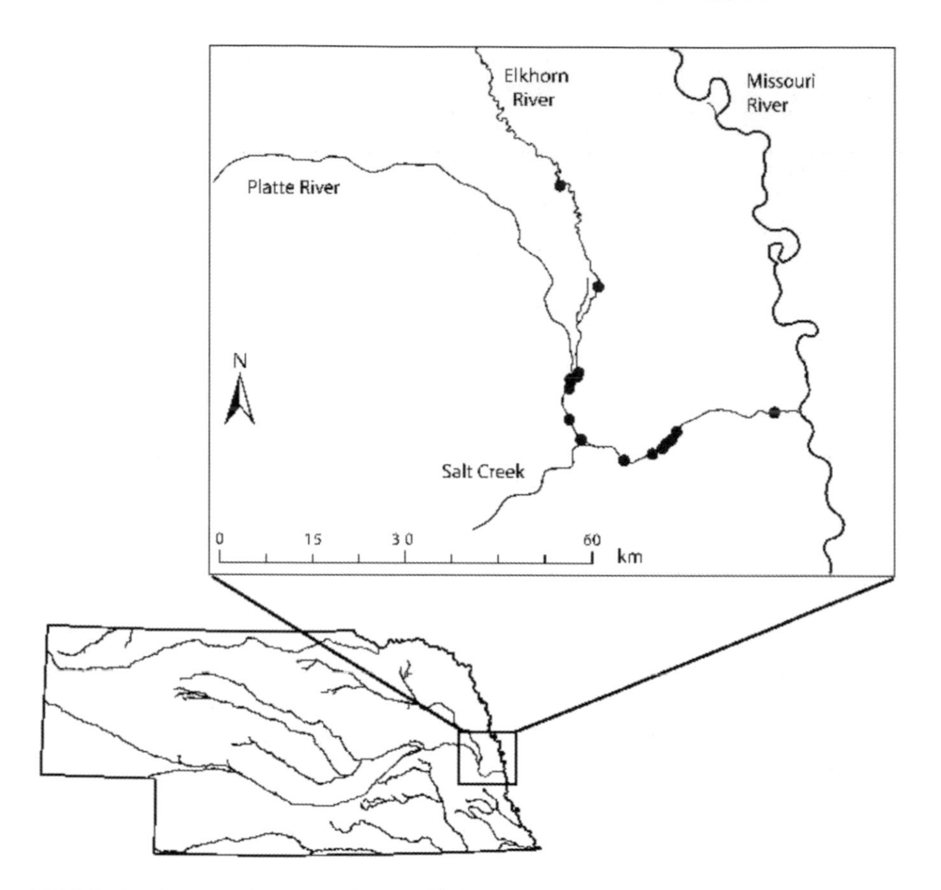

FIGURE 7-2 Map showing where pallid sturgeon had been caught in Platte River and its tributaries in 1979-2003. Each dot represents one location. Source: D. Feit, Nebraska Game and Parks Commission, unpublished data, 1979-2003.

350 sturgeon captured in the Yellowstone River. In comparison, continuing studies in the Platte River during 2000-2003 have captured four pallid and 929 shovelnose sturgeon, for a ratio of 1:232.

Factors contributing to the decline in abundance of pallid sturgeon are diverse and in some cases incompletely documented. The illegal harvest of adult sturgeon for their eggs is probably a serious drain on the breeding stock.

In the Platte River, catches of pallid sturgeon are uncommon but occur regularly. In 1979-2002, anglers reported catching pallid sturgeon in the Platte and Elkhorn Rivers. Most of the catches have been in the reach of the Platte River from the mouth upstream to about the mouth of the Elkhorn River (Table 7-1, Figure 7-2). As of 2003, no wild pallid sturgeon have been

TABLE 7-1 Angler Reports of Pallid Sturgeon Catches from Platte and Elkhorn Rivers, Nebraska

Date	Location	Length, Weight
May 10, 1979	I-80 bridge	37 in, 6.25 lb
May 25, 1993	1 mi below Elkhorn River	35 in, 6.5 lb
April 15, 1995	NE Hwy 50 bridge	36 in, 8-10 lb
May 10, 1997	Elkhorn River mouth	41 in, 6.25 lb
May 25, 1997	0.5 mi below Elkhorn River	—, 6.5 lb
June 9, 1997	US Hwy 6 bridge	36 in, —
May 25, 1998	Elkhorn River mouth	18 in, —
May 22, 1999	1 mi E. Ak-Sar-Ben Aquarium	42.5 in, —
May 30, 1999	Louisville, NE	—, —
September 5, 1999	Elkhorn River 3 mi N. NE Hwy 36	30 in, —
May 23, 2002	Elkhorn River 3 mi N. NE Hwy 91	36 in, —

Source: D. Feit, Nebraska Game and Park Commission, unpublished data, 1979-2003.

documented in the Platte River upstream of the mouth of the Elkhorn River confluence.

In addition to those fish, three wild and one hatchery-reared pallid sturgeon have been captured by researchers on the Platte River. The wild fish were caught in the vicinity of the Nebraska Highway 50 bridge at river kilometer (RK) 26 on May 3, 2001, and May 23, 2002, and near the US Highway 75 bridge at RK 10 on April 3, 2003. They were tagged with passive integrated transponder tags, implanted with radiotransmitters, and released back into the Platte River. They were later tracked. The hatchery-reared fish was captured on May 22, 2003, at RK 25; it had been stocked in the Missouri River at Bellevue, Nebraska, in 2002.

REPRODUCTION AND POPULATION TRENDS

There have been no direct observations of reproduction by pallid sturgeon in the wild, but movement patterns and collection of ripe females suggest that they spawn during June or July in South Dakota and as early as March in Louisiana. Spawning in most sturgeon is not an annual event, and current evidence suggests that pallid sturgeon males spawn on a 2- to 3-year cycle and pallid sturgeon females on a 3- to 5-year cycle (Keenlyne and Jenkins 1993).

Successful spawning by pallid sturgeon in the lower Platte River has not been recorded. No spawning areas have been documented, and no trend data exist for pallid sturgeon, but *Scaphirhynchus* sp. larvae have been collected from the lower Platte River. Until recently, collections of the larvae had been rare; until 1999, no larvae had ever been collected in the wild. In August 1999, several pallid sturgeon larvae were collected at the

Big Muddy Wildlife Refuge in Missouri; where they were spawned has not been determined. In 1996, 1998, 1999, 2000, 2001, and 2002, sturgeon larvae less than a day after hatching (protolarvae) were collected from the Platte River near Ashland, Nebraska, during late May and early June. They were in all likelihood shovelnose sturgeon larvae, but the fact that sturgeon are hatching in the Platte River provides clues to the potential sites of pallid sturgeon spawning. Although there is no direct evidence of pallid sturgeon spawning in the lower Platte River, a female carrying a large quantity of eggs was captured on May 3, 2001, and a radiotransmitter was implanted in her. She remained in the lower Platte River for more than a month before moving back to the Missouri River in early June 2001. Another pallid sturgeon tagged on May 23, 2002, showed no evidence of reproductive products; it moved consistently downstream and left the lower Platte River within 6 days. During both years, the downstream movements occurred within a week of the time when sturgeon larvae were collected in the lower Platte River. These larvae were too small for their species to be accurately identified. In spring 2003, netting on a gravel bar in the Mississippi River captured 44 adult sturgeon, including one pallid sturgeon. After that collection of adult fish, sturgeon larvae were collected at the same site (Robert Hrabik, Missouri Department of Conservation, pers. comm., March 18, 2004).

The lower Platte River seems to be the tributary of the Missouri River most likely used for spawning; no recent records of pallid sturgeon in other major tributary streams, such as the Kansas River, exist. That seems to be borne out by the observation that most collection takes place during the spring and early summer, when spawning is expected. However, the rarity of these animals, the turbid water, and the sparseness of our knowledge of their biology in the wild make the likelihood of observing their reproduction small. In addition, all wild-caught radio-tagged pallid sturgeon have been caught in April or May; they have moved out of the Platte River by early June. High water temperatures and loss of connectivity during years of low discharge may be important limiting factors.

Systematic sampling for pallid sturgeon commenced in May 2000. Before that, fish sampling was sporadic and concentrated on techniques that were more suited to collection of catfish and other species. Therefore, there are few data on which to base an assessment of population trends.

HABITAT COMPONENTS REQUIRED FOR SURVIVAL

Pallid sturgeon are described as fish of large turbid rivers (Cross and Collins 1995; Harlan and Speaker 1951; Bailey and Allum 1962; Lee 1980). The areas where they are most regularly found include areas with many islands (Bramblett and White 2001), the mouths of tributaries, and the downstream ends of islands and sand bars where currents converge (Hurley 1999; Snook et al. 2002). Pflieger (1997) describes their habitat as areas that exhibit strong current and firm substrate along sand bars and behind wing dikes. Studies in the upper Missouri River by Bramblett and White (2001), the Mississippi River by Hurley (1999), and the lower Platte River by Snook et al. (2002) describe the diversity of general conditions under which pallid sturgeon can live but indicate several common threads of habitat requirements.

Depth

Depths used by pallid sturgeon vary with river system and range from less than 1 m in the Platte River to 12 m in the Mississippi. Table 7-2 summarizes depth use for pallid sturgeon habitat by river studied. Snook (2001) and Snook et al. (2002) found that the hatchery-reared pallid sturgeon in the Platte River used depths of 0.15-2.75 m; the average was 0.84 m. Swigle (2003) found that wild pallid sturgeon caught in the Platte River used water depths that averaged 1.29 m. Bramblett (1996) and Bramblett and White (2001) found that pallid sturgeon in the upper Missouri and Yellowstone Rivers used depths that averaged 3.3 m (range, 1-7 m), and Hurley (1999) found that pallid sturgeon in the Mississippi River used depths of 6-12 m. Erickson (1992) studied habitats used by pallid sturgeon in Lake Sharpe, South Dakota, and stated that they were found in water with an average depth of over 4 m in this lake environment.

Catches of pallid sturgeon from the tail race of the Missouri River at Fort Peck Dam, Montana, came from depths of 1.2-3.7 m (Clancey 1990). Watson and Stewart (1991) captured them at 0.6-14.5 m in the upper Missouri and Yellowstone Rivers, Montana, and Constant et al. (1997) captured them at an average of 15.2 m in a constructed channel of the Atchafalaya River in Louisiana. Wild pallid sturgeon caught in the Platte

TABLE 7-2 Depth Use of Pallid Sturgeon Documented with Telemetry

Study Location	Average Depth (Range), m	Reference
Missouri and Yellowstone Rivers (Montana)	3.3 (0.6-14.5)	Bramblett (1996), Bramblett and White (2001)
Mississippi River (Illinois)	— (6-12)	Hurley (1999)
Missouri River at Lake Sharpe (South Dakota)	(<5kg) 4.62 (1.52-6.71) (>5kg) 4.66 (1.22-10.4)	Erickson (1992)
Platte River (Nebraska)	0.84 (0.15-2.75)	Snook (2001),[a] Snook et al. (2002)[a]
Platte River (Nebraska)	1.29 (0.58-2.71)	Swigle 2003

[a]Hatchery-reared fish.

River, Nebraska, during 2001 and 2002 were captured in water that was 0.7 and 1.5 m deep, respectively (Ed Peters, University of Nebraska, unpublished material, May 2, 2001, and May 23, 2002).

Water Velocity

General descriptions of current velocity often use terms like *strong* (Lee 1980) and *swift* (Carlson et al. 1985; Pflieger 1997), but these refer to the appearance of the river from the surface. In a laboratory study of swimming endurance, juvenile pallid sturgeon were able to sustain swimming at 0.25 m/s and attain burst speeds of up to 0.7 m/s (Adams et al. 1999). It seems likely that pallid sturgeon are using low-velocity microhabitats near the substrate of higher-velocity river reaches. To describe the water velocities experienced by pallid sturgeon quantitatively, both column velocities and bottom velocities have been used to characterize pallid sturgeon habitats. Column velocities tend to be more variable, as would be expected over the wide range of water depths where pallid sturgeon occur. Bottom velocities, which more accurately describe conditions of the sturgeon near the substrate, generally are less variable. Table 7-3 summarizes column velocities and bottom velocities from several studies across the range of the pallid sturgeon.

Water velocity measured during collection of pallid sturgeon was 0.46-0.96 m/s (Clancey 1990) in the Missouri River tail race of Fort Peck Reservoir in Montana. In the Platte River, Nebraska, wild pallid sturgeon were captured in habitats with a mean column velocity of 0.82 m/s in 2001 and a range of 0.38-0.65 m/s in 2002 (Ed Peters, University of Nebraska, unpublished material, May 2, 2001, and May 23, 2002). Bottom velocity at those sites was 0.31 m/s in 2001 and ranged from 0.12 to 0.36 m/s in 2002.

TABLE 7-3 Mean Column Velocities and Bottom Velocities at Pallid Sturgeon Sites in Telemetry Studies

Study Location	Column Velocity (Average), m/s[a]	Bottom Velocity (Average), m/s	Reference
Missouri and Yellowstone Rivers (Montana)	0.0-1.55 (0.90)	0.0-0.70	Bramblett (1996), Bramblett and White (2001)
Lake Sharpe (South Dakota)	0-0.73 (0.40)	0-0.55 (0.18)	Erickson (1992)
Platte River (Nebraska)	0.03-1.26 (0.69)	0.17-0.97 (0.38)	Snook (2001),[b] Snook et al. (2002)[b]
Platte River (Nebraska)	0.43-1.28 (0.86)	0.28-0.84 (0.58)	Swigle (2003)

[a]Average velocities measured in water column at point 0.6 × water depth for water less than 1 m deep or by averaging velocity measurements at 0.2 × and 0.8 × water depth for water depths greater than 1 m.
[b]Hatchery-reared fish.

Other Factors

Sand appears to be the substrate of choice for pallid sturgeon, in contrast with the preference of shovelnose sturgeon for gravel substrates (Bramblett and White 2001). The most common feature of the substrate that characterizes pallid sturgeon habitat is an irregular bottom contour (Bramblett and White 2001; Hurley 1999; Snook et al. 2002), which occurs at the downstream ends of sunken sand bars and open channels with dunes (Swigle 2003).

Pallid sturgeon seem to tolerate a wide range of temperatures, 0-33°C in the Platte River, Nebraska (temperatures averaged 24.8°C during summer 1998 and 20.6°C during summer 1999). That is consistent with the range of temperatures noted by Dryer and Sandvol (1993).

Little is known of the dissolved-oxygen requirements of pallid sturgeon, but they are generally found in areas where dissolved oxygen is considered good for most species (>5 mg/L).

Pallid sturgeon seem to prefer "excessively turbid waters" (Lee 1980) and avoid areas that lack turbidity (Bailey and Cross 1954; Erickson 1992). In the lower Platte River, suspended-solids concentrations where hatchery-reared pallid sturgeon have been found ranged from 56 to 1,000 mg/L (Snook 2001). Suspended-solids concentrations at locations where wild pallid sturgeon were captured in the lower Platte River range from 313 to 359 mg/L (Swigle 2003). Turbidity may act as a component of cover for pallid and shovelnose sturgeon. Studies in the Platte River (Snook 2001; Swigle 2003) found sturgeon away from objects, such as submerged logs, that afford cover to many species of riverine fishes (Peters et al. 1989), whereas studies in low-turbidity sections of the Missouri River have noted use of such areas by telemetry-tagged pallid sturgeon.

Rivers where pallid sturgeon are found are often high in dissolved solids and high in conductivity, but little is known of their specific requirements for these conditions. Similarly, virtually nothing is known about their tolerance of pesticides and other pollutants. Ruelle and Keenlyne (1993) examined tissues from three pallid sturgeon and found a variety of organic pesticides but documented no specific effects. Current studies by the U.S. Fish and Wildlife Service and the U.S. Geological Survey are assessing the health of shovelnose sturgeon in the Platte River.

Biologic Interactions

Pallid sturgeon occur in rivers that support a wide variety of other fish species, but the habitat conditions that they seem to prefer probably limit their direct associations to species that inhabit the same microhabitats. It is generally accepted that pallid sturgeon are piscivorous as adults (Coker 1930; Herb Bollig, U.S. Fish and Wildlife Service, pers. comm., 1998), but there have been no supporting quantitative field studies. Therefore, we surmise that smaller fishes that are found in habitats with pallid sturgeon are important food resources. In the Platte River, fishes that use the habitats where pallid sturgeon have been found include speckled chubs (*Macrhybopsis aestivalis*), sturgeon chubs *(M. gelida*), and a host of other minnow and sucker species.

Methods for Determining Habitat Components

The methods used to determine habitat use by fishes in general have centered on measuring a variety of habitat characteristics in areas where the species of interest has been collected or observed. It is particularly challenging for species, like pallid sturgeon, that seem to prefer turbid water, in which direct observation of behaviors, such as feeding and spawning, is unlikely or nearly impossible. Therefore, biologists need to depend on focused collection techniques and telemetry. Recent advances in sonar technology are affording additional methods for imaging fish in turbid river habitats.

CURRENT PLATTE RIVER CONDITIONS

The lower Platte River still maintains the braided channel pattern with abundant areas of converging currents that pallid sturgeon seem to select. The shifting sand has been noted as a preferred substrate by pallid sturgeon. Patches of gravel substrate may afford habitats for spawning during the spring warmup of the Platte River and allow spawning earlier in the season than in the Missouri River. On the basis of surveys, the lower Platte appears to afford pallid sturgeon usable habitat up to the vicinity of the Loup Power Canal confluence near Columbus, Nebraska. Flows upstream of this point are too unpredictable to provide reliable habitat, and diversion structures impose barriers to access. Even though conditions in the lower Platte may be appropriate for pallid sturgeon use, these areas of acceptable or even preferred habitat may be isolated by sections of inhospitable habitat during periods of low flow, such as those of the drought in 2002 and 2003. The section of the lower Platte downstream of the mouth of the Elkhorn River appears to retain most of the appropriate habitat conditions and the connectivity that reliably allows use by pallid sturgeon. Historically, those conditions may not have posed as much of a problem for the pallid sturgeon and other riverine fishes; today, channelization and damming of the Missouri River have depleted pallid sturgeon habitats throughout its former range. Pallid sturgeon have been captured in the Elkhorn River, but they must traverse the lower Platte to reach it. There are no other known habitats for pallid sturgeon in tributaries of the lower Platte River system or in nearby Missouri River tributaries. For those reasons, the lower Platte River may be even more important for the recovery of pallid sturgeon.

RECOVERY OF PALLID STURGEON

The pallid sturgeon recovery plan (Dryer and Sandvol 1993) has as its goal the downlisting of the species from endangered to threatened, or to delisting by 2040. Because of the large and fragmented range of the pallid sturgeon, the recovery plan identifies six recovery priority management areas (Figure 7-3). The mouth of the Platte River is included in recovery priority management area 4. Although no specific population target densities have

FIGURE 7-3 Locations of priority management areas 1-6 for recovery of pallid sturgeon. Source: Dryer and Sandvol 1993.

been identified, a population structure that includes at least 10% sexually mature females is part of the recovery goal. It is proposed that this goal be accomplished by a combined effort that uses protection from harvest, protection of habitat, and supplemental stocking of hatchery-reared fish as major features.

The purpose of the planned efforts to stock hatchery-reared pallid sturgeon is to make up for the apparent lack of reproduction of the native populations. The parentage of the released offspring is wild fish that were captured from and released back into their native areas after 1 or 2 years of spawning. The offspring are reared for the shortest time possible to retain their natural, wild behavioral characteristics and because larger fish are expensive to feed. An additional consideration in how long they are reared is whether they can accept tags that allow them to be identified as hatchery-reared fish, which would distinguish them from non-hatchery-reared stocks. Hatchery-reared pallid sturgeon have not been released into the Platte River since the late 1990s; 75 were released in 1998 (of which 10 were radio-tagged) and 25 in 1999 (of which 15 were radio-tagged). All releases since then have been in the Missouri River. Evidently, some fish stocked in the Missouri River are finding their way into the lower Platte River.

The committee views the use of hatcheries in restoration programs as a last resort for an extremely depleted population, as did the recent National Research Council report on Atlantic salmon in Maine (NRC 2004c). That report reviewed recent literature on the genetic and other hazards that can be imposed by hatcheries used to rehabilitate depressed salmonid populations; it also described protocols that can reduce some of the adverse effects, although some genetic hazards cannot be avoided by any hatchery program. Much has been learned about salmonid hatchery programs, especially in the Pacific Northwest, Canada, the United Kingdom, and Scandinavia, and research continues there. The life history of sturgeon differs from that of salmon in important ways, but much of the experience with salmon hatcheries is applicable to any diploid, heterosexual organism. Any hatchery program for sturgeon should be thoroughly informed by that experience.

The current range of conditions in the Platte River seems to attract pallid sturgeon to this river in numbers at least as great as those in the rest of its range in recovery-priority management area 4. Pallid sturgeon have been documented as moving long distances during their lives, so it seems likely (though not proved) that they are attracted to the Platte River. Because of low population densities and rather recent initiation of regular sturgeon sampling, it has not been determined what role the Platte plays in the ultimate recovery of pallid sturgeon in the middle and lower Missouri River population. Today, the lower Platte receives water from the central Platte and three main tributaries—the Loup system, the Elkhorn system, and the Salt Creek system. Discharge from the Loup system is modified by irrigation withdrawals and peaking power generation at Columbus, Nebraska. Those fluctuations and runoff events from the Elkhorn River, Salt Creek, and smaller tributaries result in irregular rises in discharge in the lower Platte that may be important for its use by pallid sturgeon; many riverine fishes require rising flows to initiate spawning activities. The tributaries also contribute fine sediments to the river.

IMPORTANCE OF THE LOWER PLATTE RIVER TO PALLID STURGEON

Critical habitat has not been designated for pallid sturgeon. However, recovery-priority management areas have included the mouth of the lower Platte River and adjacent portions of the Missouri River on the basis of the preponderance of collections of pallid sturgeon in the region. Research has documented use of the lower Platte by pallid sturgeon during April, May, and June in the last 3 years.

The question of the importance of Platte River flows for the existence of the pallid sturgeon is still problematic. On the one hand, the lower Platte River accounts for only a small fraction of the distribution of the remaining area available to the pallid sturgeon of the lower Missouri population. On the other hand, the lower Platte is the only tributary in this reach where pallid sturgeon are regularly captured, and the captures coincide with conditions and times that correspond to the presumed reproductive requirements of the pallid sturgeon. Although there is no recorded documentation of pallid sturgeon reproduction in the lower Platte River, it seems to be the tributary of the lower Missouri River most likely to have the combination of discharge, substrate, and channel morphometry necessary for spawning to occur. There is recent evidence that sturgeon are spawning on sand and gravel bars in the lower Missouri River and the Mississippi River.

As for habitat conditions, the lower Platte seems to afford pallid sturgeon the turbid, shifting sand substrate that was apparently typical of most of the Missouri River before the imposition of the Pick-Sloan impoundments that stabilized river flows and reduced sediment loads. The major habitat differences between the lower Platte and the Missouri may be in water depth and volume of flow. Those differences may limit the usefulness of the lower Platte River as overwintering habitat for pallid sturgeon. During periods of low flow and high temperatures, the Platte may exceed their thermal tolerance. However, it is important to recognize that the pallid sturgeon using the lower Platte River are part of a much larger population that may include all the Missouri River from Gavins Point Dam downstream to the Mississippi River and possibly beyond. Efforts that are under way to mitigate the impacts of the channelization that occurred in the

Missouri River over the last 60 years may eventually lessen the importance of the lower Platte River for the recovery of pallid sturgeon in this part of its range. Today, however, the lower Platte habitat is important for the recovery of the pallid sturgeon by virtue of their relative abundance in the vicinity of the mouth of the Platte River in the Missouri River.

It is clear that the lower Platte River by itself can be only a part of the overall recovery of the lower Missouri River population of the pallid sturgeon. Continued efforts to expand appropriate habitats for spawning, overwintering, development, and maturation of pallid sturgeon that will increase the overall population are urgently needed. Furthermore, recovery of pallid sturgeon and down-listing or delisting of the species will depend on population responses to the introduction of hatchery-reared fish (dependence on hatcheries to sustain populations that reproduce inadequately essentially ensures continued listing status), on control of contaminants in key habitats, and on assurance that accidental overharvest does not affect the known populations that apparently are naturally sparse.

SUMMARY AND CONCLUSIONS

The endangered population of pallid sturgeon is exceptionally small and apparently declining. Much of its original habitat in the Missouri River has been altered by dams and their reservoirs. The lower Platte River remains the habitat with a flow regime most similar to the original, unaltered habitat of pallid sturgeon. The braided channel of warm, sediment-rich waters with shifting sandbars and islands and a sandy substrate offers suitable habitat that serves as a refuge for pallid sturgeon and a possible starting point for expanding the population. For those reasons, the committee concluded that current habitat conditions in the lower Platte do not adversely affect the likelihood of survival or recovery of the pallid sturgeon. The loss of lower Platte River habitat would probably result in a catastrophic reduction in the pallid sturgeon population. Any recovery effort for the pallid sturgeon will of necessity include the lower Platte River.

Questions about the biology of the pallid sturgeon and the role of the lower Platte River in its recovery are as follows.

What is the connection between pallid sturgeon that use the lower Platte River and those found throughout the middle and lower Missouri River? Knowledge of the connection could help to answer the question of the importance of the Platte for the recovery of the pallid sturgeon in this section of its range.

How does the year-to-year variability of the discharge in the Platte River affect the reproduction of the pallid sturgeon? Concerted studies on the pallid sturgeon in the lower Platte River started in 2000, and the Platte has since experienced some of the lowest discharge volumes on record.

Continuing the current studies for another 5-10 years may allow an evaluation of how pallid sturgeon respond to higher flows. In addition, evaluation of the recovery of any animal with a maturation time of 10-plus years that reproduces on a 3- to 5-year cycle requires longer-term studies.

How do proposed modifications along the Platte River affect flow patterns and instream habitat in the river? The habitat in the lower Platte River needs to be evaluated with models that more accurately predict channel responses to altered river management. Such prediction requires intensive studies of the hydrological relations in shifting sand-bed streams.

8

CONCLUSIONS AND RECOMMENDATONS

In the previous chapters, the Committee on Endangered and Threatened Species in the Platte River Basin has explored science and its application for policy on the central and lower Platte River. The committee presents here its responses to the series of questions (reviewed in Box 1-2) included in its charge. In this chapter, for each question, we state our conclusions and the primary sources of evidence leading to them.

To reach its conclusions, the committee considered the extent of the data available for each question and whether the data was generated according to standard scientific methods that included, where feasible, empirical testing. The committee also considered whether those methods were sufficiently documented and whether and to what extent they had been replicated, whether either the data or the methods used had been published and subject to public comment or been formally peer-reviewed, whether the data were consistent with accepted understanding of how the systems function, and whether they were explained by a coherent theory or model of the system. To assess the scientific validity of the methods used to develop instream-flow recommendations, the committee applied the criteria listed above, but focused more directly on the methods. For example, the committee considered whether the methods used were in wide use or generally accepted in the relevant field and whether sources of potential error in the methods have been or can be identified and the extent of potential error estimated. The committee acknowledges that no one of the above criteria is decisive, but taken together they provide a good sense of the extent to which any conclusion or decision is supported by science. Because some of the decisions in question were made many years ago, the committee felt that it was important to ask whether they were supported by the existing science at the time they were made. For that purpose, the committee asked, in addition to the questions above, whether the decision makers had access to and made use of state-of-the-art knowledge at the time of the decision.

The population viability analysis (PVA) developed by the committee was constrained by the short study period. It did not include systematic sensitivity analyses and did not base stochastic processes and environmental variation on data from the Platte River region. A more thorough representation of environmental variation in the Platte River could be developed from regional records of climate, hydrology, disturbance events, and other stochastic environmental factors. Where records on the Platte River basin itself are not adequate, longer records on adjacent basins could be correlated with records on the Platte to develop a defensible assessment of environmental variation and stochastic processes. In addition, a sensitivity analysis could demonstrate the effects of wide ranges of environmental variation on the outcomes of PVAs. In its analysis, the committee did not consider methods and techniques that are under development by researchers such as the new SEDVEG model. SEDVEG is being developed, but is not yet completed or tested, by USBR to evaluate the interactions among hydrology, river hydraulics, sediment transport, and vegetation for application on the Platte River. The committee did not consider USGS's in-progress evaluation of the models and data used by USFWS to set flow recommendations for whooping cranes. The committee did not consider any aspects of the Environmental Impact Statement that was being drafted by U.S. Department of the Interior (DOI) agencies related to species recovery, because it was released after the committee finished its deliberations. The Central Platte River recovery implementation program proposed in the cooperative agreement by the Governance Committee also was not evaluated, because it was specifically excluded from the committee's charge.

The committee's experience with data, models, and explanations led us to the identification of three common threads throughout the issues related to threatened and endangered species. First, change across space and through time is pervasive in all natural and human systems in the central and lower Platte River. Change implies that unforeseen events may affect the survival or recovery of federally listed species. Land-use and water-use changes are likely in the central and lower Platte River region in response to market conditions, changing lifestyles, shifts in the local human population, and climate change; such changes will bring about pressures on wildlife populations that are different from those observed today. For example, riparian vegetation on the central Platte River has changed because of both natural and anthropogenic impacts. Regardless of its condition and distribution before European settlement in the middle 1800s, the riparian forest of the central Platte River was geographically limited from the middle 1800s to the first decades of the 1900s. At the time of the first aerial photography of the river in 1938, extensive sandbars, beaches, and braided channels without extensive forest cover were common in many reaches of the central Platte. Between the late 1930s and the middle to late 1960s, woodland covered increasing portions of the areas that had previously been without trees. By the late 1990s, clearing of woodlands had become a major habitat-management strategy to benefit whooping cranes that desire open roosting areas with long sight lines. Whooping cranes have used the newly cleared areas, but the overall effects of clearing on the crane population and on the structure of the river are not completely known. As with most habitat-management strategies in the central Platte River, there has been no specific monitoring to assess the success of clearing. Unintended effects remain to be investigated.

From a planning and management perspective, stable conditions are desirable so that prediction of outcomes of decisions can be simplified; but stability is rare, especially in the Platte River Basin. Explanations of existing hydrological, geomorphologic, and biological conditions and predictions of future conditions that fail to discern and accommodate change are not likely to be successful. Science can inform decision makers about expected outcomes of various choices, but prediction of the outcomes is likely to be imprecise because of ecosystem variability. Management choices therefore must include some flexibility to deal with the inevitable variability and must be adaptive, continually monitoring and adjusting. The conditions our parents would have seen in these ecosystems a half-century ago were not the conditions we see now, and present conditions are not likely to be the ones our children or grandchildren will see.

A second thread identified by the committee is that one's view of an ecosystem depends on the temporal and spatial scales on which it is examined. The variability in scale of processes in smaller drainage basins nested within larger ones is obvious, but most natural systems have a similar nested hierarchical structure. The groups of birds and fish that use the Platte River Basin are a fraction of the larger, more widely distributed population, so conditions along the river affect only a portion of each population at any time. Loss of the subpopulations that use the Platte River might not damage the entire population if there were no losses elsewhere—something that Platte River managers cannot assume. The concentration of listed species along the central Platte indicates the importance of the river, despite the fact that the birds can be found elsewhere in Nebraska during migration or nesting periods. The river is important from a management perspective because it contains all the habitat features that are included in the regulatory definitions of critical habitat.

The river supplies the needs of an assemblage of species in addition to serving the needs of single species.

Climate also operates on a series of hierarchical scales. Regional climate in the central and northern Great Plains evinces a variety of changes that depend on the time scale used for analysis. Over a period of 5 or even 10 years, we do not see the complete range of temperature and rainfall conditions likely to be experienced over a century. Decades-long drought or wet periods are likely to be important in species survival and recovery, so short-term observations of less than a few years cannot illuminate the expected conditions that a recovery effort must face.

The various scales of scientific analysis with respect to threatened and endangered species in the Platte River Basin imply that decisions based on science should also recognize scale. Decisions concerning the Platte River Basin that are based on short-term multiyear data and a local perspective are not likely to benefit the long-term (multidecadal) viability of a species that operates on a continental or intercontinental scale. The costs of efforts to recover threatened or endangered species are often most obvious on a local scale, but the benefits are much more widely distributed.

The third thread is that water links the needs of human, wildlife, and habitat more than any other ecological process. Many of the risks to threatened and endangered species, and all the comprehensive solutions to the problem of recovery, require a refined understanding of hydrological processes. The hydrological system of the Platte River is highly interconnected, so solutions to the species issues that attempt to protect commodity values of water must also be interconnected, particularly between surface water and groundwater. Climatic changes create a changing backdrop for the more important human-induced changes in the hydrology of the basin. The committee is firmly convinced that upstream storage, diversion, and distribution of the river's flow are the most important drivers of change that adversely affect species habitat along the Platte River.

COMMITTEE'S FINDINGS

1. Do current central Platte habitat conditions affect the likelihood of survival of the whooping crane? Do they limit (adversely affect) its recovery?

Conclusions: The committee concluded that, given available knowledge, current central Platte habitat conditions adversely affect the likelihood of survival of the whooping crane, but to an unknown degree. The Platte River is important to whooping cranes: about 7% of the total whooping crane population stop on the central Platte River in any one year, and many, if not all, cranes stop over on the central Platte at some point in their lifetimes. Population viability analyses show that if mortality were to increase by only 3%, the general population would likely become unstable. Thus, if the cranes using the Platte River were eliminated, population-wide effects would be likely. Resources acquired by whooping cranes during migratory stopovers contribute substantially to meeting nutrient needs and probably to ensuring survival and reproductive success. Because as much as 80% of crane mortality appears to occur during migration, and because the Platte River is in a central location for the birds' migration, the river takes on considerable importance. The committee concluded that current habitat conditions depend on river management in the central Platte River, but the population also depends on events in other areas along the migratory corridor. If habitat conditions on the central Platte River—that is, the physical circumstances and food resources required by cranes—decline substantially, recovery could be slowed or reversed. The Platte River is a consistent source of relatively well-watered habitat for whooping cranes, with its water source in distant mountain watersheds that are not subject to drought cycles that are as severe as those of the Northern Plains. There are no equally useful habitats for whooping cranes nearby: the Rainwater Basin dries completely about once a decade, and the Sandhills are inconsistent as crane habitat, while the Niobrara and other local streams are subject to the same variability as the surrounding plains. Future climatic changes may exacerbate conflicts between habitat availability and management and human land use. If the quality or quantity of other important habitats becomes less available to whooping cranes, the importance of the central Platte River could increase.

Primary Sources of Scientific Information: The basis of the above conclusion is published documents that were available to other researchers and the public including the original listing document and recovery plan for the species and a review of knowledge about the cranes by the Interstate Task Force on Endangered Species (EA Engineering, Science and Technology, Inc. 1985). Other important contributions to knowledge include Allen (1952) and Austin and Richert (2001). The committee also reviewed and discussed critical comments presented in open sessions and written testimony exemplified by Lingle (G. Lingle, unpublished material, March 22, 2000) and Czaplewski et al. (M.M. Czaplewski et al., Central Platte Natural Resource District, unpublished material, August 22, 2003) that was critical of the research conducted by DOI agencies.

2. Is the current designation of central Platte River habitat as "critical habitat" for the whooping crane supported by existing science?

Conclusions: An estimated 7% of the wild, migratory whooping crane population now uses the central Platte River on an annual basis and many, if not all, cranes stop over on the central Platte at some point in their lifetimes. The proportion of whooping cranes that use the central Platte River and the amount of time that they use it are increasing (with expected inter-annual variation). The designation of central Platte River migratory stopover habitat as critical to the species is therefore supported because the birds have specific requirements for roosting areas that include open grassy or sandy areas with few trees, separation from predators by water, and proximity to foraging areas such as wetlands or agricultural areas. The Platte River critical habitat area is the only area in Nebraska that satisfies these needs on a consistent basis. However, some habitats designated as critical in 1978 appear to be largely unused by whooping cranes in recent years, and the birds are using adjacent habitats that are not so designated (Stehn 2003).

Habitat selection (to the extent that it can be measured) on multiple geographic scales strongly suggests that Nebraska provides important habitat for whooping cranes during their spring migration. Riverine, palustrine, and wetland habitats serve as important foraging and roosting sites for whooping cranes that stop over on the central Platte River. Whooping cranes appear to be using parts of the central Platte River that have little woodland and long, open vistas, including such areas outside the zone classified as critical habitat. In some cases the cranes appear to be using areas that have been cleared of riparian woodland, perhaps partly explaining their distribution outside the critical habitat area.

Primary Sources of Scientific Information: The basis of the committee's conclusion is published documents that were available to other researchers and the public including the original listing document, recovery plan, and declaration of critical habitat; and information in Howe (1989) and Austin and Richert (2001). The committee also considered commentary that was critical of the research conducted by DOI agencies exemplified by open sessions and written testimony presented by Lingle (G. Lingle, unpublished material, March 22, 2000), EA Engineering, Science and Technology, Inc. (1985) and Czaplewski et al. (M.M. Czaplewski et al., Central Platte Natural Resources District, unpublished material, August 22, 2003).

3. Do current central Platte habitat conditions affect the likelihood of survival of the piping plover? Do they limit (adversely affect) its recovery?

Conclusions: Reliable data indicate that the northern Great Plains population of the piping plover declined by 15% from 1991 to 2001. The census population in Nebraska declined by 25% during the same period. Resident piping plovers have been virtually eliminated from natural riverine habitat on the central Platte River. No recruitment (addition of new individuals to the population by reproduction) has occurred there since 1999. The disappearance of the piping plover on the central Platte can be attributed to harassment caused by human activities, increased predation of nests, and losses of suitable habitat due to the encroachment of vegetation on previously unvegetated shorelines and gravel bars.

The committee concluded that current central Platte River habitat conditions adversely affect the likelihood of survival of the piping plover, and, on the basis of available understanding, those conditions have adversely affected the recovery of the piping plover. Changes in habitat along the river—including reductions in open, sandy areas that are not subject to flooding during crucial nesting periods—have been documented through aerial photography since the late 1930s and probably have adversely affected populations of the piping plover. Sandpits and reservoir edges with beaches may, under some circumstances, mitigate the reduction in riverine habitat areas. Because piping plovers are mobile and able to find alternative nesting sites, changes in habitat may not be as severe as they would be otherwise, but no studies have been conducted to support or reject this hypothesis.

Primary Sources of Scientific Information: Corn and Armbruster (1993) demonstrated differences (including higher river invertebrate densities and catch rates) in foraging habitat between the river and sand pit sites; this suggests that riverine habitat areas are superior to the sand mines and reservoir beaches for the piping plover. Basic information sources include the listing document and recovery plan. Higgins and Brashier (1993) provide additional information on habitat conditions, survival, and recovery. The committee also considered commentary presented in open sessions and written testimony exemplified by Lingle (G. Lingle, unpublished material, March 22, 2000) and Czaplewski et al. (M.M. Czaplewski et al., Central Platte Natural Resources District, unpublished material, August 22, 2003) that was critical of the research conducted by DOI agencies.

4. Is the current designation of central Platte River habitat as "critical habitat" for the piping plover supported by the existing science?

Conclusions: The designation of central Platte habitat as critical habitat for the piping plover is scientifically supportable. Until the last several years, the central Platte supported substantial suitable habitat for the piping plover, including all "primary constituent elements" required for successful reproduction by the species. Accordingly, the central Platte River contributed an average of more than 2 dozen nesting pairs of plovers to the average of more than 100 pairs that nested each year in the Platte River Basin during the 1980s and 1990s. The critical habitat designation for the species explicitly recognizes that not all areas so designated will provide all necessary resources in all years and be continuously suitable for the species. It is also now understood that off-stream sand mines and reservoir beaches are not an adequate substitute for natural riverine habitat.

Primary Sources of Scientific Information: Data generated according to standard scientific methods in well-defined and well-executed scientific investigations support the critical habitat designation for the piping plover—including work by Ziewitz et al. (1992), Ducey (1983), and Faanes (1983)—as does the designation in the *Federal Register* (67: 57638 [2002]). The committee also considered commentary presented in open sessions and written testimony exemplified by Lingle (G. Lingle, unpublished material, March 22, 2000) and Czaplewski et al. (M.M. Czaplewski et al., Central Platte Natural Resources District, unpublished materials, August 10, 2001, and August 22, 2003) that was critical of the research conducted by DOI agencies.

5. Do current central Platte habitat conditions affect the likelihood of survival of the interior least tern? Do they limit (adversely affect) its recovery?

Conclusions: The committee concluded that current habitat conditions on the central Platte River adversely affect the likelihood of survival of the interior least tern—in much the same fashion as they affect the likelihood of survival of the piping plover—and that on the basis of available information, current habitat conditions on the central Platte River adversely affect the likelihood of recovery of the interior least tern. Reliable population estimates indicate that the total (regional) population of interior least terns was at the recovery goal of 7,000 in 1995, but some breeding areas, including the central Platte River, were not at identified recovery levels. The central Platte subpopulation of least terns declined from 1991 to 2001. The number of terns using the Platte River is about two-thirds of the number needed to reach the interior least tern recovery goal for the Platte. The interior tern is nesting in substantial numbers on the adjacent lower Platte River, but numbers continue to decline on the central Platte, reflecting declining habitat conditions there. The decline in the tern population on the central Platte River has been coincidental with the loss of numerous bare sandbars and beaches along the river. Control of flows and diversion of water from the channel are the causes of these geomorphic changes. Woodland vegetation, unsuitable as tern habitat, has colonized some parts of the central Platte River. Alternative habitats, such as abandoned sand mines or sandy shores of Lake McConaughy, are not suitable substitutes for Platte River habitat because they are susceptible to disturbance by humans and natural predators. The shores of Lake McConaughy are available only at lower stages of the reservoir, and they disappear at high stages.

Primary Sources of Scientific Information: The scientific underpinnings of these conclusions are extensive and substantial, including work by Smith and Renken (1990), Sidle and Kirsch (1993), Ziewitz et al. (1992), and Higgins and Brashier (1993), all of whom used sound, widely accepted, standard scientific methods. The committee also considered commentary presented in open sessions and written testimony exemplified by Lingle (G. Lingle, unpublished material, March 22, 2000) and Czaplewski et al. (M.M. Czaplewski et al., Central Platte Natural Resources District, unpublished material, August 22, 2003) that was critical of the research conducted by DOI agencies.

6. Do current habitat conditions in the lower Platte (below the mouth of the Elkhorn River) affect the likelihood of survival of the pallid sturgeon? Do they limit (adversely affect) its recovery?

Conclusions: Current habitat conditions on the lower Platte River (downstream of the mouth of the Elkhorn River) do not adversely affect the likelihood of survival and recovery of the pallid sturgeon because that reach of the river appears to retain several habitat characteristics apparently preferred by the species: a braided channel of shifting sandbars and islands; a sandy substrate; relatively warm, turbid waters; and a flow regime that is similar to conditions that were found in the upper Missouri River and its tributaries before the installation of large dams on the Missouri. Alterations of discharge patterns or channel features that modify those characteristics might irreparably alter this habitat for pallid sturgeon use. In addition, the lower Platte River is connected with a long undammed reach of the Missouri River, which allows access of the pallid sturgeon in the Platte River to other segments of the existing population. Channelization and damming of the Missouri River have depleted pallid sturgeon habitats throughout its former range, so the lower Platte may be even more important for its survival and recovery. The population of pallid sturgeon is so low in numbers, and habitat such as the lower Platte River that replicates the original undisturbed habitat of the species is so rare that the lower Platte River is pivotal in the management and recovery of the species.

Primary Sources of Scientific Information: Scientific studies supporting those conclusions are reported in numerous peer-reviewed publications, as exemplified by general research on the habitat of hatchery-derived pallid sturgeon in the lower Platte River by Snook (2001) and Snook et al. (2002). Carlson et al. (1985) and Kallemeyn (1983) provided useful background information. Additional investigations in the Missouri River system by Bramblett (1996) and Bramblett and White (2001) have results that are applicable to the lower Platte River. The committee also considered commentary presented in open sessions and written testimony exemplified by Czaplewski et al. (M.M. Czaplewski et al., Central Platte Natural Resources District, unpublished material, August 22, 2003) that was critical of the research conducted by DOI agencies.

7. Were the processes and methodologies used by the USFWS in developing its central Platte River instream-flow recommendations (i.e., species, annual pulse flows, and peak flows) scientifically valid?

Conclusions: The U.S. Fish and Wildlife Service (USFWS) used methods described in an extensive body of scientific and engineering literature. Reports of interagency working groups that addressed instream-flow recommendations cite more than 80 references that were in wide use and generally accepted in the river science and engineering community. The committee reviewed that information, as well as oral and written testimony critical of the research conducted by DOI agencies, and it concluded that the methods used during the calculations in the early 1990s were the most widely accepted at that time. Revisions were made as improved knowledge became available. Although the Instream Flow Incremental Method (IFIM) and Physical Habitat Simulation System (PHABSIM) were the best available science when DOI agencies reached their recommendations regarding instream flows, there are newer developments and approaches, and they should be internalized in DOI's decision processes for determining instream flows. The new approaches, centered on the river as an ecosystem rather than focused on individual species, are embodied in the concepts of the normative flow regime. Continued credibility of DOI instream-flow recommendations will depend on including the new approach.

The instream-flow recommendations rely on empirical and model-based approaches. Surveyed cross sections along the river provided DOI investigators with specific information on the morphology of the river and vegetation associated with the river's landforms. The portions of the cross sections likely to be inundated by flows of various depths were directly observed. Model calculations to simulate the dynamic interaction of water, geomorphology, and vegetation that formed habitat for species were handled with the prevailing standard software PHABSIM, which has seen wide use in other cases and has been accepted by the scientific community. The software was used by DOI researchers in a specific standard method, IFIM, which permits observations of the results as flow depths are incrementally increased.

The continuing DOI model developments, including the emerging SEDVEG model, are needed because of the braided, complex nature of the Platte River—a configuration that is unlike other streams to which existing models are often applied. The committee did not assess the newer models, because they have not yet been completed or tested, but it recommends that they be explored for their ability to improve decision making.

The committee also recognizes that there has been no substantial testing of the predictions resulting from DOI's previous modeling work,[1] and it recommends that calibration of the models be improved. Monitoring of the effects of recommended flows should be built into a continuing program of adaptive management to help to determine whether the recommendations are valid and to indicate further adjustments to the recommendations based on observations.

Primary Sources of Scientific Information: The literature used to support USFWS's methods ranged from basic textbook sources, such as Dunne and Leopold (1978) and Darby and Simon (1999), to specific applications exemplified by Simons & Associates, Inc. (2000) and Schumm (1998). The committee also considered the interagency working reports (Hydrology Work Group 1989; M. Zallen, DOI, unpublished memo, August 11, 1994) and oral and written testimony exemplified by Parsons (2003), Payne (1995; T.R. Payne and Associates, pers. comm., June 19, 2003), Woodward (2003), and Lewis (2003).

8. Are the characteristics described in the USFWS habitat suitability guidelines for the central Platte River supported by the existing science and are they (i.e., the habitat characteristics) essential to the survival of the listed avian species? To the recovery of those species? Are there other Platte River habitats that provide the same values that are essential to the survival of the listed avian species and their recovery?

Conclusions: The committee concluded that the habitat characteristics described in USFWS's habitat suitability guidelines for the central Platte River were supported by the science of the time of the original habitat description during the 1970s and 1980s and were consistent with accepted understanding of how the systems function. New ecological knowledge has since been developed. The new knowledge, largely from information gathered over the last 20 years, has not been systematically applied to the processes of designating or revising critical habitat, and the committee recommends that it be done.

The committee also concluded that suitable habitat characteristics along the central Platte River are essential to the survival and recovery of the piping plover and the interior least tern. No alternative habitat exists in the

[1]The committee did not consider USGS's in-progress evaluation of the models and data used by USFWS to set flow recommendations for whooping cranes.

central Platte that provides the same values essential to the survival and recovery of piping plovers and least terns. Although both species use artificial habitat (such as shoreline areas of Lake McConaughy and sandpits), the quality and availability of sites are unpredictable from year to year. The committee further concluded that suitable habitat for the whooping crane along the central Platte River is essential for its survival and recovery because such alternatives as the Rainwater Basin and other, smaller rivers are used only intermittently, are not dependable from one year to the next, and appear to be inferior to habitats offered by the central Platte River.

Primary Sources of Scientific Information: The committee relied on the following sources in reaching its conclusions: for whooping cranes, the original listing document, recovery plan, and declaration of critical habitat and Howe (1989), EA Engineering, Science and Technology, Inc. (1985), Austin and Richert (2001), and Lutey (2002); for interior least terns and piping plovers, the original listing documents, recovery plans, and declaration of critical habitat for the piping plover (Fed. Regist. 67 (176): 57638 [2002]), Smith and Renken (1990), Sidle and Kirsch (1993), Ziewitz et al. (1992), Ducey (1983), Faanes (1983), Higgins and Brashier (1993), Corn and Armbruster (1993), and Kirsch and Sidle (1999). The committee also considered commentary presented in open sessions and written testimony exemplified by Lingle (G. Lingle, unpublished material, March 22, 2000) and Czaplewski et al. (M.M. Czaplewski et al., Central Platte Natural Resources District, unpublished material, August 22, 2003) that was critical of the research conducted by DOI agencies.

9. Are the conclusions of the Department of the Interior about the interrelationships of sediment, flow, vegetation, and channel morphology in the central Platte River supported by the existing science?

Conclusions: The committee concluded that DOI conclusions about the interrelationships among sediment, flow, vegetation, and channel morphology in the central Platte River were supported by scientific theory, engineering practice, and data available at the time of those decisions. By the early 1990s, when DOI was reaching its conclusions, the community of geomorphologists concerned with dryland rivers had a general understanding of the role of fluctuating discharges in arranging the land forms of the channel, and DOI included this understanding in its conclusions about the river. In the early 1990s, engineering practice, combined with geomorphology and hydrology, commonly used IFIM and PHABSIM to make predictions and recommendations for flow patterns that shaped channels, and this resulted in adjustments in vegetation and habitat. In fact, despite some criticisms, IFIM and PHABSIM are still widely used in the professional community of river restorationists in 2004. In applying scientific theory and engineering practice, the DOI agencies used the most current data and made additional measurements to bolster the calculations and recommendations. Since the early 1990s, more data have become available, and the USBR has conducted considerable cutting-edge research on a new model (SEDVEG) that should update earlier calculations but is not yet in full operation (and was not reviewed by this committee).

Primary Sources of Scientific Information: Murphy et al. (2001) outline the basic understanding of sediment and vegetation dynamics. Sediment data are obtained by sampling sediment concentrations and multiplying the concentrations by discharges and duration. For flow, gaging records on the Platte River are 50 years in duration or longer, and they are in greater density than on many American rivers; the gages provide quality data on water discharge for the Platte River. Murphy and Randle (2003) review the analyses and other sources of knowledge about the flows that provide a sound basis for DOI decisions. In addition to the review by Murphy et al. (2001) concerning vegetation, several studies over the last 20 years have provided an explanation of vegetation dynamics that the committee found to be correct and that is the basis of DOI decisions. Early work by USFWS (1981a) and Currier (1982) set the stage for an evolution of understanding of vegetation change on the river that was later expanded by Johnson (1994). For channel morphology, there is a long history of widely respected research to draw on, including early geomorphologic investigations by Williams (1978) and Eschner et al. (1983), continuing with the reviews by Simons and Associates (2000), and culminating in recent work by Murphy and Randle (2003). The committee also considered commentary presented in open sessions and written testimony exemplified by Parsons (2003) and Lewis (2003) that was critical of the research conducted by DOI agencies.

10. What were the key information and data gaps that the NAS identified in the review?

Conclusions: The committee reached its conclusions for the preceding nine questions with reasonable confidence based on the scientific evidence available. However, the committee identified the following gaps in key information related to threatened and endangered species on the central and lower Platte River, and it recommends that they be addressed to provide improved scientific support for decision making.

• *A multiple-species perspective is missing from research and management of threatened and endangered species on the central and lower Platte River.* The interactions of the protected species with each other and with unprotected species are poorly known. Efforts to enhance one species may be detrimental to another species, but these connections remain largely unknown because research has been focused on single species. One approach is to shift from the focus on single species to an ecosystem perspective that emphasizes the integration of biotic and abiotic processes supporting a natural assemblage of species and habitats.

- *There is no systemwide, integrated operation plan or data-collection plan for the combined hydrological system in the North Platte, South Platte, and central Platte Rivers that can inform researchers and managers on issues that underlie threatened and endangered species conservation.* Natural and engineered variations in flows in one part of the basin have unknown effects on other parts of the basin, especially with respect to reservoir storage, groundwater storage, and river flows.
- *A lack of a full understanding of the geographic extent of the populations of imperiled species that inhabit the central Platte River and a lack of reliable information on their population sizes and dynamics limit our ability to use demographic models to predict accurately their fates under different land-management and water-use scenarios.* Detailed population viability analyses using the most recent data would improve understanding of the dynamics of the populations of at-risk species and would allow managers to explore a variety of options to learn about the probable outcomes of decisions. Continuation of population monitoring of at-risk bird species using the best available techniques, including color-banding of prefledged chicks and application of new telemetry techniques, is recommended.
- *There is no larger regional context for the central and lower Platte River in research and management.* Most of the research and decision making regarding threatened and endangered species in the Platte River Basin have restricted analysis to the basin itself, as though species used its habitats in isolation from other habitats outside the basin. There are substantial gaps in integrative scientific understanding of the connections between species that use the habitats of the central and lower Platte River and adjacent habitat areas, such as the Rainwater Basin of southern Nebraska and the Loup, Elkhorn, and Niobrara Rivers and other smaller northern Great Plains rivers.

The committee is confident that the central Platte River and lower Platte River are essential for the survival and recovery of the listed bird species and pallid sturgeon. However, in light of the habitat it provides and the perilously low numbers of the species, there is not enough information to assess the exact degree to which the Platte contributes to their survival and recovery.

- *Water-quality data are not integrated into knowledge about species responses to reservoir and groundwater management and are not integrated into habitat suitability guidelines.* Different waters are not necessarily equal, either from a human or a wildlife perspective, but there is little integration of water-quality data with physical or biological understanding of the habitats along the Platte River.

• *The cost effectiveness of conservation actions related to threatened and endangered species on the central and lower Platte River is not well known.* Neither the cost effectiveness nor the equitable allocation of measures for the benefit of Platte River species has been evaluated. The ESA does not impose or allow the implementing agencies to impose a cost-benefit test. Listed species must be protected no matter what the cost, unless the Endangered Species Committee grants an exemption. Cost effectiveness, however, is another matter. The ESA permits consideration of relative costs and benefits when choosing recovery actions, for example. USFWS has adopted a policy that calls for minimizing the social and economic costs of recovery actions, that is, of choosing actions that will provide the greatest benefit to the species at the lowest societal cost (Fed. Regist. 59: 3472 [1994]). In addition, persons asked to make economic sacrifices for the sake of listed species understandably want assurances that their efforts will provide some tangible benefit. In the Platte, the direct economic costs of measures taken for the benefit of species appear reasonably well understood. The biological benefits are another matter. For example, the costs of channel-clearing and other river-restoration measures are readily estimated. Their precise value for cranes is more difficult to estimate, although their general use is fairly well established.

The allocation of conservation costs and responsibility also has not been systematically evaluated. USFWS has concentrated its efforts to protect listed species in the Platte system on federal actions, such as the operation of federal water projects. That focus is understandable. Water projects with a federal nexus account for a large and highly visible proportion of diversions from the system. In addition, those actions may be more readily susceptible to regulatory control than others because they are subject to ESA Section 7 consultation. But some nonfederal actions also affect the species. Water users that depend on irrigation water from the federal projects may well feel that they are being asked to bear an inordinate proportion of the costs of recovering the system. A systematic inventory of all actions contributing to the decline of the species could help the parties to the cooperative agreement channel their recovery efforts efficiently and equitably. The National Research Council committee charged with evaluating ESA actions in the Klamath River Basin recently reached a similar conclusion (NRC 2004a).

• *The effects of prescribed flows on river morphology and riparian vegetation have not been assessed.* Adaptive-management principles require that the outcomes of a management strategy be assessed and monitored and that the strategy be adjusted accordingly, but there has been no reporting of the outcomes of the 2002 prescribed flow, no analysis of vegetation effects of managed flows, no measurement of their geomorphic effects, and no assessment of their economic costs or benefits.

• *The connections between surface water and groundwater are not well accounted for in research or decision making for the central and lower Platte River.* The dynamics of and connections between surface water and groundwater remain poorly known, but they are important for understanding river behavior and economic development that uses the groundwater resource. The effects of groundwater pumping, recently accelerated, are unknown but important for understanding river flows.

• *Some of the basic facts of issues regarding threatened and endangered species in the central and lower Platte River are in dispute because of unequal access to research sites.* Free access to all data sources is a basic tenet of sound science, but DOI agencies and Nebraska corporations managing water and electric power do not enter discussions about threatened and endangered species on the central and lower Platte River with the same datasets for species and physical environmental characteristics. USFWS personnel are not permitted to collect data on some privately owned lands. As a result, there are substantial gaps between data used by DOI and data used by the companies, and resolution is impossible without improved cooperation and equal access to measurement sites.

• *Important environmental factors are not being monitored.* Monitoring, consistent from time to time and place to place, supports good science and good decision making, but monitoring of many aspects of the issues regarding threatened and endangered species on the central and lower Platte River remains haphazard or absent. Important gaps in knowledge result from a lack of adequate monitoring of sediment mobility, the pallid sturgeon population, and movement of listed birds. Responses of channel morphology and vegetation communities to prescribed flows and vegetation removal remain poorly known because the same set of river cross sections is not sampled repeatedly. Groundwater may play an important role in flows, but groundwater pumping is not monitored.

• *Long-term (multidecadal) analysis of climatic influences has not been used to generate a basis for interpretation of short-term change (change over just a few years).* The exact interactions between climate and the system are poorly known because only short-term analyses of climate factors have been accomplished so far. In addition, the relative importance of human and climatic controls remains to be explicitly defined by researchers, even though such knowledge is important in planning river restoration for habitat purposes.

• *Direct human influences are likely to be much more important than climate in determining conditions for the threatened and endangered species of the central and lower Platte River.* Potentially important localized controls on habitat for threatened and endangered species on the central and lower Platte River are likely to be related to urbanization, particularly near freeway exits and small cities and towns where housing is replacing other land uses more useful to the species. Off-road vehicle use threatens the nesting sites of piping plovers and interior least terns in many of the sandy reaches of the river. Sandy beaches and bars are inviting to both birds and recreationists. Illegal harvesting has unknown effects on the small remaining population of pallid sturgeon. In each of those cases, additional data are required to define the threats to the listed species.

SUMMARY

USFWS faces extraordinary challenges in trying to identify the habitat needs and the critical habitat for listed species on the central and lower Platte River. Lack of data, pressures of tight deadlines for research, lack of a well-defined adaptive-management strategy with effective monitoring, and competing uses for the river's water and landscape resources complicate decision making. Despite those challenges, the science that explains forms and processes of the ecosystems along the central and lower Platte River of Nebraska is sufficient to support many decisions about the management of threatened and endangered species that use the river's habitats. In all cases, enough is known about the physical environmental processes that control habitat change to make informed decisions for the survival of the whooping crane, piping plover, interior least tern, and pallid sturgeon. Our scientific knowledge is not yet adequate to contribute to decisions regarding the exact role of the central and lower Platte River in the recovery of the whooping crane and pallid sturgeon. Valid science supports critical habitat designations for the piping plover, but the scientific support of critical habitat designation for the whooping crane is weak. Valid science and engineering related to hydrology, geomorphology, sediment transport, and riparian ecology support the DOI instream-flow recommendations and explanations for the river-channel and vegetation changes. The committee found numerous gaps in knowledge that could inform management of threatened and endangered species along the central and lower Platte River, mostly focused on problems of scientific integration, overrestricted scales of analysis, lack of systemwide connections, and lack of standardized procedures for data collection.

Land, water, and life in the region surrounding the 100th meridian on the Platte River are highly changeable and precariously balanced. Human manipulations of hydrological conditions and land cover have far-reaching consequences for wildlife populations. Policy based on a desired constant, stable, and predictable set of environmental circumstances is unlikely to be successful. Policy that relies on scientific knowledge about change through time and over geographic space is the most likely avenue to success in the search for accommodation between economic vitality and diverse and sustainable populations of wildlife that are neither threatened nor endangered.

References

Adams, S.R., J.J. Hoover, and K.J. Killgore. 1999. Swimming endurance of juvenile pallid sturgeon, *Scaphirhynchus albus*. Copeia 1999(3):802-807.

Aiken, J.D. 1980. Nebraska ground water law and administration. Nebr. Law Rev. 59(4):917-1000.

Aiken, J.D. 1987. New directions in Nebraska water policy. Nebr. Law Rev. 66(1):8-75.

Aiken, J.D. 1999. Balancing endangered species protection and irrigation water rights: The Platte River Cooperative Agreement. Great Plains Nat. Resour. J. 3(2):119-158.

Aiken, J.D. 2001. Pumpkin Creek Surface-Ground Water Dispute. Cornhusker Economics, August 8, 2001. [Online]. Available: agecon.unl.edu/pub/cornhusker/08-08-01.pdf [accessed Oct. 28, 2003].

Akçakaya, H.R., J.L. Atwood, D. Breininger, C.T. Collins, and B. Duncan. 2003. Metapopulation dynamics of the California least tern. J. Wildl. Manage. 67(4):829-842.

Alisauskas, R.T., and C.D. Ankney. 1992. The cost of egg laying and its relationship to nutrient reserves in waterfowl. Pp. 30-61 in Ecology and Management of Breeding Waterfowl, B.D.J. Batt, A.D. Afton, M.G. Anderson, C.D. Ankney, D.H. Johnson, J.A. Kadlec, and G.L. Krapu, eds. Minneapolis: University of Minnesota Press.

Alldredge, J.R., and J.T. Ratti. 1992. Further comparison of some statistical techniques for analysis of resource selection. J. Wildl. Manage. 56(1):1-9.

Allen, R.P. 1952. The Whooping Crane. Research Report No. 3. New York: National Audubon Society.

Arnell, N. 1996. Global Warming, River Flows and Water Resources. Chichester: Wiley. 224 pp.

Atwood, J.L. 1999. Patterns of Juvenile Dispersal by Least Terns in Massachusetts, Final Project Report. Prepared for Massachusetts Environmental Trust, Avian Conservation Division, Manomet Center for Conservation Sciences, Manomet, MA.

Atwood, J.L., and P.R. Kelly. 1984. Fish dropped on breeding colonies as indicators of least tern food habits. Wilson Bull. 96(1):34-47.

Austin, J.E., and A.L. Richert. 2001. A Comprehensive Review of Observational and Site Evaluation Data of Migrant Whooping Cranes in the United States 1943-99. Jamestown, ND: U.S. Geological Survey, Northern Prairie Wildlife Research Center. 157 pp.

Bailey, R.M., and M.O. Allum. 1962. Fishes of South Dakota. Miscellaneous Publications No. 119. Ann Arbor, MI: Museum of Zoology, University of Michigan.

Bailey, R.M., and F.B. Cross. 1954. River sturgeons of the American genus *Scaphirhynchus*: Characters, distribution, and synonymy. Pap. Mich. Acad. Sci. Arts Lett. 39:169-208.

Barzen, J.A., and J.R. Serie. 1990. Nutrient reserve dynamics of breeding canvasbacks. Auk. 107(1):75-85.

Baxter, G.T., and M.D. Stone. 1995. Fishes of Wyoming. Wyoming: Wyoming Game and Fish Dept.

Bean, M.J., and M.J. Rowland. 1997. The Evolution of National Wildlife Law. Westport, CT: Praeger. 544 pp.

Birstein, V.J., P. Doukakis, and R. DeSalle. 2002. Molecular phylogeny of *Acipenseridae*: Nonmonophyly of *Scaphirihynchidae*. Copeia 2002(2):287-301.

Bogan, M.A., C.D. Allen, E.H. Muldavin, S.P. Platania, J.N. Stuart, G.H. Farley, P. Mehlhop, and J. Belnap. 1998. Regional trends of biological resources - Southwest. Pp. 543-592 in Status and Trends of the Nation's Biological Resources, Vol. 2., M.J. Mac, P.A. Opler, C.E. Puckett Haecker, and P.D. Doran, eds. U.S. Department of the Interior, U.S. Geological Survey, Reston, VA.

Bovee, K.D., B.L. Lamb, J.M. Bartholow, C.B. Stalnaker, J. Taylor, and J. Henriksen. 1998. Stream Habitat Analysis Using the Instream Flow Incremental Methodology. Information and Technology Report USFS/BRD-1998-0004. Fort Collins, CO: U.S. Geological Survey, Biological Resources Division. [Online]. Available: http://www.fort.usgs.gov/products/Publications/3910/3910.pdf [accessed Feb. 5, 2004].

Bowman, D.B. 1994. Instream Flow Recommendations for the Central Platte River, Nebraska. U.S. Fish and Wildlife Services, Grand Island, NE. May 23, 1994. 9 pp.

Bowman, D.B., and D.E. Carlson. 1994. Pulse Flow Requirements for the Central Platte River. U.S. Fish and Wildlife Services, Grand Island, NE. August 3, 1994. 11 pp.

Boyce, M.S., E.M. Kirsch, and C. Servheen. 2002. Bet-hedging applications for conservation. J. Bioscience 27(4):385-392.

Boyle Engineering Corporation. 1999. Water Conservation/Supply Reconnaissance Study: Platte River Research Cooperative Agreement, Final Report. Prepared for Governance Committee of the Cooperative Agreement for Platte River Research, by Boyle Engineering Corporation, in association with BBC Research & Consulting, Anderson Consulting Engineers, Lakewood, CO. December 1999.

Bramblett, R.G. 1996. Habitats and Movements of Pallid and Shovelnose Sturgeon in the Yellowstone and Missouri Rivers, Montana and North Dakota. Ph. D. Dissertation, Montana State University, Bozeman, MT.

Bramblett, R.G., and R.G. White. 2001. Habitat use and movements of pallid and shovelnose sturgeon in the Yellowstone and Missouri Rivers, Montana and North Dakota. Trans. Am. Fish. Soc. 130(6):1006-1025.

Brook, B.W., J.R. Cannon, R.C. Lacy, C. Mirande, and R. Frankham. 1999. Comparison of the population viability analysis packages GAPPS, INMAT, RAMAS and VORTEX for the whooping crane (*Grus americana*). Anim. Conserv. 2(1):23-31.

Brookes, A., and F.D. Shields, eds. 1996. River Channel Restoration: Guiding Principles for Sustainable Projects. Chichester: Wiley. 433 pp.

Brooking, A.M. 1943. The present status of the whooping crane. Nebr. Bird Rev. 11(1):5-8.

Brown, R.H., and J.R. Whitaker. 1948. Historical Geography of the United States. New York: Harcourt, Brace.

Bruntland, G., ed. 1987. Our Common Future: The World Commission on Environment and Development. Oxford: Oxford University Press.

Bult, T.P., S.C. Riley, R.L. Haedrich, R.J. Gibson, and J. Heggenes. 1999. Density-dependent habitat selection by juvenile Atlantic salmon (*Salmo salar*) in experimental riverine habitats. Can. J. Fish. Aquat. Sci. 56(7):1298-1306.

Campton, D.E., A.L. Bass, F.A. Chapman, and B.W. Bowen. 2000. Genetic distinction of pallid, shovelnose and Alabama sturgeon: Emerging species and the U.S. Endangered Species Act. Conserv. Genet. 1(1):17-32.

Carbutt, J. 1866a. The Platte River Opposite Platte City. Photo 211 in J. Carbutt Stereograph Catalog. Union Pacific Railroad Excursion to the 100th Meridian Oct. 1866. History and Photos, Union Pacific Railroad Archives. [Online]. Available: http://www.uprr.com/aboutup/photos/carbutt/jc211.shtml [accessed March 5, 2004].

Carbutt, J. 1866b. View of Camp No. 2 from Prospect Hill. Photo 214 in J. Carbutt Stereograph Catalog. Union Pacific Railroad Excursion to the 100th Meridian Oct. 1866. History and Photos, Union Pacific Railroad Archives. [Online]. Available: http://www.uprr.com/aboutup/photos/carbutt/jc214.shtml [accessed March 9, 2004].

Carlson, D.M., W.L. Pflieger, L. Trial, and P.S. Haverland. 1985. Distribution, biology, and hybridization of *Scaphirhynchus albus* and *Scaphirhynchus platorynchus* in the Missouri and Mississippi River. Environ. Biol. Fish. 14(1):51-59.

Castleberry, D.T., J.J. Cech Jr., D.C. Erman, D. Hankin, M. Healey, G.M. Kondolf, M. Mangel, M. Mohr, P.B. Moyle, J. Nielsen, T.P. Speed, and J.G. Williams. 1996. Uncertainty and instream flow standards. Fisheries 21(8):20-21.

Chadwick, J.W., S.P. Canton, D.J. Conklin, Jr., and P.L. Winkle. 1997. Fish species composition in the central Platte River, Nebraska. Southwest. Nat. 42(3):279-289.

Chang, H.H. 1998. Fluvial Processes in River Engineering. Malabar, FL: Krieger Pub. Co. 432 pp.

Chavez-Ramirez, F. 1996. Food Availability, Foraging Ecology, and Energetics of *Whooping Cranes* Wintering in Texas. Ph.D. Dissertation, Texas A & M University. 104 pp.

Chittenden, H.M. 1902. Map of the Trans-Mississippi of the United States during the period of the American fur trade as conducted from St. Louis between the years 1807 and 1843. The American Fur Trade of the Far West, Vol. 3. New York: F.P. Harper. [Online]. Available: http://memory.loc.gov [accessed Oct. 27, 2003].

Chittenden, H.M. 1935. American Fur Trade of the Far West. New York: The Press of the Pioneers, Inc.

Chow, V.T., ed. 1964. Handbook of Applied Hydrology: A Compendium of Water-Resources Technology. New York: McGraw-Hill.

Clancey, P. 1990. Fort Peck Pallid Sturgeon Study, Annual Report. Montana Fish, Wildlife and Parks, Helena, MT.

Clark, J.A., and E. Harvey. 2002. Assessing multi-species recovery plans under the Endangered Species Act. Ecol. Appl. 12(3):655-662.

COHYST. 2003. Nebraska's Cooperative Hydrology Study (COHYST). [Online]. Available: http://nrcnt3.dnr.state.ne.us/cohyst/ [accessed Oct. 28, 2003].

Coker, R.E. 1930. Studies of common fishes of the Mississippi River at Keokuk. Bull. U.S. Bur. Fish. Comm. 45:141-225.

Colt, C.J. 1997. Breeding Bird Use of Riparian Forests along the Central Platte River: A Spatial Analysis. M.S. Thesis, University of Nebraska, Lincoln, NE.

Constant, G.C., W.E. Kelso, D.A. Rutherford, and C.F. Bryan. 1997. Habitat, Movement and Reproductive Status of Pallid Sturgeon (*Scaphirhynchus albus*) in the Mississippi and Atchafalaya Rivers. Prepared for the US Army Corps of Engineers, New Orleans District, New Orleans, LA.

Corn, J.G., and M.J. Armbruster. 1993. Prey availability for foraging piping plovers along the Platte River in Nebraska. Pp. 143-149 in Proceedings of the Missouri River and its Tributaries: Piping Plover and Least Tern Symposium-Workshop, K.F. Higgins, and M.R. Brashier, eds. Brookings, SD: South Dakota State University, Dept. of Wildlife and Fisheries Sciences.

Cowardin, L.M. 1979. Classification of Wetlands and Deepwater Habitats of the United States. Report FWS/OBS-79/31. Washington, DC: Fish and Wildlife Service, U.S. Department of the Interior.

Cox, R.R., and B.E. Davis. 2003. Habitat Use, Movements, Residency Times and Survival of Female Northern Pintails during Spring Migration in Nebraska. Presentation at the Annual Rainwater Basin Joint Venture Research Symposium, 16-17 September 2003, Rowe Sanctuary near Gibbon, NE.

Cross, F.B., and J.T. Collins. 1995. Fishes in Kansas, 2nd Ed, rev. Lawrence, KS: Museum of Natural History, University of Kansas.

Currier, P.J. 1982. The Floodplain Vegetation of the Platte River: Phytosociology, Forest Development and Seedling Establishment. Ph.D. Dissertation, Iowa State University of Science and Technology, Ames, IA. 332 pp.

Currier, P.J. 1995. Relationships between vegetation, groundwater hydrology, and soils on Platte River wetland meadows. Cooperative Agreement 14-16-0006-90-917, Platte River Whooping Crane Maintenance Trust, Inc. (Pp. 172 in Proceedings of the 1995 Platte River Basin Ecosystem Symposium, February 28-March 1, 1995, Kearney, NE).

Currier, P.J. 1997. Woody vegetation expansion and continuing declines in open channel habitat on the Platte River in Nebraska. Pp. 141-152 in Proceedings of the Seventh North American Crane Workshop, January 10-13, 1996, Biloxi, MS, R.P. Urbanek, and D.W. Stahlecker, eds. Grand Island, NE: North American Crane Working Group.

Currier, P.J., and C.A. Davis. 2000. The Platte as a prairie river: A response to Johnson and Boettcher. Great Plains Res. 10(1):69-84.

Darby, S.E., and A. Simon. 1999. Incised River Channels: Processes, Forms, Engineering and Management. New York: John Wiley & Sons. 442 pp.

Davis, C.A. 2001. Nocturnal roost site selection and diurnal habitat use by sandhill cranes during spring in central Nebraska. Pp. 48-56 in Proceedings of the Eighth North American Crane Workshop, 11-14 January 2000, Albuquerque, NM, D.H. Ellis, ed. Grand Island, NE: North American Crane Working Group.

Dinan, J. 2003. Biology and Habitat Requirements of Plovers and Terns. Presentation at the First Meeting on Threatened and Endangered Species, May 6, 2003, Kearney, NE.

Dobkin, D.S. 1994. Conservation and Management of Neotropical Migrant Landbirds in the Northern Rockies and Great Plains. Moscow, ID: University of Idaho Press. 220 pp.

DOI (U.S. Department of the Interior). 1997. Cooperative Agreement for Platte River Research and Other Efforts Relating to the Endangered Species Habitats along the Central Platte River, Nebraska, Signed July 1, 1997, by the Governors of Wyoming, Colorado, and Nebraska, and the Secretary of the Interior. [Online]. Available: http://www.platteriver.org/library/CooperativeAgreement/index.htm. [accessed Oct. 27, 2003].

DOI (U.S. Department of the Interior). 2003. Plate River Recovery Implementation Program, Draft Environmental Impact Statement. Bureau of Reclamation, and U.S. Fish and Wildlife Service, U.S. Department of the Interior. December 2003. [Online]. Available: http://www.platteriver.org/ [accessed March 9, 2004].

Dryer, M.P., and A.J. Sandvol. 1993. Recovery Plan for Pallid Sturgeon (*Scaphirhynchus albus*). Bismarck, ND: U.S. Fish and Wildlife Service.

Ducey, J. 1983. Notes on the birds of the lower Niobrara Valley in 1902 as recorded by Myron H. Swenk. Nebr. Bird Rev. 51:37-44.

Ducey, J. 1988. Nest scrape characteristics of piping plover and least tern in Nebraska. Nebr. Bird Rev. 56(2):42-44.

Dugger, K.M. 1997. Foraging Ecology and Reproduction Success of Least Terns Nesting on the Lower Mississippi River. Ph.D. Dissertation, University of Missouri, Columbia. 137 pp.

Dunne, T., and L.B. Leopold. 1978. Water in Environmental Planning. San Francisco: W.H. Freeman. 818 pp.

EA Engineering, Science and Technology, Inc. 1985. Migration Dynamics of the Whooping Crane with Emphasis on Use of the Platte River in Nebraska. Prepared for Interstate Task Force on Endangered Species, Colorado Water Congress, Nebraska Water Resources Association, Wyoming Water Development Association, by EA Engineering, Science and Technology, Inc., Lincoln, NE. December 1985.

EA Engineering, Science and Technology, Inc. 1988. Status of the Interior Least Tern and Piping Plover in Nebraska (Period of Record through 1986). Prepared for Interstate Task Force on Endangered Species, Colorado Water Congress, Nebraska Water Resources Association, Wyoming Water Development Association, by EA Engineering, Science and Technology, Inc., Lincoln, NE. September 1988.

Edmonds, M. 2001. The pleasures and pitfalls of written records. Pp. 73-99 in The Historical Ecology Handbook: A Restorationist's Guide to Reference Ecosystems, D. Egan, and E.A. Howell, eds. Washington, DC: Island Press.

Ellis, M.J., and D.T. Pederson. 1986. Ground Water Levels in Nebraska, 1985. Nebraska Water Survey Paper No. 61. Lincoln, NE: Conservation and Survey Division, Institute of Agriculture and Natural Resources, the University of Nebraska.

Erickson, J.D. 1992. Habitat Selection and Movement of Pallid Sturgeon in Lake Sharpe, South Dakota. M.S. Thesis, South Dakota State University, Brookings, SD.

Eschner, T.R. 1983. Hydraulic Geometry of the Platte River near Overton, South-Central Nebraska. U.S. Geological Survey Professional Paper 1277-C. Washington, DC: U.S. Government Printing Office. 32 pp.

Eschner, T.R., R.F. Hadley, and K.D. Crowley. 1983. Hydrologic and Morphologic Changes in the Channels of the Platte River Basin in Colorado, Wyoming, and Nebraska: A Historical Perspective. U.S. Geological Survey Professional Paper 1277-A. Washington, DC: U.S. Government Printing Office. 39 pp.

Faanes, C.A. 1983. Aspects of the nesting ecology of least terns and piping plovers in central Nebraska. Prairie Nat. 15:145-154.

Faanes, C.A. 1992. Unobstructed visibility at Whooping Crane Roost Sites on the Platte River, Nebraska. Pp. 117-120 in Proceedings of the 1988 North American Crane Workshop, D.A. Wood, ed. Florida Nongame Wildlife Program Technical. Report No. 12. Tallahassee, FL: Florida Game and Fresh Water Fish Commission.

Faanes, C.A., and M.J. LeValley. 1993. Is the distribution of sandhill cranes on the Platte River changing? Great Plains Res. 3(2):297-304.

Faanes, C.A., D.H. Johnson, and G.R. Lingle. 1992. Characteristics of whooping crane roost sites in the Platte River. Pp. 90-94 in Proceedings of the Sixth North American Crane Workshop, October 3-5, 1991, Regina, Saskatchewan, D.W. Stahlecker, and R.P. Urbanek, eds. Grand Island, NE: North American Crane Working Group.

Farber, D.A., and P.P. Frickey. 1987. The jurisprudence of public choice. Tex. Law Rev. 65:873.

Fieberg, J., and S.P. Ellner. 2000. When is it meaningful to estimate an extinction probability? Ecology 81(7):2040–2047.

Forbes, S.A., and R.E. Richardson. 1905. On a new shovelnose sturgeon from the Mississippi River. Bull. Ill. State Lab. Nat. Hist. 7:37-44.

Forsberg, M. 1999. Partners in flight. NEBRASKAland (April 1999):39-45.

Franklin, J.F. 1993. Preserving biodiversity: Species, ecosystems, or landscapes? Ecol. Appl. 3(2):202-205.

Fretwell, S.D. 1972. Populations in a Seasonal Environment. Monographs in Population Biology No. 5. Princeton, NJ: Princeton University Press. 217 pp.

Friend, M. 1981. Waterfowl diseases—changing perspectives for the future. Pp. 189-196 in the Fourth International Waterfowl Symposium, January 30-February 1, 1981, New Orleans, LA. Chicago, IL: Ducks Unlimited, Inc.

Friesen, B., J. van Loh, J. Schrott, J. Butler, D. Crawford, and M. Pucherelli. 2000. Central Platte River 1998 Land Cover/Use Mapping Project, Nebraska. Technical Report of the Platte River EIS Team, U.S. Department of the Interior, Bureau of Reclamation, Fish and Wildlife Service, Denver, CO. October 20, 2000.

GAO (U.S. General Accounting Office). 2003. Endangered Species: Fish and Wildlife Service Uses Best Available Science to Make Listing Decisions, but Additional Guidance Needed for Critical Habitat Designations. GAO-03-803. U.S. General Accounting Office, Washington, DC. [Online]. Available: http://www.gao.gov/new.items/d03803.pdf [accessed Dec. 3, 2003].

Gersib, R.A., J.E. Cornley, A. Trout, J.M. Hyland, and P.J. Gabig. 1990. Concept Plan for Waterfowl Habitat Protection: Rainwater Basin Area of Nebraska, Category 25 of the North American Waterfowl Management Plan. Nebraska Game and Parks Commission, Lincoln, NE.

Gillilan, D.M., and T.C. Brown. 1997. Instream Flow Protection: Seeking a Balance in Western Water Use. Washington, DC: Island Press. 417 pp.

Glantz, M.H., ed. 1994. Drought Follows the Plow: Cultivating Marginal Areas. Cambridge, UK: Cambridge University Press.

Glenn, T.C., W. Stephan, and M.J. Braun. 1999. Effects of a population bottleneck on mitochondrial DNA variation in whooping cranes. Conserv. Biol. 13(5):1097-1107.

Glennon, R.J. 2002. Water Follies: Groundwater Pumping and the Fate of America's Fresh Waters. Washington, DC: Island Press.

Gomez, G. 1992. Whooping cranes in southwest Louisiana: History and human attitudes. Pp. 19-23 in Proceedings of the Sixth North American Crane Workshop, October 3-5, 1991, Regina, Saskatchewan, D.W. Stahlecker, and R.P. Urbanek, eds. Grand Island, NE: North American Crane Working Group.

Graf, W.L. 1988. Fluvial Processes in Dryland Rivers. New York: Springer-Verlag. 346 pp.

Haig, S.M. 1992. Piping plover (Charadrius melodus). Pp. 1-18 in Birds of North America, No. 2, A.F. Poole, P.R. Stettenheim, and F. Gill, eds. Washington, DC: American Ornithologists' Union; Philadelphia, PA: Academy of Natural Sciences.

Haig, S.M., and E. Elliott-Smith. In press. The piping plover #2. In Birds of North America, A. Poole, P. Stettenheim, and F. Gill, eds. American Ornithologists' Union, Philadelphia.

Harlan, J.R., and E.B. Speaker. 1951. Iowa Fish and Fishing. Des Moines: Iowa State Conservation Commission.

Hayes, M.A., and J.A. Barzen. 2003. Historical Breeding and Wintering Habitat Use by the Whooping Crane (*Grus americana*). Presentation at the Ninth North American Crane Workshop, January 21-25, 2003, Sacramento, CA.

Higgins, K.F., and M.R. Brashier, eds. 1993. Proceedings, the Missouri River and its Tributaries: Piping Plover and Least Tern Symposium. South Dakota State University, Brookings, SD. [Online]. Available: http://www.fs.fed.us/r2/nebraska/gpng/lt_plover/leastternandpipingploverpapersmissouri.htm [accessed Nov. 25, 2003].

Hoke, E. 1995. A survey and analysis of the unionid mollusks of the Platte Rivers of Nebraska and their minor tributaries. Trans. Nebr. Acad. Sci. 22:49-72.

Holm, C.F., J.D. Armstrong, and D.J. Gilvear. 2001. Investigating a major assumption of predictive instream habitat models: Is water velocity preference of juvenile Atlantic salmon independent of discharge? J. Fish Biol. 59(6):1653-1666.

Holmgren, P.S., M.W. Schuyler, and R. Davis. 1993. The Big Bend country of the Platte River: A history of human settlement. Pp. 58-74 in the Platte River: An Atlas of the Big Bend Region, A. Jenkins, ed. Kearney, NE: University of Nebraska at Kearney.

Howe, M.A. 1989. Migration of Radio-Marked Whooping Cranes from the Aransas-Wood Buffalo Population: Patterns of Habitat Use, Behavior, and Survival. Fish and Wildlife Technical Report No. 21. Washington, DC: U.S. Department of Interior, Fish and Wildlife Service. 33 pp.

Hurley, K.L. 1999. Habitat Use, Selection, and Movements of Mississippi River Pallid Sturgeon and Validity of Pallid Sturgeon Age Estimates from Pectoral Fin Rays. M.S. Thesis, Southern Illinois University, Carbondale, IL.

Hurr, T.R. 1983. Ground-Water Hydrology of the Mormon Island Crane Meadows Wildlife Area near Grand Island, Hall County, Nebraska. Geological Survey Professional Paper 1277-H. Washington, DC: U.S. Government Printing Office. 12 pp.

Hydrology Work Group. 1989. Evaluation Management Alternatives: Sediment, Flow, and Channel Geometry Considerations. Summary of Findings. Platte River Management Joint Study. December 1989.

Iverson, G.C., P.A. Vohs, and T.C. Tacha. 1987. Habitat use by mid-continent Sandhill Cranes during spring migration. J. Wildl. Manage. 51:448-458.

Jackman, L. 1847. Journal of Levi Jackman. The Gospel Library, the Pioneer Story. Salt Lake City: The Church of Jesus Christ of the Latter-day Saints. [Online]. Available: http://www.lds.org/gospellibrary/pioneer/15_Platte_River.html [accessed Jan. 2, 2004].

Jefferson, T. 1787. Notes on the State of Virginia, 2nd Ed. London: J. Stockdale.

Johnson, B.B., and L. Kuenning. 2002. 25th Year Edition Nebraska Farm Real Estate Market Developments 2001-02. Nebraska Cooperative Extension EC 02-809-S. University of Nebraska, Lincoln. [Online]. Available: http://agecon.unl.edu/realestate/re2002.pdf [accessed Oct. 28, 2003].

Johnson, D.H. 1980. The comparison of usage and availability measurements for evaluating resource preference. Ecology 61(1):65-71.

Johnson, K.A. 1982. Whooping crane use of the Platte River, Nebraska: History, status, and management recommendations. Pp. 33-43 in Proceedings 1981 Crane Workshop, J.C. Lewis, ed. Tavernier, FL: National Audubon Society.

Johnson, K.A., and S.A. Temple. 1980. The Migratory Ecology of the Whooping Crane (*Grus americana*). Contract No. 14-16-009-78-034. Prepared for U.S. Fish and Wildlife Service, by University of Wisconsin, Madison. 88 pp.

Johnson, W.C. 1994. Woodland expansion in the Platte River, Nebraska: Patterns and causes. Ecol. Monogr. 64(1):45-84.

Johnson, W.C. 1997. Equilibrium response of riparian vegetation to flow regulation in the Platte River, Nebraska. Regul. River. 13(5):403-415.

Johnson, W.C. 2000. Tree recruitment and survival in rivers: Influence of hydrological processes. Hydrol. Process. 14(16/17):3051-3074.

Johnson, W.C., and S.E. Boettcher. 1999. Restoration of the Platte River: What is the target? Land Water 43(3):20-23.

Johnson, W.C., and S.E. Boettcher. 2000a. The presettlement Platte: Wooded or prairie river? Great Plains Res. 10(1):39-68.

Johnson, W.C., and S.E. Boettcher. 2000b. The Platte as a wooded river: A response to Currier and Davis. Great Plains Res. 10(1):85-88.

Kallemeyn, L. 1983. Status of the pallid sturgeon, *Scaphirhynchus albus*. Fisheries 8(1):3-9.

Keenlyne, K.D. 1989. A Report on the Pallid Sturgeon. Pierre, SD: U.S. Fish and Wildlife Service.

Keenlyne, K.D., and L.G. Jenkins. 1993. Age at sexual maturity of the pallid sturgeon. Trans. Am. Fish. Soc. 122(3):393-396.

Keenlyne, K.D., C.J. Henry, A. Tews, and P. Clancey. 1994. Morphometric comparisons of upper Missouri River sturgeons. Trans. Am. Fish. Soc. 123(5):779-785.

Kirsch, E.M. 1996. Habitat Selection and Productivity of Least Terns on the Lower Platte River, Nebraska. Wildlife Monographs No. 132. Bethesda, MD: Wildlife Society.

Kirsch, E.M., and J.G. Sidle. 1999. Status of the interior population of least tern. J. Wildl. Manage. 63(2):470-483.

Kirchner, J.E. 1983. Interpretation of Sediment Data for the South Platte River in Colorado and Nebraska, and the North Platte and Platte Rivers in Nebraska. Geological Survey Professional Paper No. 1277-D. Washington, DC: U.S. Department of the Interior, Geological Survey. 37 pp.

Knighton, D. 1998. Fluvial Forms and Processes: A New Perspective. London: Arnold. 383 pp.

Knopf, F.L., R.R. Johnson, T. Rich, F.B. Samson, and R.C. Szaro. 1988. Conservation of riparian ecosystems in the United States. Wilson Bull. 100(2):272-284.

Kondolf, G.M, E.W. Larsen, and J.G. Williams. 2000. Measuring and modeling the hydraulic environment for assessing instream flows. N. Am. J. Fish. Manage. 20(4):1016-1028.

Krapu, G.L. 2003. The Role of the Platte River in Contributing to Needs of the Mid-continent Population of Sandhill Crane. Presentation at the Third Meeting on Threatened and Endangered Species, August 11, 2003, Grand Island, NE.

Krapu, G.L., and D.H. Johnson. 1990. Conditioning of sandhill cranes during fall migration. J. Wildl. Manage. 54(2):234-238.

Krapu, G.L., D.E. Facey, E K. Fritzell, and D.H. Johnson. 1984. Habitat use by migrant sandhill Cranes in Nebraska. J. Wild. Manage. 48(2):407-417.

Krapu, G.L., G.C. Iverson, K.J. Reinecke, and C.M. Boise. 1985. Fat deposition and usage by arctic-nesting sandhill cranes during spring. Auk 102:362-368.

Kuyt, E. 1979. Banding of juvenile whooping cranes and discovery of the summer habitat used by nonbreeders. Pp. 109-111 in Proceedings 1978 Crane Workshop, December 1978, Rockport, TX, J.C. Lewis, ed. Fort Collins, CO: Colorado State University Printing Service.

Kuyt, E. 1992. Aerial Radio-Tracking of Whooping Cranes Migrating between Wood Buffalo National Park and Aransas National Wildlife Refuge, 1981-1984. Occasional Paper No. 74. Ottawa: Canadian Wildlife Service. 53 pp.

Lacey, J.F. 1900. Remarks. House of Representatives Report 56-474, 56th Congress, 1st Session, Vol. 1 (1900), Congressional Record 33:4871.

Lackey, J.L. 1997. Management Implications for Least Terns and Piping Plovers in Manmade Habitats. 1997 Platte River Basin Ecosystem Symposium Feb. 18-19, 1997, Kearney, NE. [Abstract]. [Online]. Available: http://www.ianr.unl.edu/ianr/pwp/products/97abst/avian97.htm [accessed Jan. 23, 2004].

Lacy, R.C., M.B. Borbat, and J.P. Pollak. 2003. VORTEX: A Stochastic Simulation of the Extinction Process, Version 9.3. Brookfield, IL: Chicago Zoological Society.

Larson, M.A., M.R. Ryan, and B.G. Root. 2000. Piping plover survival in the Great Plains: An updated analysis. J. Field Ornithol. 71(4):721-729.

Larson, M.A., M.R. Ryan, and R.K. Murphy. 2002. Population viability of piping plovers: Effects of predator exclusion. J. Wildl. Manage. 66(2):361-371.

Lawson, M.P. 1974. The Climate of the Great American Desert: Reconstruction of the Climate of Western Interior United States, 1800-1850. Lincoln, NE: University of Nebraska Press.

Lee, D.S. 1980. Pallid sturgeon (*Scaphirhynchus albus* Forbes and Richardson). P. 43 in Atlas of North American Freshwater Fishes, D.S. Lee, C.R. Gilbert, C.H. Hocutt, R.E. Jenkins, D.E. McAllister, and J.R. Stauffer Jr., eds. Publ. No. 1980-12. Raleigh, NC: North Carolina State Museum of Natural History.

Leopold, L.B. 1994. A View of the River. Cambridge, MA: Harvard University Press. 298 pp.

Leopold, L.B. 1997. Water, Rivers and Creeks. Sausalito, CA: University Science Books. 185 pp.

Lewis, G. 2003. Platte River Channel Dynamics Investigations. Presentation at the Third Meeting on Threatened and Endangered Species, August 11, 2003, Grand Island, NE.

Lewis, J.C. 1995. Whooping crane (Grus americana). Birds of North America, No. 153, A. Poole, and F. Gill, eds. Washington, DC: American Ornithologists' Union; Philadelphia, PA: Academy of Natural Sciences.

Lewis, J.C, E. Kuyt, K.E. Schwindt, and T. Stehn. 1992. Mortality in fledged whooping cranes of the Aransas/Wood Buffalo population. Pp. 145-148 in Proceedings 1988 North American Crane Workshop, D.A. Wood, ed. Tallahassee, FL: Florida Game and Freshwater Fish Commission.

Lewis, J.C., G. Archibald, R.C. Drewien, R. Edwards, G. Gee, B. Huey, L.A. Linam, R.A. Lock, S. Nesbitt, and T. Stehn. 1994. Whooping Crane Recovery Plan. Albuquerque, NM: U.S. Fish and Wildlife Service. [Online]. Available: http://arizonaes.fws.gov/Documents/RecoveryPlans/WhoopingCrane.pdf [accessed Jan. 29, 2004].

Lingle, G.R. 1987. Status of whooping crane migration habitat within the Great Plains of North America. Pp. 331-340 in Proceedings of the 1985 Crane Workshop, J.C. Lewis, ed. Grand Island, NE: Platte River Whooping Crane Maintenance Trust.

Lingle, G.R. 1993a. Nest success and flow relationships on the central Platte River. Pp. 69-72 in Proceedings of the Missouri River and its Tributaries: Piping Plover and Least Tern Symposium-Workshop, K.F. Higgins, and M.R. Brashier, eds. Brookings, SD: South Dakota State University, Department of Wildlife and Fisheries Sciences.

Lingle, G.R. 1993b. Causes of nest failure and mortality of Least Terns and Piping Plovers along the central Platte River. Pp. 130-136 in Proceedings of the Missouri River and its Tributaries: Piping Plover and Least Tern Symposium-Workshop, K.F. Higgins, and M.R. Brashier, eds. Brookings, SD: South Dakota State University, Department of Wildlife and Fisheries Sciences.

Lingle, G.R. 1993c. Site fidelity and movements of Least Terns and Piping Plovers along the Platte River, Nebraska. Pp. 189-191 in Proceedings of the Missouri River and its Tributaries: Piping Plover and Least Tern Symposium-Workshop, K.F. Higgins, and M.R. Brashier, eds. Brookings, SD: South Dakota State University, Department of Wildlife and Fisheries Sciences.

Lingle, G.R., G.A. Wingfield, and J.W. Ziewitz. 1991. The migration ecology of whooping cranes in Nebraska, U.S.A. Pp. 395-401 in Proceedings 1987 International Crane Workshop, J.T. Harris, ed. Baraboo, WI: International Crane Foundation.

Long, S.H., and E. James. 1823. Account of An Expedition From Pittsburgh to the Rocky Mountains, Performed in the Years 1819 and 1820, by Order of the Hon. J.C. Calhoun, Sec'y of War: Under the Command of Major Stephen H. Long. From the Notes of Major Long, Mr. T. Say, and Other Gentlemen of the Exploring Party. Philadelphia: Carey and Lea.

Losos, E., J. Hayes, A. Phillips, D. Wilcove, and C. Alkire. 1995. Taxpayer-subsidized resource extraction harms species. BioScience 45(7):446-455.

Ludwig, D. 1999. Is it meaningful to estimate a probability of extinction? Ecology 80(1):298-310.

Lutey, J.M. 2002. Species Recovery Objectives for Four Target Species in the Central and Lower Platte River (Whooping Crane, Interior Least Tern, Piping Plover, Pallid Sturgeon). Denver, CO: U.S. Fish and Wildlife Service. 36 pp.

Manly, B.F.J., L.L. McDonald, D.L. Thomas, T.L. McDonald, and W.P. Erickson. 2002. Resources Selection by Animals: Statistical Design and Analysis for Field Studies, 2nd Ed. Dordrecht: Kluwer. 240 pp.

Mattes, M.J. 1969. The Great Platte River Road: The Covered Wagon Mainline via Fort Kearny to Fort Laramie. Lincoln, NE: Nebraska State Historical Society. 583 pp.

McBride, M.J. 1995. Benthic Macroinvertebrate Communities Associated with Forested and Open Riparian Areas Along the Central Platte River. M.S. Thesis, University of Nebraska, Lincoln, NE. 74 pp.

McCue, R. 2003. Overview of ESA Consultations in the Platte River Basin. Presentation at the First Meeting on Endangered and Threatened Species in the Platte River Basin, May 6-8, 2003, Kearney, NE.

McQuown, E.C., B.L. Sloss, R.J. Sheehan, J. Rodzen, G.J. Tranah, and B. May. 2000. Microsatellite analysis of genetic variation in sturgeon: New primer sequences for Scaphirhynchus and Acipenser. Trans. Am. Fish. Soc. 129(6):1380-1388.

Melvin, S.M., and S.A. Temple. 1982. Migration ecology of sandhill cranes: A review. Pp. 73-87 in Proceedings 1981 Crane Workshop, J.C. Lewis, ed. Tavernier, FL: National Audubon Society.

Meyers, C. 1966. The Colorado river. Stanford Law Rev. 17:1.

Millington, A.C., and K. Pye, eds. 1994. Environmental Change in Drylands: Biogeographical and Geomorphological Perspectives. New York: Wiley.

Millspaugh, J.J., J.R. Skalski, B.J. Kernohan, K.J. Raedeke, G.C. Brunidge, and A.B. Cooper. 1998. Some comments on spatial independence in studies of resource selection. Wildl. Soc. Bull. 26(2):232-236.

Mirande, C.M., R. Lacy, and U. Seal, eds. 1991. Whooping Crane (*Grus americana*) Conservation Viability Assessment Workshop. Apple Valley, MN: IUCN/SSC Captive Breeding Specialist Group. 115 pp.

Mirande, C.M., J.R. Cannon, K. Agzigian, R.E. Bogart, S. Christiansen, J. Dubow, A.K. Fernandez, D.K. Howarth, C. Jones, K.G. Munson, S.I. Pandya, G. Sedaghatkish, K.L. Skerl, S.A. Stenquist, and J. Wheeler. 1997. Computer simulations of possible futures for two flocks of whooping cranes. Pp. 181-200 in Proceedings of the Seventh North American Crane Workshop, January 10-13, 1996, Biloxi, MS, R.P. Urbanek, and D.W. Stahlecker, eds. Grand Island, NE: North American Crane Working Group.

Mock, C.J. 2000. Rainfall in the garden of the United States Great Plains, 1870-1889. Climatic Change 44(1-2):173-195.

Murphy, P.J., and T.J. Randle. 2001. Platte River Channel: History and Restoration, Draft. U.S. Bureau of Reclamation, Denver, CO. [Online]. Available: http://www.platteriver.org/actions/eis.htm#TECHRPT [accessed Oct.30, 2003].

Murphy, P.J., and T.J. Randle. 2003. Platte River Channel: History and Restoration, Draft. U.S. Bureau of Reclamation, Denver, CO. July 24, 2003.

Murphy, P.J., T.J. Randle, and R.K. Simons. 2001. Platte River Sediment Transport and Riparian Vegetation Model, Draft, April 7, 2001. U.S. Bureau of Reclamation, Denver, CO. 63 pp.

Nadler, C.T., and S.A. Schumm. 1981. Metamorphosis of the South Platte and Arkansas Rivers, eastern Colorado. Phys. Geogr. 2(2):95-115.

NASS (National Agricultural Statistics Service). 1999. Irrigated Corn for Grain or Seed by County: 1997. 1997 Agricultural Atlas of the United States. Census of Agriculture, Nebraska Agricultural Statistics Service, U.S. Department of Agriculture. [Online]. Available: http://www.nass.usda.gov/census/census97/atlas97/map115.htm [accessed Nov. 5, 2003].

NCDC (National Climatic Data Center). 2003. Nebraska - Average 12 Month Precipitation, 1993-2003. Climate at a Glance, Climate Monitoring Reports and Products. National Climatic Data Center. [Online]. Available: http://www.ncdc.noaa.gov/oa/climate/research/cag3/NE.html [accessed Nov. 7, 2003].

Nebraska Agricultural Statistics Service. 2002. Nebraska-Corn: Acres Harvested for Grain, 1996. Nebraska Agricultural Statistics Service, National Agricultural Statistics Service, U.S. Department of Agriculture. [Online]. Available: http://www.nass.usda.gov/ne/9596wthr/pag_009.htm [accessed Nov. 5, 2003].

Nebraska Agricultural Statistics Service. 2004. Census of Agriculture 2002, Crops County and District Data, Agricultural Statistics Data Base. Nebraska Agricultural Statistics Service, National Agricultural Statistics Service, U.S. Department of Agriculture. [Online]. Available: http://www.nass.usda.gov:81/ipedb/ [accessed Feb. 20, 2004].

Nebraska Department of Environmental Quality. 2002. 2002 Nebraska Groundwater Quality Monitoring Report. Water Quality Assessment Section, Groundwater Unit, Nebraska Department of Environmental Quality, December 2002. [Online]. Available: http://www.deq.state.ne.us/Publica.nsf/Pages/WAT039 [accessed Dec. 3, 2003].

Nebraska Department of Natural Resources. 1986. Excerpt from "Policy Issue Study on Integrated Management of Surface Water and Groundwater," April, 1986. [Online]. Available: www.dnr.state.ne.us/watertaskforce/Resourcematerials/PolicyIssueStudy.doc [accessed Dec. 3, 2003].

Nebraska Natural Resources Commission. 1994. Estimated Water Use in Nebraska, 1990. Lincoln, NE: U.S. Geological Survey and Nebraska Natural Water Resources Commission, 58 pp.

Nebraska Public Power District. 2003. A Look at NPPD's Water Resources. Nebraska Public Power District, Columbus, NE.

Nesbitt, S.A. 1982. The past, present, and future of the whooping crane in Florida. Pp. 151-154 in Proceedings 1981 Crane Workshop, J.C. Lewis, ed. Tavernier, FL: National Audubon Society.

Nesbitt, S.A., M.J. Folk, M.G. Spalding, J.A. Schmidt, S.T. Schwikert, J.M. Nicolich, M. Wellington, J.C. Lewis, and T.H. Logan. 1997. An experimental release of whooping crane in Florida - the first three years. Pp. 79-85 in Proceedings of the Seventh North American Crane Workshop, January 10-13, 1996, Biloxi, MS, R.P. Urbanek, and D.W. Stahlecker, eds. Grand Island, NE: North American Crane Working Group.

Nixon, R.M. 1972. Comments on Endangered Species. Weekly Compilation of Presidential Documents 218(February 8, 1972): 223-224.

Nordstrom, L.H., and M.R. Ryan. 1996. Invertebrate abundance at occupied and potential piping plover nesting beaches: Great Plains alkali wetlands vs. the Great Lakes. Wetlands 16(4):429-435.

NPNRD (North Plate Natural Resources District). 2003. North Plate Natural Resources District, Gering, NE. [Online]. Available: http://www.npnrd.org [accessed Oct. 29, 2003].

NRC (National Research Council). 1986. Ecological Knowledge and Environmental Problem-Solving: Concepts and Case Studies. Washington, DC: National Academy Press.

NRC (National Research Council). 1995. Science and the Endangered Species Act. Washington, DC: National Academy Press.

NRC (National Research Council). 2002a. Riparian Areas: Functions and Strategies for Management. Washington, DC: National Academy Press.

NRC (National Research Council). 2002b. Scientific Evaluation of Biological Opinions on Endangered and Threatened Fishes in the Klamath River Basin, Interim Report. Washington, DC: National Academy Press.

NRC (National Research Council). 2003. Cumulative Environmental Effects of Oil and Gas Activities on Alaska's North Slope. Washington, DC: The National Academies Press.

NRC (National Research Council). 2004a. Endangered and Threatened Fishes in the Klamath River Basin: Causes of Decline and Strategies for Recovery. Washington, DC: The National Academies Press.

NRC (National Research Council). 2004b. Adaptive Management for Water Resources Project Planning. Washington, DC: The National Academies Press.

NRC (National Research Council). 2004c. Atlantic Salmon in Maine. Washington, DC: The National Academies Press.

NRCS (Natural Resources Conservation Service). 2001. Percent of Non-Federal Area in Developed Land, Map. Results from the 2001 NRI Urbanization. Natural Resources Conservation Service, U.S. Department of Agriculture. [Online]. Available: http://www.nrcs.usda.gov/technical/land/urban.html [accessed Oct. 28, 2003].

Olsen, D.L., D.R. Blankinship, R.C. Erickson, R. Drewien, H.D. Irby, R. Lock, and L.S. Smith. 1980. Whooping Crane Recovery Plan, January 1980, S.D. Derrickson, ed. Albuquerque, NM: U.S. Fish and Wildlife Service. 206 pp.

Parsons. 2003. Platte River Channel Dynamics Investigation, Executive Summary. Technical Memorandums A, B, C, and D. Department of Natural Resources, State of Nebraska. May 2003.

Patlis, J.M. 2001. Paying tribute to Joseph Heller with the Endangered Species Act: When critical habitat isn't. Stanford Environ. Law J. 20(1):133-217.

Patten, D.T., D.A. Harpman, M.I. Voita, and T.J. Randle. 2001. A managed flood on the Colorado River: Background, objectives, design, and implementation. Ecol. Appl. 11(3):635-643.

Payne, T.R. 1995. In the Matter of Applications A-17329 through A-17333 for Water Appropriations for Instream Flow Filed by the Game and Parks Commission. Department of Water Resources, Nebraska. Hyatt Court Reporting & Video Registered Court Reporters, Denver, CO. December 19, 1995.

Pereira, J.M., and R.M. Itami. 1991. GIS-based habitat modeling using logistic multiple regression: A study of the Mt. Graham red squirrel. Photogramm. Eng. Rem. S. 57(11):1475-1486.

Peters, E.J., R.S. Holland, M.A. Callam, and D.L. Bunnell. 1989. Platte River Habitat Suitability Criteria: Habitat Utilization, Preference and Suitability Index Criteria for Fish and Aquatic Invertebrates in the lower Platte River, G. Zuerlein, ed. Nebraska Technical Series No. 17. Lincoln, NE: Nebraska Game and Parks Commission.

Petts, G.E., and P. Calow, eds. 1996a. River Flows and Channel Forms: Selected Extracts from the Rivers Handbook. London: Blackwell Science. 262 pp.

Petts, G.E., and P. Calow, eds. 1996b. River Restoration: Selected Extracts from the Rivers Handbook. London: Blackwell Science. 231 pp.

Petts, G.E., and P. Calow, eds. 1996c. River Biota: Diversity and Dynamics. Cambridge, MA: Blackwell Science. 257 pp.

Pflieger, W.L. 1997. The Fishes of Missouri. Jefferson City, MO: Missouri Department of Conservation.

Phelps, S.R., and F.W. Allendorf. 1983. Genetic identity of pallid and shovelnose sturgeon (Scaphirhynchus albus and S. platorynchus). Copeia 1983(3):696-700.

Pitlick, J., and P.R. Wilcock. 2001. Relations between streamflow, sediment transport, and aquatic habitat in regulated rivers. Pp. 185-198 in Geomorphic Processes and Riverine Habitat, J.M. Dorava, D.R. Montgomery, B.B. Palcsak, and F.A. Fitzpatrick, eds. Washington, DC: American Geophysical Union.

Platte River Whooping Crane Maintenance Trust. 1999. Crane and Waterfowl Migration Report-Spring 1999. Platte River Whooping Crane Maintenance Trust, Grand Island, NE. 8 pp.

Plettner, R.G. 1997. Reproductive Success of Least Terns and Piping Plovers on Nebraska Public Power District's Nesting Sites. 1997 Platte River Basin Ecosystem Symposium, Feb. 18-19, 1997, Kearney, NE. [Abstract]. [Online]. Available: http://www.ianr.unl.edu/ianr/pwp/products/97abst/avian97.htm [accessed Jan. 23, 2004].

Plissner, J.H., and S.M. Haig. 2000. Viability of piping plovers (*Charadrius melodus*) metapopulations. Biol. Conserv. 92(2):163-173.

Poff, N.L., J.D. Allan, M.B. Bain, J.R. Karr, K.L. Prestegaard, B.D. Richter, R.E. Sparks, and J.C. Stromberg. 1997. The natural flow regime: A paradigm for river conservation and restoration. BioScience 47(11):769-784.

Poff, N.L., J.D. Allan, M.A. Palmer, D.D. Hart, B.D. Richter, A.H. Arthington, K.H. Rogers, J.L. Meyer, and J.A. Stanford. 2003. River flows and water wars: Emerging science for environmental decision making. Front. Ecology Environ. 1(6)298-306.

Postel, S., and B.D. Richter. 2003. Rivers for Life: Managing Water for People and Nature. Washington, DC: Island Press.

Powell, J.W., G.K. Gilbert, C.E. Dutton, A.H. Thompson, and W. Drummond, Jr. 1878. Report on the Lands of the Arid Region of the United States. Washington, DC: U.S. Government Printing Office. 195 pp.

Prindiville-Gaines, E., and M.R. Ryan. 1988. Piping Plover habitat use and reproductive success in North Dakota. J. Wildl. Manage. 52(2):266-273.

Raisz, E., and W.W. Atwood. 1957. Landforms of the United States, 6th Ed. Cambridge, MA: E. Raisz.

Randle, T., and P. Murphy. 2003. Geomorphology of the Platte River. Presentation at the First Meeting on Threatened and Endangered Species, May 6, 2003, Kearney, NE.

Randle, T.J., and M.A. Samad. 2003. Platte River Flow and Sediment Transport Between North Platte and Grand Island, Nebraska (1895-1999), Draft. Bureau of Reclamation, U.S. Department of the Interior, Denver, CO. July 2003.

Reed, J.M. 2003. Variability Issues for Target Species in the Platte River. Report to the National Academy of Sciences, National Research Council, by J.M. Reed, Department of Biology, Tufts University, Medford, MA. October 2003.

Regan, H.M., M. Colyvan, and M.A. Burgman. 2002. A taxonomy of treatment of uncertainty for ecology and conservation biology. Ecol. Appl. 12(2):618-628.

Rhoads, B.L. 1994. Fluvial geomorphology. Prog. Phys. Geog. 18(1):103-123.

Ricciardi, A., and J.B. Rasmussen. 1999. Extinction rates of North American freshwater fauna. Conserv. Biol. 13(5):1220-1222.

Richards, K.S. 1982. Rivers: Form and Process in Alluvial Channels. London: Methuen. 358 pp.

Richert, A.L.D. 1999. Multiple Scale Analyses of Whooping Crane Habitat in Nebraska. Ph.D. Dissertation, University of Nebraska, Lincoln, NE. 175 pp.

Richter, B.D., J.V. Baumgartner, J. Powell, and D.P. Braun. 1996. A method for assessing hydrologic alteration within ecosystems. Conserv. Biol. 10(4):1163-1174.

Richter, B.D., J.V. Baumgartner, R. Wigington, and D.P. Braun. 1997. How much water does a river need? Freshwater Biol. 37(1):231-249.

Richter, B.D., R. Mathews, D.L. Harrison, and R. Wigington. 2003. Ecologically sustainable water management: Managing river flows for ecological integrity. Ecol. Appl. 13(1):206-224.

Robinson, S.K., F.R. Thompson III, T.M. Donovan, D.R. Whitehead, and J. Faaborg. 1995. Regional forest fragmentation and the nesting success of migratory birds. Science 267(5206):1987-1990.

Ruelle, R., and K.D. Keenlyne. 1993. Contaminants in Missouri River pallid sturgeon. Bull. Environ. Contam. Toxicol. 50(6):898-906.

Ryan, M.R., B.G. Root, and P.M. Mayer. 1993. Status of piping plovers in the Great Plains of North America: A demographic simulation model. Conserv. Biol. 7(3):581-585.

Saarinen, T.F. 1966. Perception of the Drought Hazard on the Great Plains. Research Paper No. 106. Chicago: University of Chicago.

Sax, J.L. 2000. Environmental law at the turn of the century: A reportorial fragment of contemporary history. Calif. Law Rev. 88(6):2375-2402.

Schainost, S., and M.D. Koneya. 1999. Fishes of the Platte River Basin. Nebraska Game and Parks Commission, Lincoln, NE.

Scharf, W.C. 2003. The Avifauna of Three Habitats at Cottonwood Ranch and the Jeffrey Island Habitat Area, Nebraska 2001 and 2002. Report to Central Nebraska Public Power and Irrigation District and Nebraska Public Power District. Submitted to the FERC as part of the 2002 Wildlife Monitoring Report for Projects #1417 and #1835.

Schumm, S.A. 1977. The Fluvial System. New York: Wiley. 338 pp.

Schumm, S.A. 1998. Geomorphology of North Platte River and Platte River in Nebraska. Ayres Associates, Ft. Collins, CO. 53 pp.

Schwalbach, M.J. 1988. Conservation of Least Terns and Piping Plovers along the Missouri River and Its Major Western Tributaries in South Dakota. M.S. Thesis. South Dakota State University, Brookings, SD. 104 pp. [Online]. Available: LTPPMissouriSchwalbach1988b.pdf [accessed Jan. 28, 2003].

Shaffer, M.L., L. Hood-Watchman, W.J. Snape, and I.K. Latchis. 2002. Population viability analysis and conservation policy. Pp. 123-146 in Population Viability Analysis, S.R. Beissinger, and D.R. McCullough, eds. Chicago, IL: University of Chicago Press.

Sharp, D.E., J.A. Dubovsky, and K.L. Kruse. 2003. Status and Harvests of the Mid-Continent and Rocky Mountain Populations of Sandhill Cranes. U.S. Fish and Wildlife Service, Denver, CO. 9 pp. [Online]. Available: http://migratorybirds.fws.gov/reports/status03/crane.pdf [accessed Oct. 28, 2003].

Sheehan, R.J., R.C. Heidinger, P.S. Wills, M.A. Schmidt, G.A. Conover, and K.L. Hurley. 1999. Guide to the Pallid Sturgeon Shovelnose Sturgeon Character Index (CI) and Morphometric Character Index (mCI). SIUC Fisheries Bulletin No. 14. Fisheries Research Laboratory, Southern Illinois University, Carbondale, IL. [Online]. Available: http://ws3.coopfish.siu.edu/pallid_guide/ [accessed Nov. 3, 2003].

Sheldon, A.E. 1913. History and Stories of Nebraska. Chicago: University Pub. Co.

Shen, H.W., ed. 1971. River Mechanics. Fort Collins, CO: H.W. Shen.

Sidle, J.G. 1993. Least tern and piping plover use of sand and gravel pits along the Platte and Loup rivers, Nebraska. Pp. 91 in Proceedings of the Missouri River and its Tributaries: Piping Plover and Least Tern Symposium-Workshop, K.F. Higgins, and M.R. Brashier, eds. Brookings, SD: South Dakota State University, Department of Wildlife and Fisheries Sciences.

Sidle, J.G., and E.M. Kirsch. 1993. Least terns and piping plovers nesting at sandpits in Nebraska. Colon. Waterbird 16(2):139-148.

Simons, A.M., R.M. Wood, L.S. Heath, B.R. Kuhajda, and R.L. Mayden. 2001. Phylogenetics of *Scaphirhynchus* based on mitochondrial DNA sequences. Trans. Am. Fish. Soc. 130(3):359-366.

Simons, D.B., and F. Sentürk. 1992. Sediment Transport Technology: Water and Sediment Dynamics. Littleton, CO: Water Resources Publications. 897 pp.

Simons, R.K., and D.B. Simons. 1994. An analysis of Platte River channel changes. Pp. 341-361 in the Variability of Large Alluvial Rivers, S.A. Schumm, and B.R. Winkley, eds. New York: American Society of Civil Engineers Press.

Simons and Associates, Inc. 2000. Physical History of the Platte River in Nebraska: Focusing Upon Flow, Sediment Transport, Geomorphology, and Vegetation. Ft. Collins, CO: Simons and Associates, Inc. [Online]. Available: http://www.platteriver.org/actions/eis.htm#TECHRPT [accessed Dec. 12, 2003].

Simpson, H. 2003. Hydrology and Water Use in the South Platte River Basin. Presentation at the Third Meeting on Threatened and Endangered Species, August 11, 2003, Grand Island, NE.

Smith, L.S. (Whooping Crane Recovery Team). 1986. Whooping Crane Recovery Plan 1986. Albuquerque, NM: U.S. Fish and Wildlife Service. 283 pp.

Smith, J.W., and R.B. Renken. 1990. Improving the Status of Endangered Species in Missouri (Least Tern Investigations). Final Report. Endangered Species Project No. SE-01-19. Missouri Department of Conservation, Columbia. 33 pp.

Smith, J.W., and R.B. Renken. 1993. Reproductive success of least terns (*Sterna antillarum*) in the Mississippi River Valley. Colon. Waterbird. 16(1):39-44.

Smith, L.S. 1986. Whooping Crane Recovery Plan 1986. Albuquerque, NM: U.S. Fish and Wildlife Service. 283 pp.

Snook, V.A. 2001. Movements and Habitat Use by Hatchery-Reared Pallid Sturgeon in the Lower Platte River, Nebraska. M.S. Thesis, University of Nebraska, Lincoln, NE.

Snook, V.A., E.J. Peters, and L.J. Young. 2002. Movements and habitat use by hatchery-reared pallid sturgeon in the lower Platte River, Nebraska. Pp. 161-173 in Biology, Management and Protection of North American Sturgeon, W. Van Winkle, P.J. Anders, D.H. Secor, and D.A. Dixon, eds. American Fisheries Society Symposium 28. Bethesda, MD: American Fisheries Society.

Solberg, J.W. 2002. Coordinated Spring Mid-Continent *Sandhill Crane* Survey. U.S. Fish and Wildlife Service, Bismarck, ND. 10 pp.

SPNRD (South Platte Natural Resources District). 2003. South Platte Natural Resources District. [Online]. Available: http://www.spnrd.org [accessed Oct. 29, 2003].

Stahlecker, D.W. 1997. Availability of stopover habitat for migrant whooping cranes in Nebraska. Pp. 132-140 in Proceedings of the Seventh North American Crane Workshop, January 10-13, 1996, Biloxi, MS, R.P. Urbanek, and D.W. Stahlecker, eds. Grand Island, NE: North American Crane Working Group.

Stalnaker, C.B., B.L. Lamb, J. Henriksen, K. Bovee, and J. Bartholow. 1995. The Instream Flow Incremental Methodology: A Primer for IFIM. Biological Report 29. Washington, DC: National Biological Service, U.S. Department of the Interior. 45 pp.

Stanford, J.A., J.V. Ward, W.J. Liss, C.A. Frissell, R.N. Williams, J.A. Lichatowich, and C.C. Coutant. 1996. A general protocol for restoration of regulated rivers. Regul. River. 12(4/5):391-413.

Stearns, F.W. 1949. Ninety years change in a northern hardwood forest in Wisconsin. Ecology 30(3):350-358.

Steed, T. 1850. Journal of Tomas Steed. The Gospel Library, the Pioneer Story. Salt Lake City: The Church of Jesus Christ of the Latter-day Saints. [Online]. Available: http://www.lds.org/gospellibrary/pioneer/15_Platte_River.html [accessed Jan. 02. 2004].

Stehn, T. 2001. Relation between Inflows, Crabs, Salinities, and Whooping Cranes. Aransas National Wildlife Refuge. November 26, 2001. 4 pp. [Online]. Available: http://www.learner.org/jnorth/tm/crane/Stehn_CrabDocument.html [accessed Jan. 30, 2004].

Stehn, T. 2003. Importance of the Platte River to Whooping Cranes. Presentation at the First Meeting on Endangered and Threatened Species in the Platte River Basin, May 6-8, 2003, Kearney, NE.

Stroup, D., M. Rodney, and D. Anderson. 2001. Flow Characterization for the Platte River Basin in Colorado, Wyoming, and Nebraska, Draft. Platte River Recovery Program EIS Office, Lakewood, CO. 21 pp plus appendices.

Su, L. 2003. Habitat Selection by Sandhill Cranes, *Grus Canadensis tabida*, at Multiple Geographic Scales in Wisconsin. Ph. D. Dissertation, University of Wisconsin, Madison, WI.

Swenk, M.H. 1933. The present status of the whooping crane. Nebr. Bird Rev. 11(4):111-129.

Swigle, B.D. 2003. Movement and Habitat Use by Shovelnose and Pallid Sturgeon in the Lower Platte River, Nebraska. M.S. Thesis, University of Nebraska, Lincoln, NE.

Thompson, B.C., J.A. Jackson, J. Burger, L.A. Hill, E.M. Kirsch, and J.L. Atwood. 1997. Least Tern (Sterna antillarum). The Birds of North America, No. 290, A. Poole, and F. Gill, eds. Philadelphia, PA: Academy of Natural Sciences; Washington, DC: American Ornithologists' Union. [Online]. Available: http://www.birds.cornell.edu/birdsofna/excerpts/lsttern.html [accessed Nov. 3, 2003].

USACE (U.S. Army Corps of Engineers). 1996. The National Water Control Infrastructure: National Inventory of Dams. CD-ROM. Washington DC: U.S. Army Corps of Engineers, Federal Emergency Management Agency.

USBR (U.S. Bureau of Reclamation). 1998. The Platte River Program. [Online]. Available: http://mcmcweb.er.usgs.gov/platte [accessed Feb. 4, 2004].

USFWS (U.S. Fish and Wildlife Service). 1980. Habitat Evaluation Procedures (HEP). Ecological Service Manual ESM 103. Fort Collins, CO: U.S. Fish and Wildlife Service.

USFWS (U.S. Fish and Wildlife Service). 1981a. The Platte River Ecology Study: Special Research Report. Jamestown, ND: U.S. Fish and Wildlife Service, Northern Prairie Wildlife Research Center. [Online]. Available: http://www.npwrc.usgs.gov/resource/othrdata/platteco/platteco.htm [accessed Oct. 28, 2003].

USFWS (U.S. Fish and Wildlife Service). 1981b. Standards for the Development of Habitat Suitability Index Models. EMS 103. Washington, DC: Division of Ecological Services, U.S. Fish and Wildlife Service, Department of the Interior.

USFWS (U.S. Fish and Wildlife Service). 1988. Recovery Plan for Piping Plovers, *Charadrius melodus*, of the Great Lakes and Northern Great Plains, Great Lakes/Northern Great Plains Piping Plover Recovery Team. Twin Cities, MN: Department of the Interior, U.S. Fish and Wildlife Service.

USFWS (U.S. Fish and Wildlife Service). 1990. Recovery Plan for the Interior Population of the Least Tern (*Sterna antillarum*). Twin Cities, MN: Department of the Interior, U.S. Fish and Wildlife Service. 90 pp.

USFWS (U.S. Fish and Wildlife Service). 1994. Endangered Species Listing Handbook: Procedural Guidance for the Preparation and Processing of Rules and Notices Pursuant to the Endangered Species Act, 4th Ed. Washington, DC: U.S. Fish and Wildlife Service, Division of Endangered Species.

USFWS (U.S. Fish and Wildlife Service). 2002a. Revised Intra-Service Section 7 Consultation for Federal Agency Actions Resulting in Minor Water Depletions to the Platte River system. Memorandum to Assistant Regional Director, Ecological Services, Region 6, from Regional Director, Region 6, U.S. Fish and Wildlife Service, Mountain-Prairie Region, Denver, CO. March 4, 2002.

USFWS (U.S. Fish and Wildlife Service). 2002b. USFWS Instream Flow Recommendations: Proposed Definitions and Usage for the Platte River Recovery Implementation Program, Draft. U.S. Fish and Wildlife Service, Mountain-Prairie Region, Region 6, Fort Collins, CO. December 23, 2002. 25 pp.

USFWS (U.S. Fish and Wildlife Service). 2003. Waterfowl Population Status, 2003. U.S. Department of the Interior, Washington, DC. 53 pp. [Online]. Available: http://migratorybirds.fws.gov/reports/status03/statusofwaterfowl03.pdf [accessed Oct. 28, 2003].

USFWS (U.S. Fish and Wildlife Service). 2004a. Piping Plover. Mountain-Prairie Region, U.S. Fish and Wildlife Service. [Online]. Available: http://mountain-prairie.fws.gov/pipingplover/ [accessed Feb. 5, 2004].

USFWS (U.S. Fish and Wildlife Service). 2004b. The Interior Least Tern. U.S. Fish and Wildlife Service. [Online]. Available: http://images.amrivers.org/content/images/photo.acs?object_id=117&photo_id=916 [accessed Feb. 5, 2004].

USFWS and NMFS (U.S. Fish and Wildlife Service/ National Marine Fisheries Service). 1998. Endangered Species Consultation Handbook: Procedures for Conducting Section 7 Consultations and Conferences. Washington, DC: U.S. Fish and Wildlife Service and National Marine Fisheries Service.

USGCRP (U.S. Global Change Research Program). 2000a. Climate Change Impacts on the United States. The Potential Consequences of Climate Variability and Change Overview: Water Sector. National Assessment Synthesis Team, U.S. Global Change Research Program. [Online]. Available: http://www.usgcrp.gov/usgcrp/Library/nationalassessment/overviewwater.htm [accessed Oct. 28, 2003].

USGCRP (U.S. Global Change Research Program). 2000b. Climate Change Impacts on the United States. The Potential Consequences of Climate Variability and Change Overview: Great Plains. National Assessment Synthesis Team, U.S. Global Change Research Program. [Online]. Available: http://www.usgcrp.gov/usgcrp/Library/nationalassessment/overviewgreatplains.htm [accessed Oct. 28, 2003].

USGS (U.S. Geological Survey). 2003. Daily Streamflow for Nebraska. USGS 06774000 Platte River near Duncan, NE. Water Resources of the United States, U.S. Geological Survey. [Online]. Available: http://nwis.waterdata.usgs.gov/ne/nwis/discharge/?site_no=06774000&agency_cd=USGS. [accessed Oct. 29, 2003].

Van Horne, B. 1983. Density as a misleading indicator of habitat quality. J. Wildl. Manage. 47:893-901.

Waddle, T.J., ed. 2001. PHABSIM for Windows: User's Manual and Exercises. USGS Open-File Report 01-340. U.S. Geological Survey, Fort Collins, CO. 288 pp. [Online]. Available: http://www.fort.usgs.gov/products/publications/15000/preface.html [accessed Nov. 12, 2003].

Watson, J.H., and P.A. Stewart. 1991. Lower Yellowstone River Pallid Sturgeon Study. Miles City, MT: Montana Department of Fish, Wildlife and Parks.

Webb, W.P. 1931. The Great Plains. Boston: Ginn and Co. (Reprint 1981, University of Nebraska Press, Lincoln, NE).

Weller, M.W., and B.D.J. Batt. 1988. Waterfowl in winter: Past, present, and future. Pp. 3-8 in Waterfowl in Winter: Selected Papers from Symposium and Workshop held in Galveston, Texas, 7-10 January 1985, M.W. Weller, ed. Minneapolis: University of Minnesota Press.

Wemmer, L.C. 2000. Conservation of the Piping Plover in the Great Lakes Region: A Landscape-ecosystem Approach. Ph.D. Dissertation, University of Minnesota, Twin Cities, MN.

Westerskov, K. 1950. Methods for determining the age of game bird eggs. J. Wildl. Manage. 14:56-67.

Whitney, G.G., and J.P. DeCant. 2001. Government land office surveys and other early land surveys. Pp. 147-172 in the Historical Ecology Handbook, D. Egan, and E.A. Howell, eds. Washington, DC: Island Press.

Williams, G.P. 1978. The Case of the Shrinking Channels: The North Platte and Platte Rivers in Nebraska. Geological Survey Circular 781. Washington, DC: U.S. Geological Survey. 48 pp.

Wilson, A.A.G. 1988. Width of firebreak that is necessary to stop grassfires: Some field experiments. Can. J. For. Res. 18:682-687.

Wilson, E.C. 1991. Nesting and Foraging Ecology of Interior Least Terns on Sandpits in Central Nebraska. M.S. Thesis, University of Wyoming, Laramie.

Wilson, E.C., W.A. Hubert, and S.H. Anderson. 1993. Foraging and diet of least terns nesting at sand pits in the central Platte Valley, Nebraska. Pp. 142 in Proceedings of the Missouri River and its Tributaries: Piping Plover and Least Tern Symposium-Workshop, K.F. Higgins, and M.R. Brashier, eds. Brookings, SD: South Dakota State University, Dept. of Wildlife and Fisheries Sciences.

Wingfield, G.A. 1993. Least tern and piping plover use of Lake McConaughy, Nebraska. Pp. 64-65 in Proceedings of the Missouri River and its Tributaries: Piping Plover and Least Tern Symposium-Workshop, K.F. Higgins, and M.R. Brashier, eds. Brookings, SD: South Dakota State University, Dept. of Wildlife and Fisheries Sciences.

Winter, T.C., J.W. Harvey, O.L. Franke, and W.M. Alley. 1998. Groundwater and Surface Water - A Single Resource. Circular 1139. Denver, CO: U.S. Geological Survey. 79 pp.

Woodward, D. 2003. Platte River Surface Water and Ground Water Hydrology. Presentation at the Third Meeting on Threatened and Endangered Species, August 11, 2003, Grand Island, NE.

Yang, T. 2003. Some Comments on the Review by Parsons. Presentation at the Third Meeting on Threatened and Endangered Species, August 11, 2003, Grand Island, NE.

Ziewitz, J.W. 1992. Whooping crane riverine roosting habitat suitability model. Pp. 71-81 in Proceedings of the 1988 North American Crane Workshop, D.A. Wood, ed. Technical Report 12. Tallahassee: Florida Game and Freshwater Fish Commission.

Ziewitz, J.W., J.G. Sidle, and J.J. Dinan. 1992. Habitat conservation for nesting least terns and piping plovers on the Platte River, Nebraska. Prairie Nat. 24(1):1-20.

Appendix A

Biographical Information on Committee Members

Front (left to right) Edwin E. Herricks, William L. Graf, W. Carter Johnson, Hsieh Wen Shen
Back (left to right) Edward J. Peters, Dennis D. Murphy, Richard N. Palmer, Frank Lupi, Holly Doremus, Francesca Cuthbert, Lisa M. Butler Harrington, John "Jeb" A. Barzen, Katharine L. Jacobs, James Anthony Thompson

GEOMORPHOLOGY

William L. Graf (Chair) is Educational Foundation University Professor and professor of geography at the University of South Carolina. He earned a PhD from the University of Wisconsin, Madison in 1974 in physical geography with a minor in water resources management. His specialties include fluvial geomorphology and policy for public land and water, with emphasis on river channel change, human influences on river processes and morphology, contaminant transport and storage in river sediments, and the downstream effects of large dams. Much of his early work focused on dryland rivers; for the last several years, his work has been national in scope. He has served as an officer in the Geological Society of America and is past president of the Association of American Geographers.

In public-policy work, he has emphasized the interaction of science and decision making and resolution of the conflict between economic development and environmental preservation. Dr. Graf's work has been funded by 52 grants and contracts from federal, state, and local agencies. He has given more than 100 professional presentations and published 130 papers, articles, book chapters, and reports on geomorphology, riparian ecology, river management, and the interaction between science and public policy. His eight books include *Geomorphic Systems of North America, The Colorado River: Basin Stability and Management, Fluvial Processes in Dryland Rivers, Wilderness Preservation and the Sagebrush Rebellions,* and *Plutonium and the Rio Grande*, and he is the primary author and editor of *New Strategies for America's Watersheds* and *Research Opportunities in Geography at the U.S. Geological Survey*. He is principal author of *Dam Removal: Science and Decision Making* and editor of *Science for Dam Removal*, and he is working on *Dam the Consequences: The Effects of Dams on America's Rivers*. His work has produced awards from the Association of American Geographers, the Geological Society of America, and the British Geomorphological Research Group, and he has received a Guggenheim Fellowship, a Fulbright Senior Scholarship, and the Founders' Medal awarded by Queen Elizabeth II of Great Britain and the Royal Geographical Society.

Dr. Graf has served as a science and policy adviser in numerous capacities for federal, state, and local agencies and organizations. He is a national associate of the National Academy of Sciences, and at the National Research Council he has been a member of the Board on Earth Sciences and Resources, the Water and Science Technology Board, the Committee on Glen Canyon Environmental Studies, the Panel to Review the Critical Ecosystem Studies Initiative for Everglades National Park, Committee on the Restoration of the Greater Everglades Ecosystem, and the Committee on Rediscovering Geography. He has also chaired the Research Council Committee on Innovative Watershed Management, the Workshop to Advise the President's Council on Sustainable Development, and a committee to advise the U.S. Geological Survey on research priorities in geography. He chairs the Heinz Center's committee on the Social, Economic, and Environmental Outcomes of Dam Removal and has been the river specialist on teams to advise Costa Rica on dam and river management and a member of a recovery team of the U.S. Fish and Wildlife Service for endangered riparian birds. He also serves on the Committee on Research and Exploration of the National Geographic Society. President Clinton appointed Dr. Graf to the

Presidential Commission on American Heritage Rivers to advise the White House on river management.

RIVER ECOLOGY

W. Carter Johnson is professor of ecology at South Dakota State University in Brookings. Dr. Johnson earned a BS in biology from Augustana College (Sioux Falls) in 1968 and a PhD in botany (plant ecology) from North Dakota State University in 1971. Dr. Johnson began his professional career as research associate and research staff member at Oak Ridge National Laboratory (1971-1977), followed by 12 years in the Department of Biology at Virginia Polytechnic Institute. From 1989 to 1995, he served as head of the Department of Horticulture, Forestry, Landscape, and Parks at South Dakota State University. His research interests include river regulation and riparian forest ecology, climate change and prairie wetlands, seed dispersal in fragmented landscapes, and paleoecology (climate reconstruction using tree rings and Holocene seed dispersal and plant migration). His research program is strongly multidisciplinary and interinstitutional. He was the recipient of the William S. Cooper Award from the Ecological Society of America in 1996 and the Best Paper Award from the International Association of Landscape Ecology in 1995 for his 1994 ecological monograph, *Woodland Expansion in the Platte River, Nebraska: Patterns and Causes.* He is a life member of the Ecological Society of America and has served on the editorial boards of *Landscape Ecology* and *Wetlands.* He was previously appointed to two National Research Council panels: the Committee on Water Resources Management, Instream Flows, and Salmon Survival in the Columbia River (2002-2004) and the Committee on Missouri River Basin Ecosystem Science (1999-2002).

AVIAN BIOLOGY

Whooping Crane

John "Jeb" A. Barzen has been director of field ecology at the International Crane Foundation (ICF) for 15 years. He has a BS in wildlife biology from the University of Minnesota and an MS in biology from the University of North Dakota. He has worked to implement conservation activities on private lands; worked to restore prairie, savanna, and wetland ecosystems in southern Wisconsin and in Asia; and conducted research on cranes and other species of waterbirds. Specifically, he has directed a long-term research project on sandhill cranes, with a focus on crop damage and other habitat selection issues; Siberian cranes as they interact with a variable and hidden food resource on winter habitats; the creation of wetland reserves in

Southeast Asia to prevent the extinction of the eastern sarus crane; and factors that influence the re-establishment of prairie plant communities as they change over time.

Piping Plover and Interior Least Tern

Francesca Cuthbert is professor in the Department of Fisheries and Wildlife at the University of Minnesota, Twin Cities campus. She is also co-director of the university's Conservation Biology Graduate Program. She earned a PhD at the University of Minnesota. Dr. Cuthbert is also an adjunct professor at the University of Michigan Biological Station. Her research interests include the biology and conservation of small avian populations (especially colonial waterbirds and shorebirds), and the recovery of endangered populations using an ecosystem perspective. A current research focus is the recovery of the endangered Great Lakes piping plover population, including studies on demography, captive rearing and reintroduction, and winter and breeding ecology, all in the context of management of coastal shore ecosystems.

FISH BIOLOGY

Edward J. Peters is professor of fisheries in the School of Natural Resource Sciences at the University of Nebraska. Dr. Peters earned a BS in conservation and biology from Wisconsin State University and an MS and a PhD in zoology from Brigham Young University. His primary research has been in habitat use and ecology of riverine and reservoir fisheries. Dr. Peters has published and presented papers on habitat use by riverine fish and invertebrates and pallid sturgeon in the Platte River. He has served multiple terms as president of the Nebraska Chapter of the American Fisheries Society and has been an active member of the Middle Basin Pallid Sturgeon Recovery Work Group.

CONSERVATION ECOLOGY

Dennis D. Murphy is research professor in the Biology Department and director of the graduate program in ecology, evolution, and conservation biology at the University of Nevada, Reno. He received a BS at the University of California, Berkeley and a doctorate from Stanford University. Until recently, he served as director and then as President of the Center for Conservation Biology at Stanford. Author of more than 170 published papers and book chapters on the biology of butterflies and on key issues in the conservation of imperiled species, Dr. Murphy has worked in conflict resolution in land-use planning on private property since the first federal Habitat Conservation Plan (HCP) on San Bruno Mountain, including HCPs in the Pacific Northwest, southern California, and Nevada. He won the industry's oldest and most respected prize in conservation, the Chevron Conservation Award; has been named a Pew Scholar in Conservation and the Environment; and has received the California Governor's Leadership Award in Economics and the Environment.

Dr. Murphy has served a number of scientific societies and environmental organizations and is past president of the Society for Conservation Biology. His professional activities outside academia include service on the Interagency Spotted Owl Scientific Advisory Committee, enjoined by Congress to develop a solution to that planning crisis in the Pacific Northwest, as chair of the National Park Service's Scientific Advisory Committee on Bighorn Sheep, as cochair of the Department of State's American-Russian Young Investigators Program in Biodiversity and Ecology, as codirector of the statewide Nevada Biodiversity Initiative based at the University of Nevada at Reno, and as chair of the Scientific Review Panel of the first Natural Community Conservation Planning Program in southern California's coastal sage scrub ecosystem. He served the National Research Council on its Committee on Scientific Issues in the Endangered Species Act and in its contribution to the recent General Accounting Office review of desert tortoise management and recovery. He has been a member of both the Applied Science Panel and the Interagency Working Group of the federal-state Coastal Salmon Initiative in northern California.

Dr. Murphy's continuing activities in conservation planning and adaptive management include service on the Science Board of the Cal-Fed Ecosystem Restoration Planning Program for the Sacramento and San Joaquin River systems, development of a conservation strategy for the imperiled Tahoe Yellow Cress for the U.S. Fish and Wildlife Service, development of a watershed-based ecosystem-management framework for the Truckee, Carson, and Walker hydrological units in the Humboldt-Toiyabe National Forest, and adaptive-management design for the nation's largest HCP under the Endangered Species Act in Clark County, Nevada. Dr. Murphy recently served as team leader for the committee of scientists carrying out the Lake Tahoe Watershed Assessment, a presidential deliverable to the Tahoe Federal Interagency Partnership via the U.S. Forest Service.

ENVIRONMENTAL ENGINEERING

Edwin E. Herricks is professor of environmental biology in the Environmental Engineering and Science Program in the Department of Civil and Environmental Engineering at the University of Illinois. He also holds affiliate appointments in the Departments of Animal Biology (College of Liberal Arts and Sciences) and Natural Resources and Environmental Sciences (College of Agriculture, Community, Consumer, and Environmental Sciences) and on the Environmental Council. Dr. Herricks earned a BA in zoology and English from the University of Kansas in 1968, an MS in sanitary and environmental engineering from The Johns Hopkins University in 1970, and a PhD in biology from Virginia Polytechnic Institute and State University in 1973. He was a research and field biologist with Union Carbide Corporation from 1973 to 1975.

While at the University of Illinois, Dr. Herrick's research, teaching, and professional activities were related to a program that couples understanding of the environment with engineering-based approaches to management and regulation. As a biologist, he focused this program on the analysis and interpretation of the effects of contaminants and other environmental alterations on communities of organisms (in aquatic and terrestrial ecosystems). His research analyzes and interprets the effects of environmental change on species, populations, and communities of organisms, with emphasis on the development of methods to improve environmental decision making and ecologically relevant engineering design. Specific research subjects include development of biological monitoring procedures for environmental decision making; time-related consequences of environmental change ranging from stormwater runoff to climate-change effects; analysis of organism-habitat relationships in streams and wetlands directed to restoration and naturalization; systems analysis of interactions between human and natural systems, including transportation-system interactions with wildlife; and development of engineering design approaches that minimize environmental and ecological impact.

FLUVIAL HYDRAULICS AND HYDROLOGY

Hsieh Wen Shen, emeritus professor of civil and environmental engineering at the University of California, Berkeley, taught graduate courses on erosion, sedimentation, environmental river mechanics, and river engineering and undergraduate courses in fluid mechanics and basic hydrology. He earned a BS and an MS in civil engineering from the University of Michigan and a PhD in hydraulics from the University of California, Berkeley. He has investigated behavior of numerous rivers, including the Nile in Egypt for the United Nations, the Cauca River in Colombia for the World Bank, and the Mississippi in the United States. In recognition of his contributions in various fields, Dr. Shen received the Horton Award in hydrology from the American Geophysical Union, the Einstein Award in fluvial hydraulics from the American Society of Civil Engineers, the Special Creative Awards in sediment mechanics from the U.S. National Science Foundation, the Humboldt Foundation Senior Distinguished U.S. Scientist Award in fluvial hydraulics from the German government, and the annual Joan Hodges Queneau Award in environment conservation jointly from the U.S. National Audubon Society and the American Association of Engineering Societies. Dr. Shen was elected to the National Academy of Engineering in 1993.

WATER-RESOURCES MANAGEMENT

Katharine L. Jacobs recently joined the faculty of the University of Arizona as an associate professor in soil, water, and environmental science and a water-management specialist in the Water Resources Research Center. She is also affiliated with the Institute for the Study of the Planet Earth and with the NSF Science and Technology Center on Sustainability of Semi-Arid Hydrology and Riparian Areas (SAHRA). She previously was special assistant for policy and planning for the Arizona Department of Water Resources (ADWR), where she worked on rural water-resources issues and a drought plan for the state. She was the director of the Tucson Active Management Area (AMA) of the ADWR from 1988 through 2001. In 2001-2002, she worked on a special project at the National Oceanic and Atmospheric Administration focused on the interface between scientific information, policy, and decision making. Ms. Jacobs earned her MLA in environmental planning from the University of California, Berkeley. Her expertise is in groundwater management and developing practical, appropriate solutions to difficult public-policy issues. She has been involved in all aspects of implementation of the 1980 Groundwater Management Act, including establishing water rights and permits; developing mandatory conservation requirements for municipal, agricultural, and industrial water users; developing plans for artificial recharge; and writing the Assured Water Supply Rules that require new subdivisions in AMAs to prove a 100-year supply of water. She served on the Synthesis Team for the U.S. National Assessment of the Consequences of Climate Variability and Change and on two other National Research Council panels, the Committee on Valuing Groundwater (1994) and the Committee on the U.S. Climate Change Science Program Strategic Plan (2003).

CIVIL ENGINEERING

Richard N. Palmer is a professor in the Department of Civil and Environmental Engineering at the University of Washington. He earned a BS in civil engineering from Lamar University in 1972, an MS in environmental engineering from Stanford University in 1973, and a PhD in environmental engineering from The Johns Hopkins University in 1979. He is the author of over 80 refereed papers, conference papers, and technical reports. He is a member of the American Society of Civil Engineers (ASCE) and is a registered professional engineer in the state of Washington. Dr. Palmer received the Service to the Professional Award from the Water Resources Planning and Management Division of ASCE in 1998. He was awarded the Certificate of Recognition for his editorial services to the *Journal of Water Resources Planning and Management* of ASCE in 1997, for which he was editor from 1993 to1997. Dr. Palmer was awarded the Huber Award for Research Excellence by ASCE in 1992; this honor was based on his innovative application of simulation and optimization techniques to issues in water-resources management. His paper "Operational Guidance During Droughts: An Expert System Approach" was awarded the Prize for Best Practice-Oriented Paper of the Year in the *Journal of Water Resources Planning and Management* by the ASCE in 1989. Dr. Palmer's primary interests are in the application of structured planning approaches to water resources, including reservoir management, the application of decision support and expert systems to civil engineering management problems, and real-time water-resources management, particularly as applied to drought. He has also developed the field of "shared-vision models" in water-resources planning, a technique that incorporates stakeholders into the model building process. A primary theme in his work has been the development and application of tools that have application in nondeterministic settings, that is, those in which uncertainty and forecasts play an important role in solving the problems.

ENDANGERED SPECIES LAW AND ENVIRONMENTAL POLICY

Holly Doremus is professor of law at the University of California, Davis, where she teaches environmental law, land-use planning law, public-lands management, and property law. Dr. Doremus earned a BS in biology from Trinity College in 1981, a PhD in plant biology from Cornell University in 1986, and a JD from the University of California, Berkeley, in 1991. She has written and presented extensively on protection and restoration of endangered species, biological diversity, adaptive management, and the effective use of science in environmental policy.

AGRICULTURAL ECONOMICS

Frank Lupi has a joint appointment at Michigan State University as associate professor of environmental and natural-resource economics in the Departments of Agricultural Economics and Fisheries and Wildlife. He earned a BS and an MS from the University of Illinois and a PhD from the University of Minnesota in 1997. His applied research focuses on fisheries, wildlife, and natural-resources management issues. Dr. Lupi serves as the economist in the Partnership for Ecosystem Research and Management, a cooperative venture between university-based faculty and various state and federal fish, wildlife, and natural-resources management agencies in the Great Lakes region. He has extensive experience in modeling the demand for and value of natural resources. Dr. Lupi is nationally recognized for his work with travel-cost and stated-preference methods of demand estimation, nonmarket valuation, and choice modeling. He has served as president of U.S. Department of Agriculture Multistate Research Project W133: Benefits and Costs of Resource Policies Affecting Public and Private Land, a national project that deals with benefit-cost analysis and nonmarket valuation for natural-resources planning. Dr. Lupi has served as an associate editor of the *North American Journal of Fisheries Management* and has served on the Great Lakes Panel of Environmental Economists, sponsored by the National Oceanic and Atmospheric Administration's Coastal Oceans Program and the Northeast-Midwest Institute.

ENVIRONMENTAL AGRICULTURE

James Anthony Thompson is a farmer who operates the Willow Lake Farm near Windom, Minnesota. He earned a BS in agronomy and continued with graduate studies in plant-community ecology at Montana State University, Bozeman. With help from the Minnesota Department of Agriculture's Energy and Sustainable Agriculture Grant Program, he has been increasing the biodiversity of his cropping system, planting understory forage grasses and legumes in his corn and soybean fields. Over a period of many years, he has preserved a large expanse of native prairie and now harvests seed for ecological restoration from these sites. The Willow Lake Farm hosts an Annual Agroecology Summit with support from neighbors, and many local, regional and national agencies and organizations. This locally focused event brings together farmers, ecologists, crop consultants, agency representatives, academics, consumer advocates, and citizens to discuss the merits and implementation of agroecological practices and concepts. Mr. Thompson is a board member of the Craighead Environmental Research Institute.

LAND USE

Lisa M. Butler Harrington is associate professor of geography at Kansas State University. Dr. Harrington earned a BS from Colorado State University in 1979, an MS from Clemson University in 1982, and a PhD from the University of Oklahoma in 1986. Her research interests include natural resources and land management, environmental change, biotic resources, human-environment relations, rural landscapes, and public lands. She has been involved in the multiuniversity Global Change in Local Places and Human-Environment Regional Observatory research projects, with funding from the National Science Foundation (NSF), the National Aeronautics and Space Administration, and the Department of Energy's National Institute for Global Environmental Change. Her professional service includes membership on NSF Doctoral Dissertation Research Improvement panels and the Association of American Geographers' (AAG) research committee, chairmanship of an AAG specialty group, and board membership for the Natural Resources and Environmental Science secondary major at Kansas State University.

Appendix B

Bird Species of Conservation Concern in Nebraska[a]

Species	Status[b]	Occurrence in Central Platte[c]	Habitat[d]
Bald eagle	1	B,W	Partial
Whooping crane	1	M	Open
Piping plover	1	B	Open
Mountain plover	1	—	—
Eskimo curlew	1	M	Open
Least tern	1	B	Open
Trumpeter swan	2	M,W	Open
Ferruginous hawk	2	M	Partial
Greater prairie-chicken	2	B,W	Open
Sandhill crane	2	M	Open
Willet	2	M	Open
Long-billed curlew	2	—	—
Buff-breasted sandpiper	2	—	—
Wilson's phalarope	2	B,M	Open
Yellow-billed cuckoo	2	B	Partial
Burrowing owl	2	B	Open
Whip-poor-will	2	—	—
Red-headed woodpecker	2	B	Partial
Loggerhead shrike	2	B	Open

Species	Status[b]	Occurrence in Central Platte[c]	Habitat[d]
Bell's vireo	2	B	Open
Cerulean warbler	2	—	—
Lark bunting	2	—	—
Henslow's sparrow	2	B	Open
Harris's sparrow	2	W	Open
McCown's longspur	2	—	—
Chestnut-collared longspur	2	—	—
Dickcissel	2	B	Open
Bobolink	2	B	Open
Northern harrier	3	B,M,W	Open
Swainson's hawk	3	B	Open
Sharp-tailed grouse	3	—	—
Northern bobwhite	3	B,W	Partial
Virginia rail	3	B	Open
American avocet	3	—	—
American bittern	3	B	Open
Upland sandpiper	3	B	Open
American woodcock	3	B	Closed
Black tern	3	M	Open
Black-billed cuckoo	3	B	Closed
Barn owl	3	—	—
Short-eared owl	3	W	Open
Chuck-will's widow	3	—	—
Chimney swift	3	B	Partial
Great crested flycatcher	3	B	Closed
Cassin's kingbird	3	—	—
Long-eared owl	3	B	Closed
Sedge wren	3	B	Open
Ovenbird	3	B	Closed
Yellow-breasted chat	3	—	—
Lark sparrow	3	B	Open
Grasshopper sparrow	3	B	Open
Eastern meadowlark	3	B	Open
Orchard oriole	3	B	Partial
Baltimore oriole	3	B	Closed
Bullock's oriole	3	—	—
Clark's grebe	4	—	—

Species	Status[b]	Occurrence in Central Platte[c]	Habitat[d]
Mississippi kite	4	M	Partial
White-throated swift	4	—	—
Lewis's woodpecker	4	—	—
Pileated woodpecker	4	—	—
Acadian flycatcher	4	—	—
Cordilleran flycatcher	4	—	—
Scissor-tailed flycatcher	4	—	—
Yellow-throated vireo	4	—	—
Pygmy nuthatch	4	—	—
Mountain bluebird	4	—	—
Townsend's solitaire	4	—	—
Yellow-throated warbler	4	—	—
Prothonotary warbler	4	—	—
Louisiana waterthrush	4	—	—
Kentucky warbler	4	—	—
Brewer's sparrow	4	—	—

[a]Information from Science Advisory Workgroup and Nebraska Partnership for All Bird Conservation, except as noted.
[b]1 = threatened or endangered; 2 = high concern (PIF score, ≥ 24); 3 = moderate concern (PIF score, 20-23); 4 = concern or edge of range.
[c]Information from J. Dinan, NGPC. B = breeding; M = migrant; W = wintering.
[d]Open habitat is characterized by absence of trees. Partial habitat is characterized by widely scattered trees where canopies do not often touch and shrubs may be present but are not impenetrable. In closed habitat, tree canopies frequently touch, and understory vegetation can be very thick.

Appendix C

Confirmed Whooping Crane Sightings in Central Platte River Study Area, 1942-2003

Compiled by the U.S. Fish and Wildlife Service

Group ID	Number of Cranes	Year	Confirmed River Use	Dates of Use[a]
42A-1	3	1942	Y	4/5
43A-1	1	1943	Y	4/4
50A-1	1	1950	N	5/4
59B-2	2	1959	N	10/26
66B-2	5	1966	Y	10/21-22
74B-5	2	1974	Y	10/31-11/1
75A-6	7	1975	Y	4/18-19
75A-7	5	1975	Y	4/19-20
77A-4	1	1977	Y	3/29
80A-11	2	1980	Y	4/16-18
80A-12	2	1980	Y	4/18
83B-21	5	1983	Y	10/27-28
83B-22	3	1983	N	10/28
85A-6	4	1985	N	4/8
85B-18	3	1985	Y	10/20-21
86B-58	3	1986	Y	11/4-5
86B-88	3	1986	N	11/7
87A-1	1	1987	Y	3/17-4/19
87A-13	2	1987	N	4/16
87B-2	2	1987	Y	10/21-22
88A-1	2	1988	Y	3/24-26
88A-2	2	1988	Y	3/26-4/7
88A-3	1	1988	Y	4/1-11
88A-4	1	1988	Y	4/3-4
89A-1	1	1989	Y	3/24-4/14
89A-13	1	1989	Y	4/16-17
89B-5	4	1989	Y	10/13-14
89B-18	2	1989	Y	11/4
90A-4	3	1990	Y	4/14-15
90B-15	3	1990	Y	10/26-27
92A-3	2	1992	Y	4/10
92A-4	4	1992	Y	4/10-11
92A-5	6	1992	Y	4/10-11
92A-10	5	1992	Y	4/14
93A-2	5	1993	Y	4/9-10
93A-3	2	1993	Y	4/9-10
93A-4	2	1993	Y	4/9-10
93B-16	1	1993	N	10/23
94A-1	6	1994	N	4/4

Location, miles[b]	Bridge Segment[c]	River Miles[d]	Comment
3 E Lexington Bridge	W	—	
4 W Kearney Bridge	9	—	
SE of Overton	11	NA	
6 S, 5 E of Overton	11	NA	
6 NE of Philips	1	158	
2 E Minden (#10) Bridge	7	206	
0.5 E Odessa Bridge	9	223.5	Some of 9 birds hazed from Funk basin
0.5 E Odessa Bridge	9	223.5	
2 E Minden (#10) Bridge	7	206	
Audubon Sanctuary	6,7	207	Assumed same birds as 80A-12
1 E Minden/I-80 Interchange	7	—	
6 SE of Shelton	5	191.1	
7 SE of Shelton	5	NA	Joined 83B-21
6 S, 2 W of Shelton	6	NA	Early AM; probable river use
1.5 E Minden (#10) Bridge	7	206.4	
2 E Kearney Bridge	8	213	Assumed same birds as 86B-88
1.5 S, 1 E of Hwy 44/I-80 Interchange	8	NA	
W of Gibbon Bridge (Lowell area)	6,7	202.9-206.4	
Mormon Island Crane Meadows	3	NA	Brief stop during day's migration
1.25 E Gibbon Bridge	6	201.2	
0.75 W Minden (#10) Bridge	8	208.6, 209.6	Group moved to Odessa (88A-2)
3 W Odessa Bridge	9,10	226.3-227.6	See 88A-1
3 W Wood River Bridge	5	189.5-189.8	
1 W Alda Bridge	4	183	Possibly same bird as 88A-3
2.5 W Wood River Bridge	5	188-191.1	
3 W Minden (#10) Bridge	8	211	
2.5 W Wood River Bridge	5		
2 E Alda Bridge	3-4	180	2nd site 2 W Alda Bridge
0.5 W Minden (#10) Bridge	7	207.5	
2.25 E Shelton Bridge	5	193.5	Originally 2.5 W Wood River
0.5 E Gibbon Bridge	6	201.5	Joined 92A-4 to roost
2 W Gibbon Bridge	7	204.3	Includes 92A-3 birds
0.25 E Overton Bridge	11	239.0	
2.5 W Gibbon Bridge	7	204.7	Probably roosted 4/13-14
1.5 W Gibbon Bridge	7	206.5	
E of Alda Bridge	3	180.2	
1 W Wood River Bridge	5	188.1	
1 S, 1 E Elm Creek	10	NA	
3.5 S of Grand Island	2	NA	Corn stubble, middle to late morning

Group ID	Number of Cranes	Year	Confirmed River Use	Dates of Use[a]
94A-3	2	1994	Y	4/4
94A-4	2	1994	Y	4/4-6
94A-8	2	1994	N	4/5
94A-14	1	1994	Y	4/14
94A-15	1	1994	Y	4/14
95A-8	2	1995	Y	4/14
95A-21	1	1995	Y	5/8-14
95B-5	5	1995	Y	10/25
96A-3	1	1996	N	3/30
96A-9	3	1996	N	4/7-9
96A-10	2	1996	Y	4/10
96A-11	3	1996	Y	4/11
96A-21	1	1996	N	4/20
96A-22	3	1996	Y	4/20-21
96A-31	2	1996	Y	5/8-13
97A-1	1	1997	Y	3/9
97A-2	1	1997	Y	3/10-4/17
97A-3	1	1997	Y	3/19-4/7
97A-6	1	1997	N	4/14
97A-7	2	1997	Y	4/14-16
97A-17	3	1997	Y	4/15-17
97A-18	1	1997	N	4/16-19
98A-1	1	1998	Y	2/15-3/25
98A-3	1	1998	Y	3/23-4/9
98A-4	1	1998	N	3/27-28
98A-6	1	1998	Y	3/30
98A-7	1	1998	N	4/3
98A-16	2	1998	N	4/14
98B-8	3	1998	Y	10/22-11/9
99A-1	1	1999	Y	3/4-23
99A-3	1	1999	N	3/23
99A-4	1	1999	N	3/21
99A-7	2	1999	N	4/3
00A-2	1	2000	Y	3/2-4/4
00A-4	1	2000	Y	3/8-16
00B-30	2	2000	Y	11/3
01A-07	3	2001	N	4/11
01B-13	1	2001	Y	10/23

Location, miles[b]	Bridge Segment[c]	River Miles[d]	Comment
1 E Gibbon Bridge	6		
2.5 E of Minden Bridge	7	205.6, 206.6	2nd site 1.25 E of Minden Bridge
1 N, 1 W Hwy 10/I-80 Exit	8		
1 downstream I-80 Bridge	2	171.5	
1.5 upstream Gibbon Bridge	7	203.8	Rowe Sanctuary
5 S, 1 W Gibbon	7	203.1	
0.5 N, 5 W Doniphan	3	180.6	
2.75 E Hwy 10 Bridge	7	205.25	
2 E Elm Creek	10	NA	
3.5 S, 1.5 Odessa	9		
2.5 upstream Wood River Bridge	5	191.1	
0.5 upstream Wood River Bridge	4,5	187.8	
2 N, 1 W Doniphan	3	NA	Wet meadow, midday
1 W, 2 N Doniphan	3	176.6	
2 S, 2 E Elm Creek	10	228.6	
4.5 S, 1.75 W Gibbon	7	203.9	
3 W Doniphan	3,4	183.1, 180.7, 178.4	
4.5 S, 4 W Gibbon	7	205.7, 203.0, 204.0	
6 S Gibbon	7	NA	
2 S Gibbon	7	203.4	
5 S, 2 W Gibbon	6,7	204.1	
4.5 S, 1 W Overton	11,12		
3 W Doniphan	3,4	NA	
5.5 N, 4 E Doniphan	2	NA	
4 S I-80 Alda Interchange	3,4	NA	Likely roost on river
1.25 N, 1.75 W Doniphan	3	NA	Adult with sandhills
2 N, 3 E I-80 Alda Exit	3	NA	5:00 PM, feeding with sandhills
6.5 S, 2.5 W Wood River	5	NA	8-11:15 AM, flooded pasture
1.25 E, Hwy 10 Bridge	7	204, 207.6	Rowe Sanctuary
3 S, E Gibbon	2,3,4,6		
2 S, 2 W Wood River	5	NA	Assumed same birds as 99A-2
1.75 S, 2 W Shelton	6		
2.5 S, 1.5 E Overton/I-80	11	NA	
2.5 W Doniphan	3		
2.5 W Doniphan	3		
1.5 W Gibbon Bridge	7		
1.5 S, 2 E Gibbon Road/I-80 interchange	6		
Platte River, 3 E (downstream) Hwy 10 Bridge	7		

Group ID	Number of Cranes	Year	Confirmed River Use	Dates of Use[a]
01B-37	3	2001	Y	11/4
02A-1	1	2002	Y	3/19-4/13
02B-1	1	2002	Y	10/13-14
02B-3	1	2002	Y	10/14
02B-40	8	2002	Y	11/1-10
02B-41	2	2002	Y	11/1-3
02B-42	3	2002	Y	11/1-10
02B-43	1	2002	Y	11/5-10
03A-3	2	2003	Y	4/3-4
03A-6	2	2003	Y	4/8

[a]When initial sighting occurred in early morning, arrival was assumed to be previous day.
[b]Only two Platte River Valley sightings west of Lexington.
[c]Locations are distances in miles from identified landmark.
[d]Corps of Engineers river miles.
Sources: USFWS, Grand Island, unpublished data, 1975-2003.

Location, miles[b]	Bridge Segment[c]	River Miles[d]	Comment
Platte River, 2.75 E (downstream Hwy 10)	7		
2 S, 3 W Wood River Road /I-80 interchange	5		
3 W Doniphan; Platte River west Alda Bridge	3,5		
1 W of Rowe Sanctuary office	7		
1.5 S, 0.75 W Doniphan	3		
Near Odessa Bridge and upstream and downstream	9		
Kearney Bridge			
Upstream and downstream Alda Bridge	3,4		
5 S, 3 W Chapman	1		
0.25 E Minden Bridge	7		
2.6 E Elm Creek Bridge	10		

Appendix D

Input Data for Figures 5-6A and 5-6B

"Use-days" is combined number of days that each crane spent on the Platte River in fall and spring of that year, "sightings" are numbers of cranes in fall and spring of that year seen on the Platte River, "population" is the total number of cranes in the Aransas-Wood Buffalo population during that year, "ratio" is the number of cranes divided by the number of birds in population, and "use ratio" is the number of use-days divided by total population in that year. Double counts of crane sightings in Platte River were omitted. Data collected prior to 1975 were not used in these analyses because the data collection effort was extremely variable before 1975. Data for 2003 were not yet complete during committee's deliberations.

Year	Use-Days	Sightings	Population[a]	Ratio	Use Ratio
1950	1	1	31		
1959	2	2	33		
1966	10	2	43		
1974	4	2	49		
1975	36	12	57		
1976	0	0	69	0	0
1977	1	1	71	0.01408	0.01408
1978	0	0	75	0	0
1979	0	0	76	0	0
1980	6	2	78	0.02564	0.07692
1981	0	0	73	0	0
1982	0	0	73	0	0

Year	Use-Days	Sightings	Population[a]	Ratio	Use Ratio
1983	13	8	75	0.10667	0.17333
1984	0	0	86	0	0
1985	10	7	97	0.07216	0.10309
1986	12	3	110	0.02727	0.10909
1987	40	5	134	0.03731	0.06716
1988	41	3	138	0.02174	0.2971
1989	34	8	146	0.05479	0.23288
1990	12	6	146	0.0411	0.08219
1991	0	0	132	0	0
1992	25	15	136	0.11029	0.18382
1993	19	10	143	0.06993	0.13287
1994	18	14	133	0.10526	0.1203
1995	14	8	158	0.05063	0.08861
1996	34	15	160	0.09375	0.2125
1997	80	10	182	0.05495	0.43956
1998	120	10	183	0.05464	0.65574
1999	24	5	188	0.0266	0.12766
2000	45	4	180	0.02222	0.19444
2001	7	7	176	0.04545	0.1875
2002	151	17	185	0.08649	0.67568

[a]Population data are from Lewis et al. (1994) and Canadian Wildlife Service and U.S. Fish and Wildlife Service, unpublished report (2003).